Automobile Electronics

Automobile Electronics

ERIC CHOWANIETZ

Newnes
An imprint of Butterworth-Heinemann Ltd
Linacre House, Jordan Hill, Oxford OX2 8DP

A member of the Reed Elsevier plc group

OXFORD LONDON BOSTON
MUNICH NEW DELHI SINGAPORE SYDNEY
TOKYO TORONTO WELLINGTON

First published 1995

British Library Cataloguing in Publication Data
A catalogue record for this book
is available from the British Library

ISBN 0 7506 1878 7

Typeset by Graphicraft Typesetters Ltd, Hong Kong
Printed in Great Britain by Scotprint Ltd., Musselburgh

Contents

Preface

The stringent exhaust emission regulations introduced in the USA during the early 1970s meant that the all-mechanical automobile faced an uncertain future. Engine designers were forced to adopt electronic engine control systems as a solution to their problems and so crucial early links between the automobile and electronics industries were soon formed. Subsequently, developments in microelectronics, and particularly the availability of powerful low-cost microprocessors, resulted in the development of a whole host of automotive electronic systems. Most of these may be described as 'mechatronic' systems, meaning that they integrate mechanical, electronic and microcomputer technologies.

These changes, together with increasing environmental and social pressures, have led to a revolution in the concept of the automobile, and have placed great demands on those engineers and technicians involved in designing and maintaining cars.

This book is one of the few available that explain the operation of contemporary automobile electronic systems in simple terms. Although starting with the premise that the reader has a basic understanding of the mechanical workings of the automobile, no prior knowledge of electrical technology is assumed. The first four chapters of this book therefore aim to provide the detailed understanding of electrical, electronic and microcomputer fundamentals that is so essential when seeking to understand the operation of the modern car. Subsequent chapters are then used to explore specific aspects of automotive electronics such as engine management and pollution control, transmission control, chassis control and the operation of body electronic systems. The final chapter is devoted to the all-important topic of fault detection and diagnosis.

Each theme is illustrated by taking a 'case study' approach, examining the operation of a number of typical systems, from a wide variety of manufacturers, in great depth. Where appropriate, sufficient non-technical background information is given to place the subject in its correct historical and social context. In this way it is anticipated that the already knowledgeable reader will be quickly updated, whilst the novice will not be overcome by the complexity of the subject. Moreover, with a good grounding in microprocessor-based control systems and exposure to a broad range of example applications, the reader should be able to understand the functioning of new and different automotive electronic systems as they appear on the market.

Although the readership of this book will predominantly be students on City and Guilds and NVQ courses, it is also suitable for qualified engineers and technicians who wish to keep up-to-date in this rapidly changing area, as well as those enthusiastic motorists who want to know more about how their car works. Students pursuing an electrical, mechanical or automobile engineering degree will also discover much that is relevant to their course, whilst those studying mechatronics at either undergraduate or postgraduate level will find the book an excellent source of material for 'case study' and 'applications' modules.

Eric Chowanietz

Acknowledgements

Grateful acknowledgement is given to the following companies and organizations who have kindly given permission for certain illustrations to be reproduced in this book:

Automotive Products plc., BMW (GB) Ltd., Robert Bosch Ltd., Citroën UK Ltd., Delco Remy Division of General Motors Corp., Federation of Automatic Transmission Engineers (FATE), FKI Crypton Ltd., Ford Motor Co. Ltd., Honda (UK) Ltd., Jaguar Cars Ltd., Lucas Rists Wiring Systems Ltd., Lucas Automotive, Magneti Marelli UK Ltd., Mazda Cars (UK) Ltd., The Colt Car Co. Ltd. (Mitsubishi), Motor Industry Research Association (MIRA), Motorola, NGK Spark Plugs (UK) Ltd., Nissan Motor (GB) Ltd., Peugeot-Talbot Motor Co. Ltd., Rover Cars Ltd., RS Components Ltd., SAAB (Great Britain) Ltd., Sachs UK Ltd., Subaru (UK) Ltd., Telefunken Microelectronic GmbH., Toyota (GB) Ltd., Valeo Clutches Ltd.

1

Introduction

1.1 THE EARLY YEARS

Automotive history began in the middle of the eighteenth century when a Frenchman, Nicholas Cugnot, produced the world's first automobile. It used the power of the industrial revolution – the steam engine – to open up the possibility of moving people and goods quickly and conveniently from door to door. In the early part of the twentieth century this dream turned to reality for millions of people when Henry Ford developed the technique of the production line – the 'second industrial revolution'. Mass-production enabled the production of automobiles at low cost and spawned the world's greatest manufacturing industry.

With advances in technology, the performance of the motor car has increased beyond even the most exaggerated predictions of those early years, but so too has the number of vehicles on the road and their impact on the global environment.

Electronics and automobiles

Although car radios, using valve circuits, had been manufactured as early as the 1930s, it was the invention of the transistor in 1948 and the integrated circuit (IC or 'silicon chip') in 1959 which enabled automobile electronics to become a reality.

One of the earliest automotive applications for these new electronic devices was in transistorized ignition systems, the first of which was designed in 1962 by General Motors. Progress was comparatively swift and by 1967 Bosch had a simple electronic fuel injection control system in series production. Further new products followed in the late 1960s, including cruise control and anti-lock braking systems (ABS). Unfortunately, although novel, all of these added electronic features were expensive and were based around the relatively unreliable analogue circuits of the time, making them costly to maintain. Not surprisingly, they gained little popularity with the motoring public.

It was the environmental impact of automobiles which ultimately stimulated one of the most significant changes in car design – the widespread adoption of microcomputer-based engine control technology. The US Clean Air Act of 1971 required that harmful automobile exhaust emissions be drastically reduced and, by fortunate coincidence, 1971 was also the year that the microprocessor, the heart of all modern microcomputer-based automobile control systems, was first manufactured. This 'third industrial revolution' – the application of microcomputer technology – ultimately led to a complete change in the concept of the motor car. Engine designers quickly seized upon the microprocessor as the solution to their emission control problems and vital links between computer technology and automobiles were soon forged. The first automotive application of the microprocessor was in GM's MISAR ignition timing control system, introduced in 1976. It enabled very precise control of spark timing, leading to increased engine output and efficiency, together with much lower exhaust emissions. Other car makers soon followed GM's lead and ignition systems are now almost universally of this type.

Pressure to install more advanced microprocessor engine control systems arose in the late 1970s, when a host of US federal and state government exhaust emission and fuel economy requirements were introduced. These regulations posed a unique problem; to meet all of the requirements simultaneously and yet still maintain good driveability required interactive engine controls. Mechanical controls of the necessary sophistication were either not possible or not cost-effective. This gave rise to 'acceptance-through-necessity' of the microprocessor by the US motor industry. For example, by 1981 microcomputer-based engine controls were incorporated into the entire US petrol-engined car production of General Motors, requiring the manufacture of 3.7 million electronic control units (ECUs) per year for that company alone (a rate of 22 000 ECUs per day).

Europe has traditionally lagged well behind the USA and Japan in exhaust emission legislation and it was not until January 1993 that US-style emission control systems were made a mandatory requirement for all cars sold in EC countries. This means that almost the entire passenger car production of the advanced nations, about forty million vehicles per year, is now fitted with microcomputer-based engine controls.

Enhanced vehicle performance

The 1980s saw an explosion in the manufacture of 'high-tech' microprocessor-based consumer products, ranging from washing machines through to video cassette recorders. In parallel with this growth came ever-increasing consumer expectations of performance, functionality and reliability improvements in motor cars. Vehicle manufacturers, by now confident and proficient in the application of microprocessors to engine control, began to diversify into new areas of automobile electronics.

Microprocessor control of automatic transmissions (introduced by Toyota in 1981) provides smoother shifting and more fuel-efficient gearboxes. Traction control systems (TCS) can assist when accelerating on slippery road surfaces. ABS is an invaluable braking aid when driving conditions are poor. Ever more sophisticated chassis control is possible via electronically-controlled 4-wheel steering (E4WS), which precisely steers the rear wheels in sympathy with those at the front to increase stability when cornering or changing lanes. Electronic control of the suspension system is also increasingly utilized as a means of improving vehicle handling without compromising comfort.

Within the passenger cabin, electronics have been used to drastically improve comfort; as well as the ubiquitous in-car stereo system, many vehicles now feature such luxuries as electrically-operated seats, mirrors, sunroof and windows. Electronically-controlled air-conditioning systems are commonly found on 'high-line' vehicles and are gradually being introduced onto smaller vehicles as purchaser expectations rise.

Safety and security are the latest beneficiaries of electronic technology. Air-bag systems are a proven safety feature and are now a mandatory fitment on US cars and a standard fitment on many European cars. Side-impact air-bags have also been introduced by some manufacturers, thanks to recent developments in sensor technology.

Car security is a particularly important issue in the UK, where the level of car crime is the highest in the world. Most UK-market vehicles are now supplied with a factory-fitted burglar alarm and engine immobilization system; a trend which is helping to reduce the cost of insurance for late-model vehicles.

1.2 CURRENT AND FUTURE TRENDS IN AUTOMOBILE ELECTRONICS

The enormous advances in electronic technology throughout the 1980s and early 1990s have brought about great changes in the status of automobile electronics. Reliability has improved greatly and costs have been reduced. Electronic components are now much smaller, drastically reducing weight, space and electrical power requirements. Thanks to the availability of powerful and inexpensive microprocessors, computing power is no longer a limitation to the development of electronic control systems. Future developments are therefore likely to centre on the refinement of existing automotive electronic systems, coupled with advances in sensor and actuator technology. These changes will have an impact in the four main areas of vehicle operation:

1 environmental;
2 safety;
3 ergonomics;
4 social infrastructure.

Environmental considerations

With approximately four hundred million vehicles on the world's roads, the environmental impact of the automobile is awesome. For example, German government research has shown that, from 'cradle to grave', a typical car produces 59.7 tonnes of carbon dioxide (the 'greenhouse gas'), 2 040 million cubic metres of polluted air, and 26.5 tonnes of solid rubbish to add to the problems of waste disposal experienced by most Western countries.

If valuable fuel reserves are to be preserved and the global ecology is to be maintained, it is imperative that electronic systems are developed to improve engine efficiency. Currently, most governments insist that cars are fitted with a 'catalytic converter' which cleans up the exhaust gasses as they leave the engine. Unfortunately, in order to function properly, the catalyst requires that the engine be operated in a relatively inefficient manner – a situation that may be aptly described as 'the tail wagging the dog'. Around the world, automobile engineers are working on solutions which allow more efficient combustion to take place. Improving combustion efficiency achieves two important aims; it maximizes fuel economy, thereby conserving valuable hydrocarbon resources, and it reduces the emission of carbon dioxide and other pollutants.

Many vehicle manufacturers have already developed so-called 'lean-burn' engines which, together with advanced electronic control of the spark and fuel-injection systems, provide fuel efficiency gains of up to 25% over conventional engines, as well as lower pollution levels.

A reduced environmental impact may also be attained through the use of 'alternative fuels' and as we enter the twenty-first century it is likely that interest will increase. Methanol, electric and hybrid (hydrocarbon/electric) propulsion systems may become increasingly viable as petrol prices rise through a combination of scarcity and 'carbon tax'. Research will concentrate on making the required motors, sensors, actuators and controllers smaller, faster and smarter.

Safety

Annual road accident statistics have relatively little impact in the UK, where death or injury through car accidents is accepted almost as a fact of life. However, when these statistics are compared with vehicle use and hours of life lost, road traffic accidents emerge as a major cause of death, particularly for young people. Every car produced, on average over its lifetime, is responsible for about 800 hours of life lost through a fatality and about 3 000 hours of life damaged through injury. Statistically, about one person in every 100 will be killed in a road traffic accident. Thankfully, over the past few years safety has taken a rising profile in the marketing of automobiles. The application of advanced electronic technology focuses on two main areas: (i) *active safety*, assisting the driver in avoiding an accident and promoting safer driving; and (ii) *passive safety*, protecting the vehicle occupants once a collision has become inevitable.

Active safety

As electronic control systems have been incorporated into automobiles the driver 'workload' has been reduced and driving has become less fatiguing and safer. The extreme swiftness with which electronic systems can process data and intervene in chassis control means that in an emergency situation they can be invaluable. Currently, traction control systems (TCS) and anti-lock braking systems (ABS) can rescue a driver from a situation that would otherwise result in an accident. Improved sensing technology, such as the use of radar and infrared detectors, can enable obstacle recognition systems to be linked to speed control, TCS and ABS, in order to provide even greater levels of safety.

Electronically-controlled four-wheel steering has been used as an aid to chassis stability since the late 1980s. In future, such systems will be enhanced by the use of additional sensors to monitor rate of yaw and the cornering forces generated by each tyre. Using this information a powerful electronic controller will be able to individually steer each end of the vehicle to maintain the desired path through a curve. Other important developments could include driver monitoring systems, which can warn a driver if his behaviour or reactions become abnormal, and improved warning and display systems which present information to the driver without requiring him to look away from the road.

Passive safety

Seatbelts have been fitted to UK-market cars since the 1960s and have had a proven and dramatic impact on preventing injury when an accident occurs. Air-bag systems were developed to supplement seatbelts and were originally mechanically triggered. Fast-acting electronics have considerably enhanced their performance however, and modern air-bag controllers can detect the onset of an impact and initiate air-bag inflation in less than one-hundredth of a second. In future, air-bag controllers could be designed to predict a collision before it actually occurs, possibly using radar-based speed and distance measurement systems. Such technology would enable the bag to be already inflated at the time of impact.

Ergonomics

A primary requirement of any passenger car is that it should be comfortable and easy to operate; in other words 'ergonomically designed'. Controls need to be light and precise in operation, driver information should be presented in a clear and logical fashion.

Electronic systems can improve vehicle ergonomics by providing power-assisted and 'intelligent' controls. For example, some manufacturers have produced vehicles in which the steering wheel and column automatically swing towards the dashboard to give easier access when the driver enters or leaves the car. Others have introduced cars with electronic 'keys', which not only operate the door and ignition locks but also hold data to automatically adjust the seat, mirrors and steering column to a personalized position. Further common examples include 'speed-sensitive' power steering and cruise-control, which reduces driver fatigue on long journeys. In future, such systems will be enhanced by using additional microcomputers and sensors to assess road surface and traffic flow conditions, enabling control system behaviour to be modified accordingly.

As cars have become more sophisticated, the increase in the amount of information available to the driver has led to a proliferation in the number of indicators and warning lights on the dashboard. Some instrument panels have become very complex indeed. Electronics are now being used to simplify the presentation of this data and improve legibility through the use of liquid crystal displays (LCDs) and vacuum fluorescent displays (VFDs). Many manufacturers are currently researching novel display systems which project an image of the instrument onto the inside of the windscreen, just below the driver's line of sight. These so-called 'head-up displays' (HUDs) are proving easy to read in all lighting conditions, without requiring the driver to look away from the road. Further developments may include the use of audible warnings to augment visual information, as in aircraft cockpits.

Social infrastructure

As society becomes more information oriented, cars will increasingly be fitted with equipment to provide drivers and passengers with information from extra-vehicular sources. Future on-board communication systems will enable drivers to avoid traffic jams and accidents, and to take advantage of information relating to the availability of parking spaces or the location of particular stores and shops.

Currently, most vehicles are fitted with an AM/FM radio which provides entertainment and a basic level of traffic information from 'traffic bulletins', read by station announcers. In the early 1990s the highly successful Radio Data System (RDS) was introduced in Europe as a means of enhancing the information-carrying ability of FM broadcasts by transmitting digital data along with the radio signal. Many vehicles are now supplied with RDS radios as a standard fitment.

Future traffic data systems will be far more comprehensive; road information systems have already been demonstrated and incorporate a large-area display screen to provide navigation and route guidance facilities. The navigation system is able to display the current location of the vehicle and an ideal route to follow to reach the destination in a minimum time. Vehicle position is continuously updated using data from vehicle speed sensors, an in-car gyroscope and signals received from the GPS (Global Positioning System) satellite network. The predicted position can then be compared with a road map, held on a CD-ROM data disc.

Other navigation devices include AutoGuide, a low-cost system which relies on roadside beacons to undertake two-way communication with the vehicle's navigation electronics. The beacons are linked to a central traffic-control computer which tries to keep traffic flowing smoothly and safely by guiding drivers away from troublespots.

Since all of these systems depend upon the provision of a large-scale communications infrastructure for their operation, governments around the world are supporting developments in this area. In Europe, the EU countries are promoting the DRIVE and PROMETHEUS initiatives. In the USA, the federal government is working with universities to develop the IVHS (Intelligent Vehicle-Highway System), and in Japan VICS (Vehicle Information and Communication System) is being developed.

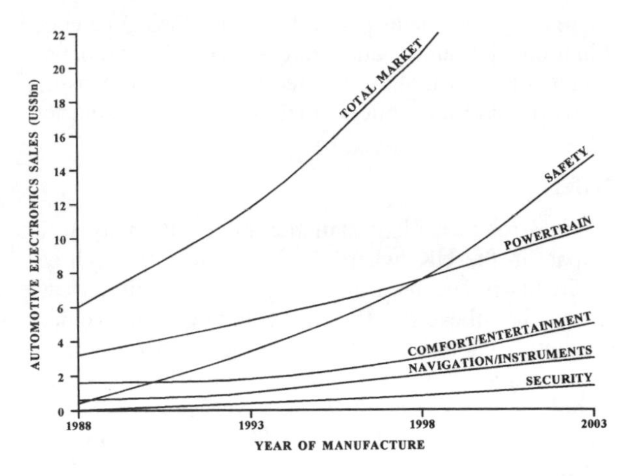

Figure 1.1 Trends in automotive electronics – automotive electronics sales in the North American market, 1988–2003

1.3 CONCLUSIONS

The 1970s brought stringent exhaust emission regulations and the first oil crisis, stimulating a demand for clean, energy efficient automobiles which was satisfied only through the application of electronic engine controls. Initial difficulties in using electronics in the harsh automotive environment were quickly overcome.

During the 1980s, the development of microprocessor control systems in tandem with precision mechanical systems made cars faster, safer, more comfortable and still more fuel efficient. Excellence in powertrain performance is now taken for granted. In the early 1990s, consumer demands became even more sophisticated. Familiarity with domestic electronic products, such as video cassette recorders, home computers and satellite television receivers has led to increased expectations of automobile comfort, safety and convenience features to the extent that CD players, mobile telephones, ABS and air-bags have become the norm on luxury vehicles.

The key issues in automobile design for the next century will be safety and the environment. Fortunately, rapid developments in electronics will help in these areas. The time delay between the development of new automotive electronic devices and their introduction into series production is continuously reducing. Each year, as the capabilities of electronic systems improve, their size, weight, cost and power consumption is reduced. Predictions by the large automotive electronics suppliers suggest that the demand for electronic systems will grow at a rate of about 12% per year through the 1990s (Figure 1.1). By the end of the 1990s it is estimated that over one-third of a typical car's components will be electrical or electronic and a typical vehicle will be fitted with as many as 50 electronic sensors. Automotive electronics will continue to grow even further in importance in the next century, requiring all those involved in designing, selling and maintaining vehicles to have a sound knowledge of these systems.

2

Electrical and electronic principles

2.1 INTRODUCTION

To study and understand automobile electronic systems a firm foundation in elementary electronic and electrical engineering is required. This chapter is therefore designed to provide a knowledge of the basic principles of operation of all electrical and electronic circuits, from the most elementary to the most complex.

Atoms and electricity

All material in the universe is made of tiny particles called *atoms*. Every atom has a compact and relatively heavy *nucleus*, which carries a positive charge and about which orbit *electrons*, each of which carries a small negative charge. The nucleus is itself composed of two types of elementary particles called *neutrons* and *protons*. Neutrons and protons each have a mass of only about 1.7×10^{-24}g, which is tiny by the standards of our everyday experience but still nearly 2 000 times greater than that of the electron, and thus they account for almost all of the mass of the atom. Neutrons and protons are almost identical, apart from the fact that protons carry a positive charge and neutrons carry no charge at all (they just add mass to the atom). All protons are identical and all carry the same amount of positive charge. All electrons are identical and all carry the same amount of negative charge which is of equal magnitude but opposite *polarity* to the positive charge of the proton.

In its normal condition an atom has just as many protons in its nucleus as there are electrons outside the nucleus and so the positive and negative charges balance out; the atom is said to be *electrically neutral*. Figure 2.1 shows a diagram of an atom of the gas helium, notice how the positive charge of the two protons is exactly balanced by the negative charge of the two orbiting electrons.

If one of the orbiting electrons were to break free from the atom, there would be an excess of positive charge remaining on the atom. The atom would then be described as a *positive ion*. Conversely, if the *free electron* where to attach itself to another (neutral) atom, then that atom would then have an excess of negative charge and so would be called a *negative ion*. It is the existence of free electrons that gives rise to the phenomenon of electricity.

Conductors, semiconductors and insulators

The process of removing electrons from atoms, called *ionization*, can be made to occur in any material, but it often requires the input of a lot of energy – usually in the form of heat or a strong electric field. Materials that are difficult to ionize have very few free electrons and so do not readily allow electric charge to move about within them; these materials are termed *insulators*.

Although in normal circumstances the flow of electricity in an insulator is so tiny that it can be ignored, most will eventually break down and ionize under a sufficiently high voltage, allowing current to flow in the form of a *spark*. Most non-metals are good insulators, although in automobile electronics, plastics and ceramics are the main insulating materials used.

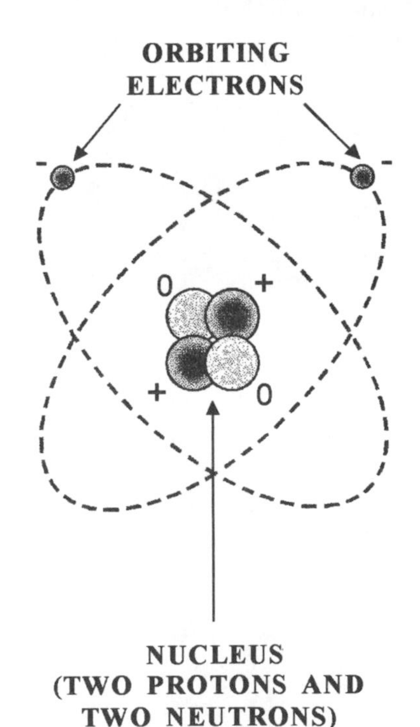

Figure 2.1 Atomic structure of helium

Conductors are materials in which atoms normally exist in an ionized state, surrounded by many free electrons. Most metals belong to this class of materials. Silver has the most free electrons of any metal and so is the best conductor of electricity; copper is almost as good and is much cheaper – so electrical *cables* are usually made from copper. A cable is simply a length of flexible conductor used to transport electrons from one point to another. An insulating layer (usually plastic) surrounds the conductor, preventing the electrons from escaping. The insulation has to be strong enough to prevent the leakage of electrons at the working voltage, for example insulation that is adequate for a car's 12 V circuits would break down if used at the high voltages encountered in spark-plug leads.

Between insulators and conductors, there exists a class of materials in which ionization occurs to just a small extent at room temperature. These materials conduct electricity much better than insulators, but not quite as well as conductors, and so are known as *semiconductors*. Semiconductor materials tend to conduct electricity better as their temperature is raised, releasing more free electrons through temperature induced ionization. Semiconductor materials are of great importance to the electronics industry since they are the raw materials from which devices such as transistors and diodes can be manufactured. Silicon is perhaps the best known of the semiconductors, but there are many others.

2.2 VOLTAGE, CURRENT AND RESISTANCE

In a metal such as copper, the heavy atomic nuclei (each containing 29 protons and 34 neutrons) occupy a fixed position in space and form a stable *lattice* structure. The outer electrons of each atom are very weakly bound to their nuclei, enabling them to easily break free and 'wander' from atom to atom. This wandering motion is normally random, with no overall direction. If, however, a force is applied to the free electrons in the copper, then they can all be made to flow in the same direction. This flow of electrons is called a *current* (symbol I), and the force that causes the current is known as the *electromotive force* (usually abbreviated to *emf*). The emf is supplied by a generator or battery and its strength is measured in units of *volts* (symbol V). In conductors, a large current will be produced by a relatively small emf, whereas in insulating materials a very large emf is required to produce even a very tiny current.

When a length of conductor is connected across the terminals of a source of emf, such as a battery, the chemical reaction within the battery produces an excess of positive ions on one terminal and an excess of negative ions on the other. The negatively charged free electrons in the conductor are attracted to the positive ions at the battery's positive terminal and so drift towards it, eventually entering the battery. The supply of free electrons in the conductor is constantly replenished by the excess of electrons at the negative terminal. This electrical current, caused by the battery's emf, is measured in units of *amperes* (often shortened to amps, symbol A), where a current of 1 A corresponds to 6.25 million million million (6.25×10^{18}) electrons passing a point in the conductor each second. This enormous number of electrons constitutes an amount of charge called a *coulomb* (C) and an ampere is therefore defined as a rate of charge flow of one coulomb per second. Note that current does not 'use up' electrons, they simply move from one battery terminal to the other; electrons cannot be created or destroyed – they can only be moved or redistributed to other parts of the electrical system.

The amount of current that the emf produces in a conductor depends on the electrical *resistance* (R) of the conductor. Resistance can be thought of as opposition to electron flow. It gives rise to a voltage difference (more correctly, a *potential difference*) between two points along a conductor. Resistance arises because drifting electrons will occasionally collide with lattice atoms and give up some of their energy by making the lattice atoms vibrate. This increase in atomic vibration is detected as additional *heat* in the conductor. Since the atoms in a hot conductor have a greater extent of vibration, they tend to be struck by more drifting electrons and so it is observed that the resistance of a conductor rises with its temperature.

Ohm's Law

In 1826 a German maths teacher, Georg Ohm, made a very important observation about the relationship between the current in a conductor and the voltage across it; he discovered that they are *linearly proportional*. This means that for a given constant resistance, if one electrical quantity is increased or decreased by a certain percentage then the other quantity will increase or decrease by the same percentage. For example, if the voltage across a conductor is doubled, the current through it will double. It is therefore possible to define the resistance of a conductor as the ratio V/I, where V is the potential difference across the conductor and I is the current in it. Mathematically,

$$V/I = R \qquad (2.1)$$

[which can also be written $V = I \times R$ or $I = V/R$]

Voltage (volts) divided by current (amps) equals resistance (ohms). This very important mathematical relationship is known as *Ohm's Law* and it is one of the most important and widely used mathematical relationships in electrical engineering. It states that the current in a conductor increases in direct proportion to the applied voltage, with resistance being the constant of proportionality.

The unit of resistance is the *ohm* (symbol Ω). One ohm (1 Ω) is thus the resistance of a conductor through which a current of one ampere flows when a potential difference of one volt is maintained across it.

Ohm's Law indicates that electrical conductors (wires, cables and so on) must be of adequate size to carry the required current without significant resistance to the current. In practice this means that the higher the current and, to some extent the longer the circuit, the thicker the conductors have to be. If a conductor is too thin for the current it carries, it will have too much resistance and the voltage at the exit from the conductor will be much less than at the entrance, in other words there will be a *voltage drop* along the conductor.

Resistors

Copper is a very good conductor and a metre length of copper wire has a resistance of much less than 1 Ω. If there is need to put large amounts of resistance into a circuit a specially made electronic component called a *resistor* is used. Most resistors found in electronic circuits are made of either a rod of powdered carbon or a very thin layer of metal; copper connecting leads are attached to the resistor body and the whole assembly is coated with an insulating material to protect it. Resistors are either manufactured with fixed values, from a fraction of an ohm to about a million ohms, or they can be made to have adjustable values – determined by a rotating shaft or slider. This type of resistor is usually known as a *variable resistor* or *potentiometer*.

The value of a resistor is always indicated on its package. It is often simply printed or embossed on a variable resistor, but fixed resistors usually have their values indicated by a series of coloured bars painted onto the encapsulation. Table 2.1 shows how to interpret the standard colour code that is used.

Electrical measurements

The electrical measurements most often required are those of current, voltage and resistance.

Current is measured with an *ammeter*, which is connected in series with a component to measure the current through the component. It is important that the ammeter does not restrict the flow of current in the circuit and so a good ammeter has a very small resistance.

Voltage is the difference in potential between two points in a circuit and so a *voltmeter* needs to be connected across the ends of the component whose voltage is to be measured. In order that a true voltage can be measured it is very important that the voltmeter draws no current. It must therefore have a very high resistance.

The resistance of a component is measured with an *ohmmeter*. An ohmmeter applies a voltage to a component and measures the ensuing current, displaying the result in ohms. Ohmmeters can only give an accurate result if the component under test is disconnected from the rest of the circuit.

Today, it is usual to find an ammeter, voltmeter and ohmmeter incorporated into one instrument, called a *digital multimeter* (DMM). As its name implies, a DMM has a digital display and incorporates push-buttons or rotary switches to select the particular type of measurement required (volts, amps, ohms, etc.). There are numerous models of DMM available, some specifically designed for automobile use.

One of the most versatile instruments for electrical measurement is the *oscilloscope*. An oscilloscope can display a rapidly changing voltage, in graphical form, on the face of a cathode ray tube (CRT) – rather like a small TV tube. Oscilloscopes are very useful for examining fast changing signals such as the voltages applied to fuel injectors or spark plugs. Very sophisticated oscilloscopes, specially designed for use on automotive electronic systems, are now available to the motor trade. These 'scopes are generally linked to a microcomputer system and form part of a complete engine analysis/diagnosis system. Specialist operator training is required if they are to be used effectively.

2.3 ELECTRICAL CIRCUITS

An electrical circuit is a conducting path around which electrons can flow. The path starts and finishes at the source of the emf (the battery, in the case of a vehicle), but between the battery terminals it may subdivide into branches that involve many components. Normally the circuit path is *closed* (i.e. unbroken) and a current flows. If the circuit path is interrupted, then current ceases and the circuit is said to be *open*. An open circuit may arise intentionally (e.g. through the action of a switch or fuse) or may be the result of

Table 2.1 Fixed resister colour codes (RS components)

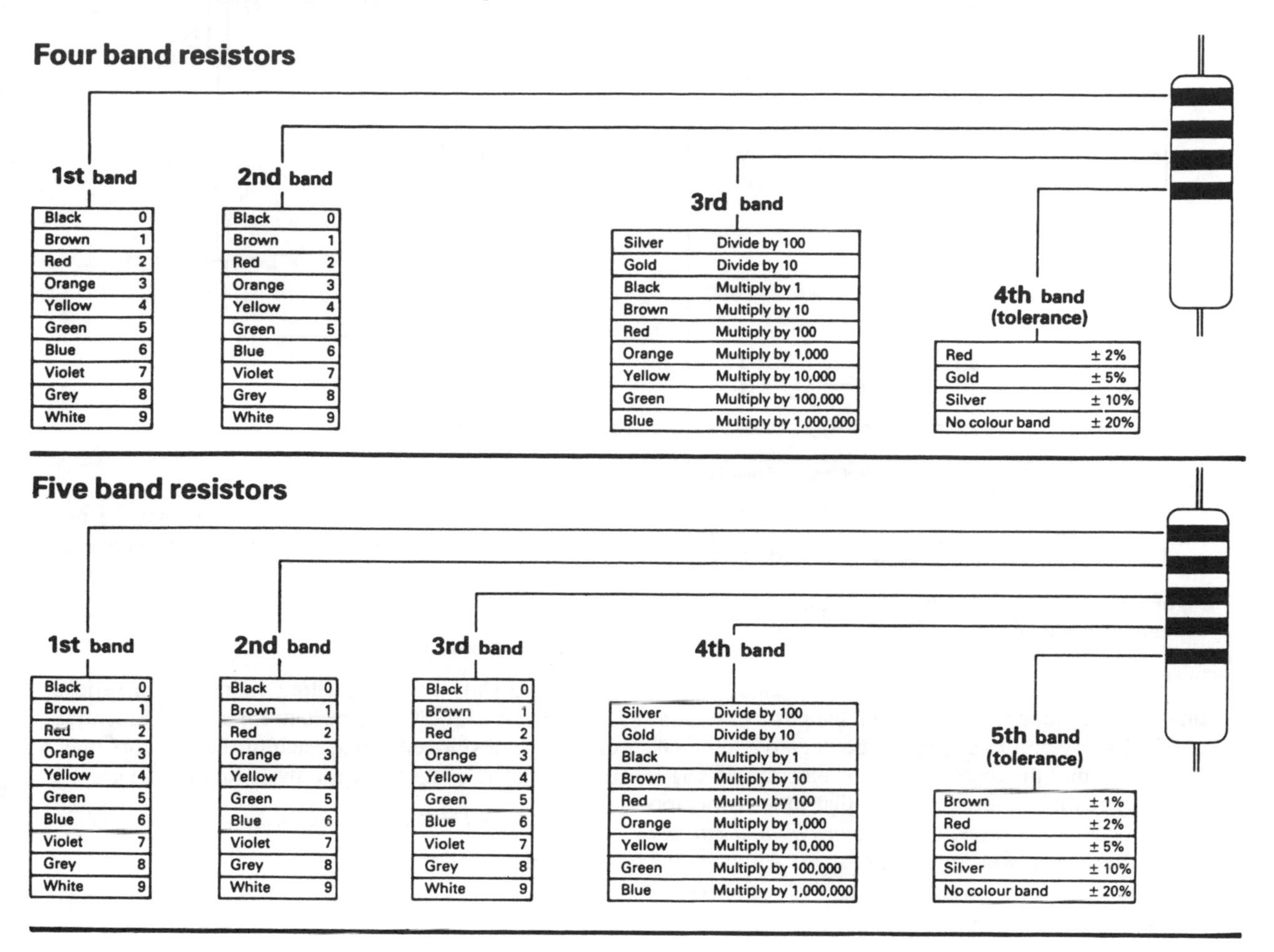

Four band resistors

1st band

Colour	Value
Black	0
Brown	1
Red	2
Orange	3
Yellow	4
Green	5
Blue	6
Violet	7
Grey	8
White	9

2nd band

Colour	Value
Black	0
Brown	1
Red	2
Orange	3
Yellow	4
Green	5
Blue	6
Violet	7
Grey	8
White	9

3rd band

Colour	Multiplier
Silver	Divide by 100
Gold	Divide by 10
Black	Multiply by 1
Brown	Multiply by 10
Red	Multiply by 100
Orange	Multiply by 1,000
Yellow	Multiply by 10,000
Green	Multiply by 100,000
Blue	Multiply by 1,000,000

4th band (tolerance)

Colour	Tolerance
Red	± 2%
Gold	± 5%
Silver	± 10%
No colour band	± 20%

Five band resistors

1st band

Colour	Value
Black	0
Brown	1
Red	2
Orange	3
Yellow	4
Green	5
Blue	6
Violet	7
Grey	8
White	9

2nd band

Colour	Value
Black	0
Brown	1
Red	2
Orange	3
Yellow	4
Green	5
Blue	6
Violet	7
Grey	8
White	9

3rd band

Colour	Value
Black	0
Brown	1
Red	2
Orange	3
Yellow	4
Green	5
Blue	6
Violet	7
Grey	8
White	9

4th band

Colour	Multiplier
Silver	Divide by 100
Gold	Divide by 10
Black	Multiply by 1
Brown	Multiply by 10
Red	Multiply by 100
Orange	Multiply by 1,000
Yellow	Multiply by 10,000
Green	Multiply by 100,000
Blue	Multiply by 1,000,000

5th band (tolerance)

Colour	Tolerance
Brown	± 1%
Red	± 2%
Gold	± 5%
Silver	± 10%
No colour band	± 20%

a poor connection. A further possibility is that the current may find a short-cut around some of the components in the circuit, the current will then be much greater than intended – a situation known as a *short circuit*. Short circuits invariably arise through a fault condition; for example, an unintended connection or the failure of insulation.

Circuit diagrams

In order that electrical engineers can easily understand how a particular circuit operates they draw *circuit diagrams*. A circuit diagram can be thought of as a kind of 'map' of the electrical system, indicating all the possible routes that the current can take, with small pictorial symbols showing all of the different components in the system and how they connect together.

It is an unfortunate fact that there is considerable variation between different motor vehicle manufacturers in their use of electrical symbols. UK manufacturers tend to use the symbols recommended by the British Standards Institution (BS3939); in much of Europe the German DIN standard (Deutsche Industrie Norm) is commonly used, and in the USA and the Far East yet other symbols are used. In all cases, when trying to interpret a vehicle's wiring diagram, it is wise to refer to the symbol key provided by the manufacturer. Some common circuit symbols used in this book are illustrated in Figure 2.2.

When drawing circuit diagrams it is always the convention to assume that the current is from *positive to negative*, i.e. opposite to the direction of electron flow. This rule arose because early electrical engineers had wrongly assumed that current was due to

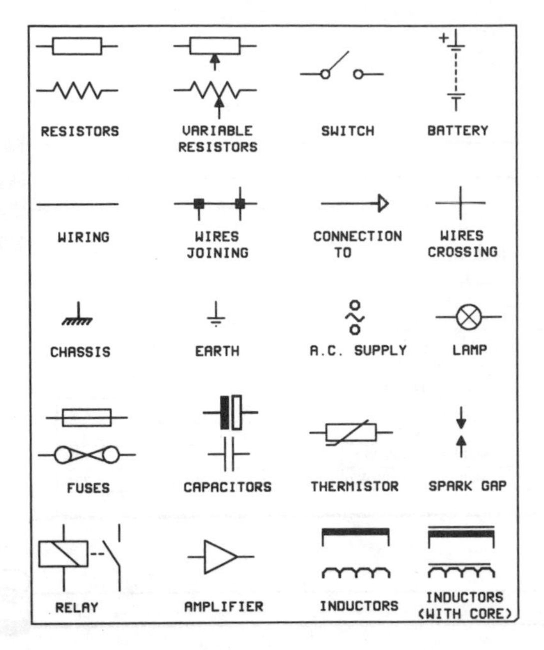

Figure 2.2 Common circuit symbols used in automotive electronics

tiny positive charges flowing from the positive terminal of a battery and returning to the negative terminal. It was not until the discovery of the electron, in 1917, that they realized their error. By that time it was too late to change the rules, and so *conventional current* is still used when working with circuits. Although the use of conventional current is technically incorrect, it rarely causes problems.

When reading a vehicle's circuit diagram remember that on a metal bodied vehicle, the bodyshell is nearly always connected to the battery's negative terminal. This is done in order that it can be used as the return conductor for the circuit, considerably simplifying the wiring and reducing the cost, weight and complexity of the electrical system. Electrical systems on a car are thus invariably connected between the battery positive terminal and the body (the 'earth' or 'ground' connection).

2.4 COMPONENTS IN SERIES AND PARALLEL

All electrical circuits in an automobile consist of a number of components connected between the vehicle's battery terminals to make a circuit.

Components in the circuit can be connected in *series* (end-to-end in a 'string'), so that the current is the

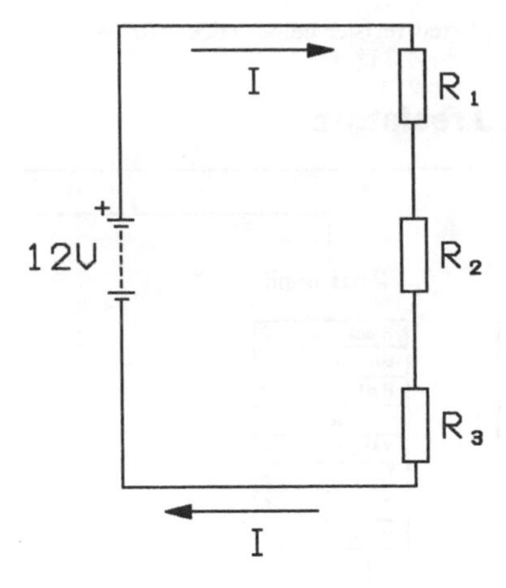

Figure 2.3 Resistors connected in series

same in each, as in Figure 2.3, or they may be arranged in *parallel* so that the current divides between them as in Figure 2.4.

Series connection

Figure 2.3 shows three resistors connected in series, so all must carry the same current I. The potential difference developed across each of the resistors may be calculated using Ohm's Law (equation 2.1),

$$V_1 = I \times R_1,\ V_2 = I \times R_2,\ V_3 = I \times R_3 \tag{2.2}$$

and so the total applied voltage V_T is,

$$V_T = V_1 + V_2 + V_3 \tag{2.3}$$

$$= (I \times R_1) + (I \times R_2) + (I \times R_3) \tag{2.4}$$

$$= I \times (R_1 + R_2 + R_3) \tag{2.5}$$

and the total resistance, R_T, of the series circuit is thus,

$$R_T = V_T/I = R_1 + R_2 + R_3 \tag{2.6}$$

In summary, for a series circuit:

(i) The current is the same through all resistors.
(ii) The total potential difference is equal to the sum of all the individual potential differences.
(iii) The individual potential differences are directly proportional to the individual resistances.
(iv) The total resistance is larger than the largest individual resistance.
(v) The total resistance is the sum of the individual resistances.

Example: In the circuit of Figure 2.3, $R_1 = 10\ \Omega$, $R_2 = 8\ \Omega$ and $R_3 = 6\ \Omega$. Calculate the current, I, in the

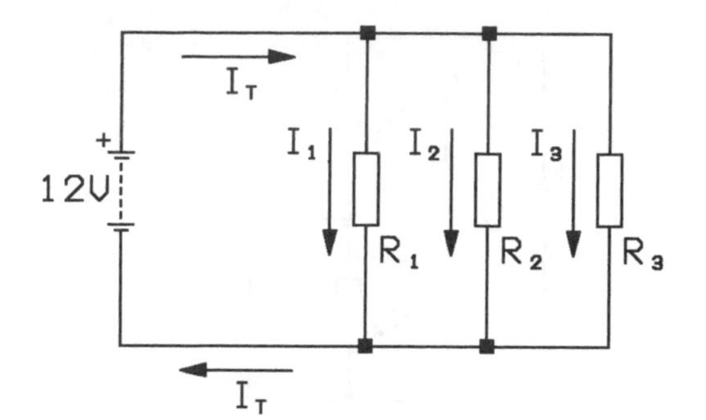

Figure 2.4 Resistors connected in parallel

circuit and the potential differences V_1, V_2 and V_3 across each resistor.

Solution: Remember, the total resistance of the series circuit is the sum of all the individual resistances in the circuit. So the total resistance, R_T, is,

$$R_T = 10\ \Omega + 8\ \Omega + 6\ \Omega$$
$$= \underline{24\ \Omega}$$

Using Ohm's Law the current, I, in the circuit can be calculated,

$$I = V/R_T \qquad \text{(Ohm's Law)}$$
$$\text{so} \quad I = 12\ \text{V}/24\ \Omega$$
$$= \underline{0.5\ \text{A}}$$

It is now possible to calculate the potential difference across each resistor, again using Ohm's Law,

$$V_1 = 0.5\ \text{A} \times 10\ \Omega$$
$$\text{so} \quad V_1 = \underline{5\ \text{V}}$$

$$V_2 = 0.5\ \text{A} \times 8\ \Omega$$
$$\text{so} \quad V_2 = \underline{4\ \text{V}}$$

$$V_3 = 0.5\ \text{A} \times 6\ \Omega$$
$$\text{so} \quad V_3 = \underline{3\ \text{V}}$$

Parallel connection

Figure 2.4 shows three resistors connected in parallel. All of the resistors are subject to the same applied voltage, V_T, but the current, I_T, divides three ways into I_1, I_2 and I_3, through the three resistor branches,

$$I_T = I_1 + I_2 + I_3 \qquad (2.7)$$

and using Ohm's Law,

$$I_1 = V_T/R_1,\ I_2 = V_T/R_2,\ I_3 = V_T/R_3 \qquad (2.8)$$

and so the current I_T can be calculated as,

$$I_T = V_T \times (1/R_1 + 1/R_2 + 1/R_3) \qquad (2.9)$$

therefore,

$$I_T/V_T = 1/R_{EQ} = 1/R_1 + 1/R_2 + 1/R_3 \qquad (2.10)$$

where R_{EQ} is the *equivalent resistance* (V_T/I_T) of the three resistors in parallel. Equation 2.10 is known as the *reciprocal resistance formula* and is used to work out the equivalent resistance of a parallel circuit.

In summary, for a parallel circuit:

(i) The potential difference is the same across each resistor.
(ii) The total current in the circuit is the sum of the individual branch currents.
(iii) The individual branch currents are inversely proportional to the individual resistances.
(iv) The equivalent resistance of the circuit is smaller than the smallest individual resistance.

Example: In the parallel circuit shown in Figure 2.4, $R_1 = 12\ \Omega$, $R_2 = 6\ \Omega$ and $R_3 = 4\ \Omega$. Calculate the total current, I_T, flowing from the battery and the branch currents I_1, I_2 and I_3 in each resistor.

Solution: In the parallel circuit the single-resistor equivalent value, R_{EQ}, of the parallel combination of R_1, R_2 and R_3 is obtained by using the reciprocal resistance formula of equation 2.10,

$$1/R_{EQ} = 1/R_1 + 1/R_2 + 1/R_3$$
$$\text{so} \quad 1/R_{EQ} = 1/12 + 1/6 + 1/4$$
$$= 1/12 + 2/12 + 3/12$$
$$= 6/12$$
$$\text{so} \quad R_{EQ} = 12/6\ \Omega = \underline{2\ \Omega}$$

The total current, I_T, can now be found using Ohm's Law,

$$I_T = V/R_{EQ}$$
$$\text{so} \quad I_T = 12/2$$
$$= \underline{6\ \text{A}}$$

The branch currents can easily be found by applying Ohm's Law to each branch in turn,

$$\text{For branch 1, } I_1 = V/R_1 = 12/12 = \underline{1\ \text{A}}$$

$$\text{For branch 2, } I_2 = V/R_2 = 12/6 = \underline{2\ \text{A}}$$

$$\text{For branch 3, } I_3 = V/R_3 = 12/4 = \underline{3\ \text{A}}$$

As a check on the calculation, note that the sum of I_1, I_2 and I_3 is equal to I_T.

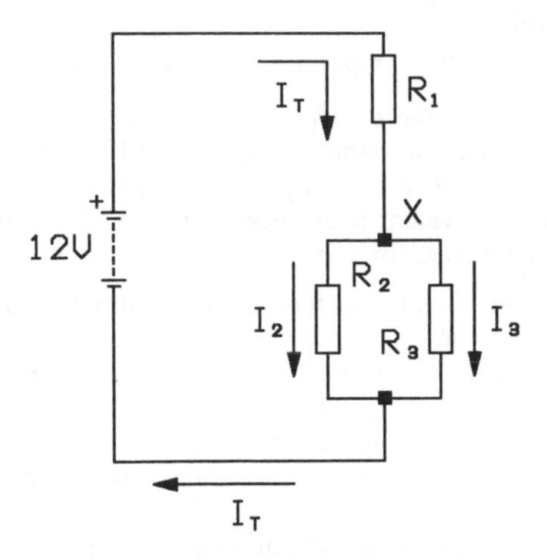

Figure 2.5 Series-parallel connected resistors

Series–parallel circuit

A series–parallel circuit is a circuit in which there is a combination of both series and parallel current paths. Figure 2.5 shows a simple series–parallel resistor circuit. All of the circuit current (I_T) passes through the resistor R_1, but divides at the point X where it flows through the resistors R_2 and R_3 in inverse proportion to their values. In order to perform calculations on series–parallel circuits it is first necessary to replace the parallel components with their single-resistor equivalent value, R_{EQ}, and then proceed as for a series circuit.

Example: In the circuit shown in Figure 2.5, $R_1 = 8\ \Omega$, $R_2 = 6\ \Omega$ and $R_3 = 12\ \Omega$. Calculate the total current drawn from the battery and also the potential difference across the resistor R_1.

Solution: Using equation 2.10, first calculate the single-resistor equivalent value of the parallel combination R_2,R_3,

$$\begin{aligned} 1/R_{EQ} &= 1/R_2 + 1/R_3 \\ &= 1/6 + 1/12 \\ &= 2/12 + 1/12 \end{aligned}$$

$$\begin{aligned} \text{so, } 1/R_{EQ} &= 3/12 \\ \text{and so } R_{EQ} &= 12/3 = \underline{4\ \Omega} \end{aligned}$$

The circuit will behave like a series circuit that has an 8 Ω resistor in series with a 4 Ω resistor, and so Ohm's Law can be used to calculate the current, I_T, in the circuit,

$$\begin{aligned} \text{where } R_T &= R_1 + R_{EQ} \\ &= 8\ \Omega + 4\ \Omega \\ &= \underline{12\ \Omega} \end{aligned}$$

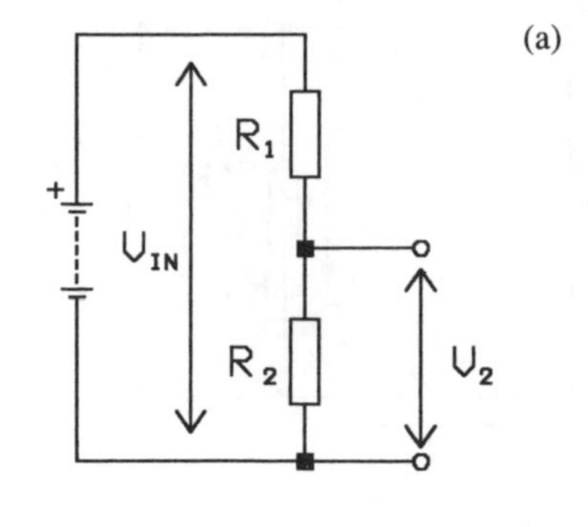

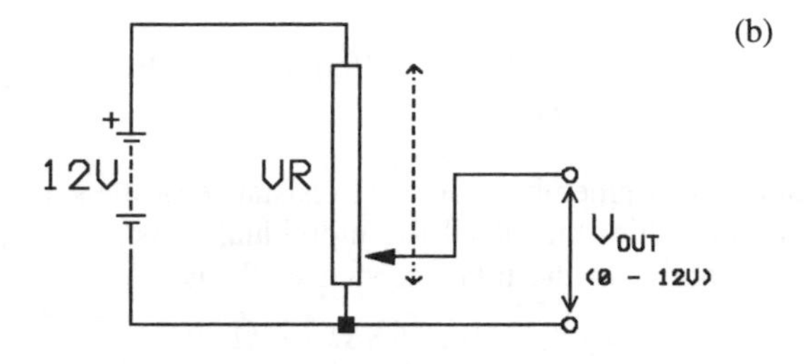

Figure 2.6 The potential divider: (a) simple two-resistor divider; (b) the potentiometer

$$\begin{aligned} \text{so} \quad I_T &= V/R_T \\ &= 12/12 \\ &= \underline{1\ \text{A}} \end{aligned}$$

The potential difference, V_1, across resistor R_1 is also calculated by using Ohm's Law,

$$\begin{aligned} V_1 &= I_T \times R_1 \\ &= 1 \times 8 \\ &= \underline{8\ \text{V}} \end{aligned}$$

This means that the remainder of the battery voltage must be dropped across the parallel resistors R_2 and R_3, and so the voltage at point X, measured with respect to the battery negative terminal, will be,

$$12\ \text{V} - 8\ \text{V} = \underline{4\ \text{V}}$$

The potential divider

The *potential divider* circuit (sometimes called a voltage divider) is very important in electronics; most electronic products have at least one potential divider within them. Figure 2.6(a) shows the arrangement of the potential divider; it is simply two resistors connected in series. The current in the circuit, I, is calculated using Ohm's Law,

$$I = V_{IN}/(R_1 + R_2) \qquad (2.11)$$

The voltage, V_2, dropped across R_2 is thus given by,

$$V_2 = I \times R_2 = [V_{IN}/(R_1 + R_2)] \times R_2. \qquad (2.12)$$

Thus the potential divider shown in Figure 2.6(a) gives a fixed fraction, V_2, of the input voltage V_{IN}.

Instead of using fixed resistors, a long resistive track with a sliding contact, called a *wiper*, can be used to give a continuously variable potential difference, as shown in Figure 2.6(b). These components are known as potentiometers (sometimes shortened to 'pots') and find very wide application in electronics. The resistive track may be either straight, to give linear adjustment, or curved, to give a rotary adjustment of wiper position over an angle of some 100–200°C. In automobile electronics, potentiometers are widely used to indicate the angular position of a shaft. As an example, Figure 2.7(a) shows the construction of a rotary potentiometer used to sense the position of a throttle shaft. The potentiometer is connected in the manner shown in Figure 2.7(b); when the throttle is closed, and the engine at idle, the wiper is at a position close to the 0 V terminal and so the output voltage sent to the engine management electronic control unit (ECU) is very small (approximately 0.5 V). As the throttle is opened the shaft rotates, moving the wiper along the resistive track toward the supply voltage terminal of the potentiometer and so the output voltage sent to the engine management system rises. Finally, at wide-open throttle (WOT), the output voltage is nearly the same as the supply voltage, in this case about 4.3 V. The output voltage from the throttle potentiometer is thus an indication of throttle opening, and hence the driver's power demand on the engine, as shown in Figure 2.7(c).

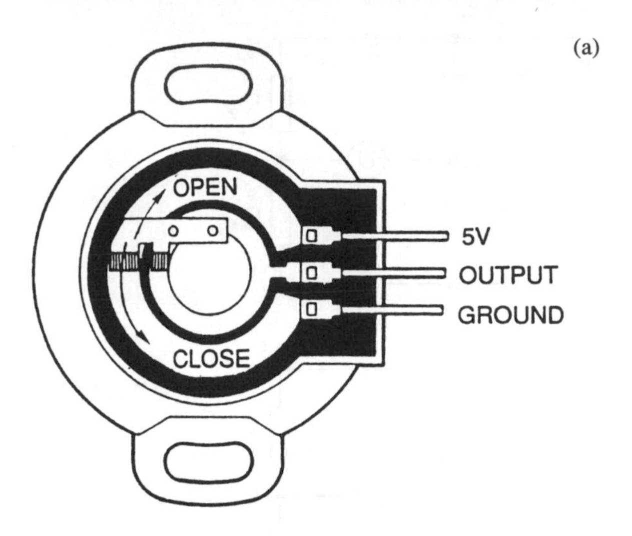

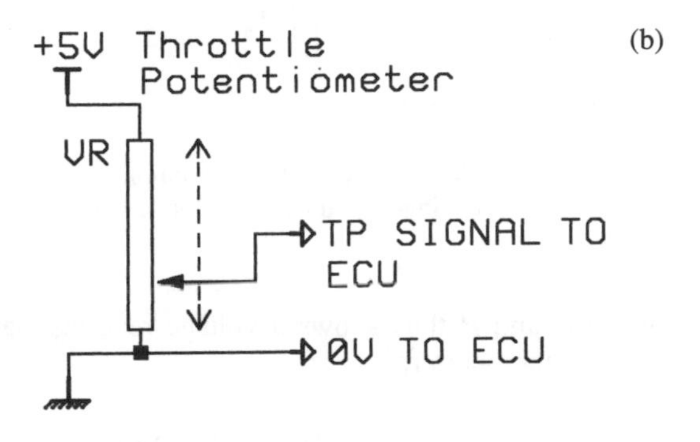

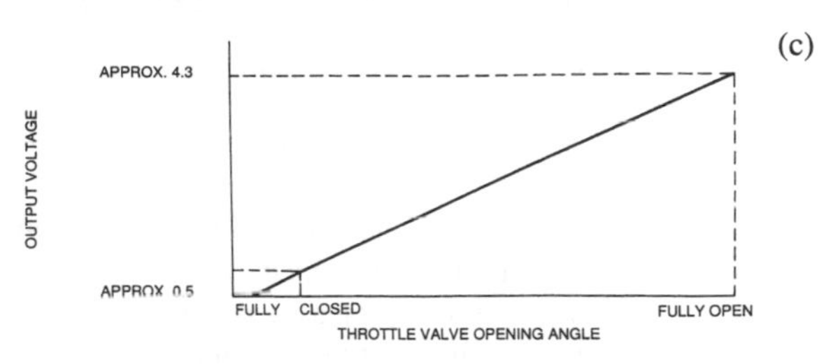

Figure 2.7 Sensing throttle position using a potentiometer: (a) construction of the throttle potentiometer (*Mazda*); (b) circuit connection; (c) output signal according to throttle valve opening

The Wheatstone bridge

Circuits using two potential dividers, known as *bridge circuits*, are frequently used in automotive measurement systems. The most common type of bridge circuit is the *Wheatstone bridge*, illustrated in Figure 2.8(a). A constant input voltage, V_{IN}, is applied to the two potential divider pairs, composed of resistors R_1,R_3 and R_2,R_4. If the ratios of R_1:R_3 and R_2.R_4 are the same, then the voltages V_A and V_B, at terminals A and B respectively, will be the same. A voltmeter connected between the two terminals records no potential difference and the bridge is said to be *balanced*. If, however, one of the resistor values is changed slightly, then the two potential dividers no longer provide exactly the same output voltage. Consequently the voltmeter shows a small potential difference that is in proportion to the *imbalance* of the bridge.

One of the most common applications of the bridge circuit in automobile electronics is in temperature sensing. Figure 2.8(b) shows a bridge circuit in which the resistor R_2 has been swapped for a thermistor (a component whose resistance is highly temperature dependent). As the temperature of the thermistor rises its resistance falls and so the potential difference between point B and battery negative rises. The resistor ratio R_1:R_3 does not vary, however, and so the voltage at point A remains constant. A voltmeter connected

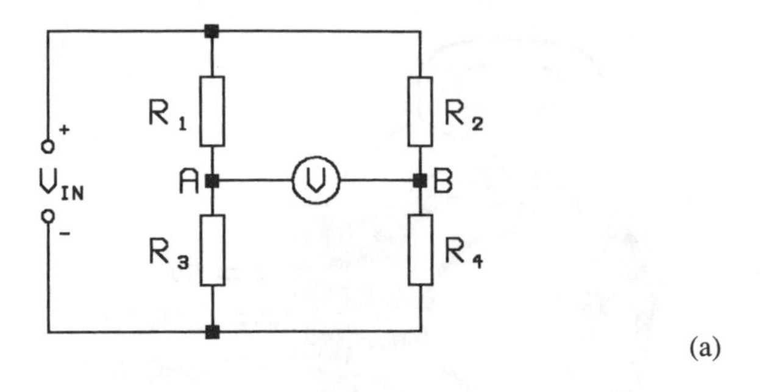

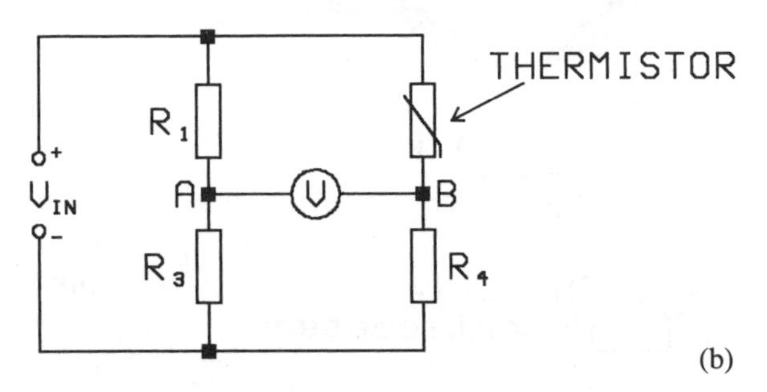

Figure 2.8 The Wheatstone bridge: (a) simple bridge circuit (b) using a bridge circuit to monitor temperature

between A and B thus shows a voltage reading that rises with temperature.

2.5 ELECTRICAL ENERGY AND POWER

In an electrical circuit, electrons gain electrical energy from a source of electromotive force (a battery or generator) and dissipate it into other forms (heat, light, mechanical energy) in various circuit components. Since electrons cannot be created or destroyed no electricity is actually 'consumed' in the components, the electrons simply act as a medium for the transfer of energy from one place to another. When considering the automobile electrical system it is important to know the electrical *power* needs of the components. Power is an indication of the rate of conversion of energy from one form to another and since the converted *energy* (symbol W) is measured in *joules* (symbol J) and time is measured in seconds, it follows that power is measured in joules per second (Js^{-1}) or *watts* (symbol W) where,

$$1 \text{ watt (W)} = 1 \text{ joule per second } (Js^{-1})$$

The electrical power (P) delivered to, or consumed by, an electrical component is calculated in terms of the current in the component and the voltage across it, by using the equation,

$$P = I \times V \qquad (2.13)$$

Watts equals amps multiplied by volts.

As an example of a power calculation, consider a 12 V headlight bulb in which there is a 4.5 A current. The power of the bulb is calculated as,

$$P = 12\text{ V} \times 4.5\text{ A}$$
$$= \underline{54\text{ W}}$$

In other words, the bulb filament is changing 54 joules of electrical energy into heat and light energy each second. The vehicle's battery is supplying the electrical energy and so must convert 54 joules of chemical energy into electrical energy each second.

Electrical heating

When current flows in a resistance *all* of the dissipated electrical energy is converted into heat energy, and since Ohm's Law applies (i.e. since $V = I \times R$) equation 2.13 can be altered to give,

$$P = I \times V = I \times I \times R = I^2R \qquad (2.14)$$

Thus, if the resistance and current in a circuit is known, it is possible to calculate the rate of production of heat in the components. Note that because of the 'squared' relationship between power and current, a doubling of current will produce a four-fold rise in heat production. This is why large currents produce significant heating, even in conductors of very low resistance. The 'square-law' of heat production is used to good effect in vehicle fuses.

A fuse is an electrical safety device consisting of a short length of resistance wire which is designed to heat-up and melt when the current through it exceeds a pre-determined value. In general, the melting temperature is obtained when the current in the fuse is about twice the maximum safe continuous current, i.e. when the rate of heat production in the fuse wire is about four times the safe continuous maximum.

2.6 DIRECT CURRENT AND ALTERNATING CURRENT

Direct current (dc) is current in which the electrons flow steadily in only one direction within the circuit; all of the circuits looked at so far have been dc circuits. Batteries produce only dc, and in order to reverse the direction of the current in a battery-powered dc circuit it would be necessary to physically reverse the battery connections.

In an *alternating current* (ac) circuit, the electrons first flow in one direction for a short time and then reverse and flow in the opposite direction, i.e. the direction of electron flow alternates on a regular basis.

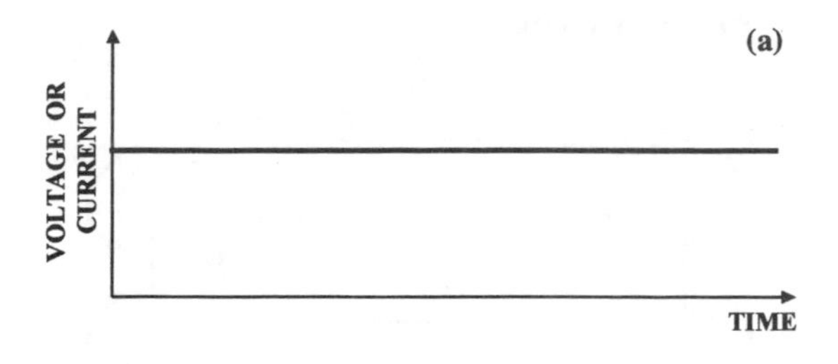

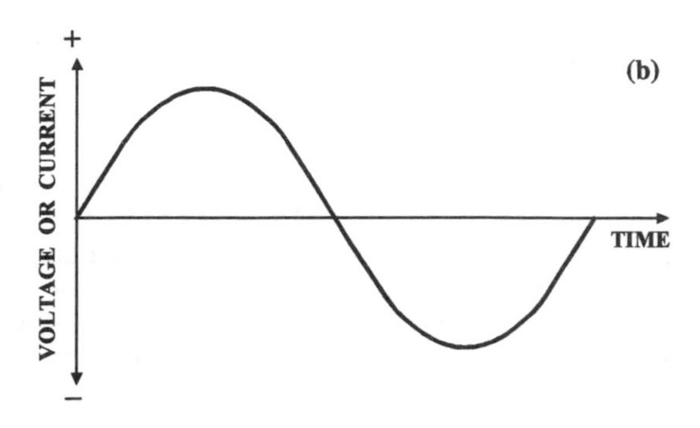

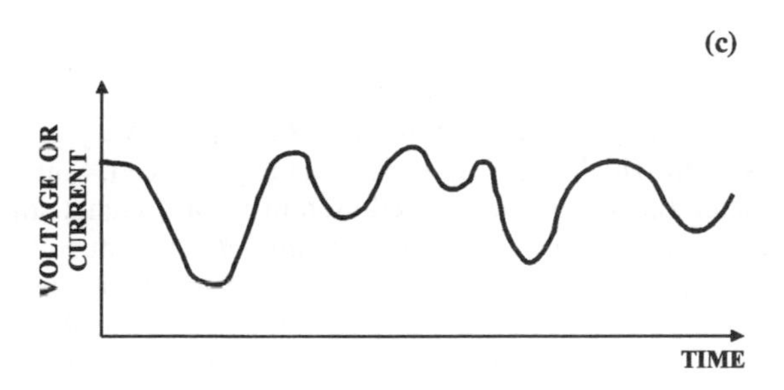

Figure 2.9 Types of waveforms: (a) dc signal; (b) ac signal; (c) varying dc signal

An understanding of ac theory is important when studying automobile electronics since although car electrical systems usually operate from a 12 V dc supply, many sensors and actuators operate with ac, or varying dc, signals (a varying dc signal is simply a steady dc signal to which an ac signal has been added).

Frequency and period of an ac waveform

The manner in which a current or voltage varies with time is illustrated by drawing a graphical plot called a *waveform*. Figure 2.9 shows waveforms for different types of direct and alternating currents.

The ac waveform drawn in Figure 2.9(b) is that for the mathematically simplest kind of current alternation and is called a sinusoidal (or sine) wave. The current starts at zero, rises to a positive peak value, declines again through zero, reverses to a negative peak value and then rises back to zero. This pattern of current alternation (called a *cycle*) may be repeated over and over again to give a continuous waveform. The number of cycles completed each second is then known as the *frequency* of the waveform.

The frequency (symbol f) of the ac is measured in units of *hertz* (Hz), where one hertz is one cycle per second ($1\ \text{Hz} = 1\ \text{cs}^{-1}$). The frequency of ac may vary enormously depending upon the application. For example, mains electricity is normally supplied as 50 Hz ac in Europe, FM radio signals, on the other hand, are produced by ac with a frequency of around 100 million hertz (100 MHz) and satellite television signals are transmitted at a frequency of about 10 billion hertz (10 GHz). The length of time taken for each individual cycle of the ac waveform to be completed is known as the *period* (T) of the waveform. Period (T) and frequency (f) have a simple mathematical relationship,

$$T = 1/f \qquad (2.15)$$

$$\text{and so } f = 1/T \qquad (2.16)$$

Knowing either f or T allows the other to be calculated. A simple calculation reveals that a cycle of 50 Hz ac mains has a period of 0.02 seconds (20 ms), whereas a cycle of 100 MHz FM radio signal has a period of only 10ns (10×10^{-9}s).

Voltage and current values

The voltage and current magnitudes of an ac waveform can be expressed in a seemingly bewildering variety of ways. The most important are the *instantaneous*, *peak*, *peak-to-peak* and *rms* values of the waveform, as illustrated in Figure 2.10.

The *instantaneous* value of the waveform is simply the voltage or current value at a particular instant of time. Instantaneous values are positive on the positive part of the waveform and negative on the negative part.

The *peak* value of the waveform (often called the *amplitude* of the waveform) is the value of the voltage or current, measured from zero to either the positive or negative peak. Peak values are usually denoted as V_p or I_p.

The *peak-to-peak* value of the waveform is the voltage or current, measured from the positive peak to the negative peak. It is the most frequently quoted parameter, ac mains is 240 V peak-to-peak, for example. For a sine wave, the peak-to-peak value of voltage or current is always twice the peak value and is denoted V_{pp} or I_{pp}.

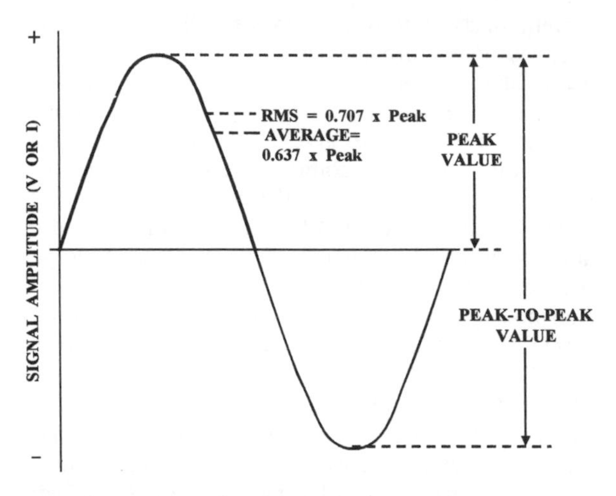

Figure 2.10 Instantaneous, peak, peak-to-peak, rms and average values of an ac sine wave

The *rms* (root mean square) value of a waveform is arrived at by mathematical derivation. The rms value for a sine wave is a measure of the heating effect of the ac in a resistor and is defined as that value of voltage (or current) which is equal to the dc voltage (or current) that will produce the same amount of heating. Peak values for a sine wave can easily be converted into rms values by using the mathematical relations,

$$V_{rms} = 0.707 \times V_p \quad (2.17)$$

$$I_{rms} = 0.707 \times I_p \quad (2.18)$$

Ohm's Law for ac circuits

Provided that a circuit contains only resistive components, Ohm's Law may be used in ac circuits just as for dc circuits, but the manner in which the voltage and current is expressed must be consistent. For example, if a peak value of voltage is used in a calculation then a peak value of current will be the result. In calculations of ac power, rms values must always be used.

2.7 MAGNETISM AND ELECTROMAGNETISM

Although natural magnets were known thousands of years ago and Chinese sailors were using magnetic compasses in the eleventh century, it was not until the early nineteenth century that a relationship between electricity and magnetism was discovered. An appreciation of this interaction is very important when studying automobile electronic systems since many components on a vehicle (solenoids, relays, ignition coils, sensors) rely on electromagnetic effects. Furthermore, the interaction of vehicle electrical components with each other and with the external electromagnetic environment (*electromagnetic compatibility* or EMC) is itself an area of considerable significance for the motor industry.

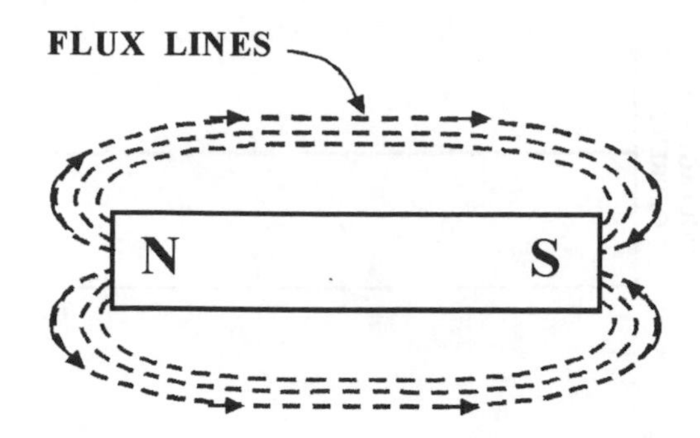

Figure 2.11 Flux lines around a bar magnet

Magnetic fields

A magnet produces a region around itself, a *magnetic field*, in which a magnetic force is experienced. Some materials, once magnetized, retain their magnetism and so are known as *permanent magnets*. A frequently used alloy for permanent magnets is alnico which has a composition of 54% iron, 18% nickel, 12% cobalt, 6% copper and 10% aluminium. Alnico retains its magnetism extremely well and so is often used to make the magnets found in small motors.

The appearance of the field around a magnet can easily be seen by covering the magnet with a sheet of paper and scattering iron filings onto it; Figure 2.11 shows the sort of pattern that would be obtained from a common bar magnet. Lines of *flux* emerge from the north pole and re-enter the magnet at the south pole; the lines show the direction of the magnetic force and the density of lines shows the concentration of the magnetism around the magnet. Since the direction of the magnetic force is opposite at the north and south poles it is observed that when two magnets are brought close together the like magnetic poles repel and the unlike magnetic poles attract.

Magnetic effects of current

The first observations of a connection between electricity and magnetism were made in 1820 by the Danish scientist Hans Oersted. He discovered that a

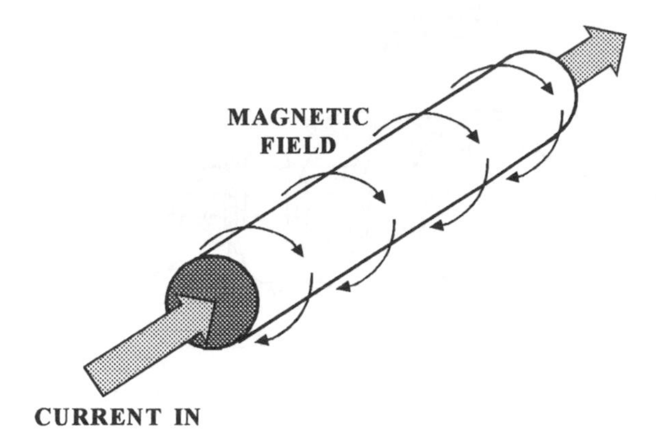

Figure 2.12 Magnet field due to a current through a wire

wire carrying a current, if moved close to a compass needle, would make the needle deflect. The reason for Oersted's observation can be seen by scattering iron filings on a piece of card that has a wire passing through it. If the wire carries a sufficiently large current, a magnetic field pattern of concentric circles will be seen. Looking along the wire in the direction of conventional current, the field lines circle the wire in a clockwise direction, as in Figure 2.12 (this is known as *Maxwell's corkscrew rule* – imagine driving a corkscrew in the direction of the current, the corkscrew would rotate clockwise).

The solenoid

If a length of wire is wound into a long cylindrical *coil*, the lines of magnetic force associated with each turn of wire add together to produce a strong magnetic field within the coil, creating north and south poles at each end, as illustrated in Figure 2.13(a). A coil designed to produce a strong magnetic field is called a *solenoid* and it has a field pattern very similar to that of a bar magnet. Adding a *core* material inside the coil, as shown in Figure 2.13(b), can increase the density of the magnetic flux and so make a stronger magnet. Some core materials, such as glass, wood and plastic have a negligible effect on the flux density, other materials, such as iron, steel or cobalt greatly increase the flux density. By increasing the field strength with a core an *electromagnet* is created; a magnet that has all the properties of a permanent magnet, but whose magnetism quickly disappears when the current is switched off.

The flux density associated with an electromagnet is proportional (within limits) to the number of turns of wire that there are in a specific length of the coil and the current in the coil. Many turns of thin wire can make a compact and powerful electromagnet, however care must be taken to ensure that the resistance of the wire does not impose an inadequate current limit on the device.

2.8 APPLICATIONS OF ELECTROMAGNETISM

There are a large number of automotive applications for electromagnetism; the most common ones are the *relay* and the *plunger solenoid.*

The relay

Relays are a relatively cheap way of enabling a small control current to switch a much larger current, such as a motor supply, in a separate external circuit. Figure 2.13(c) illustrates a typical relay design.

The main body of the relay consists of an iron-cored electromagnet, wound with many turns of thin wire and mounted on an iron frame. The switching action is accomplished by a pair of heavy-duty contacts, one of which is fixed and the other attached to the moving armature. With no operating current applied, a strong spring keeps the contacts apart and so there is no current in the external circuit. When the operating current (typically 50–100 mA) is applied to the electromagnet, the armature is pulled down and the contacts close, thus completing the heavy current circuit.

In electronic terms relays are comparatively slow-acting and unreliable devices and so are only suitable

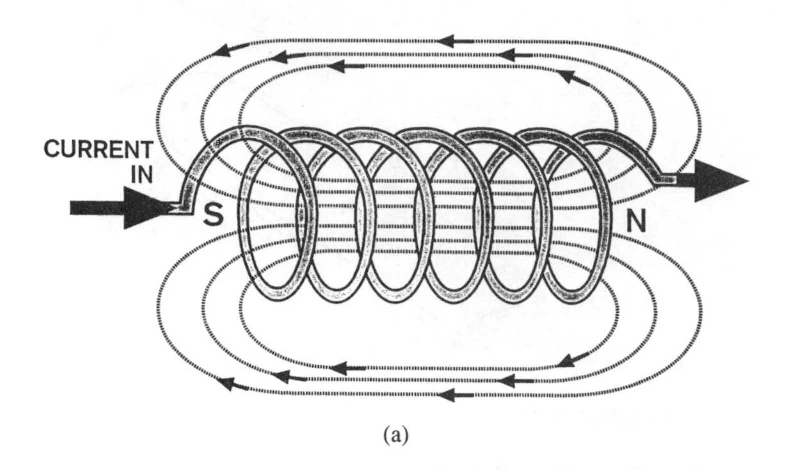
CURRENT
IN
S
N

(a)

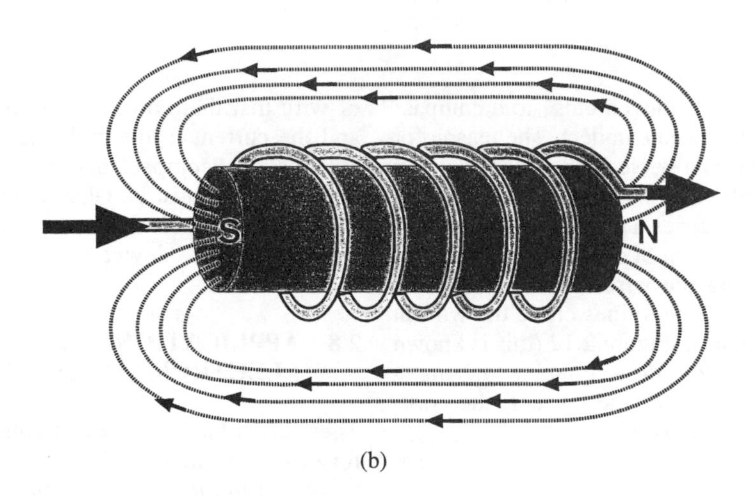
S
N

(b)

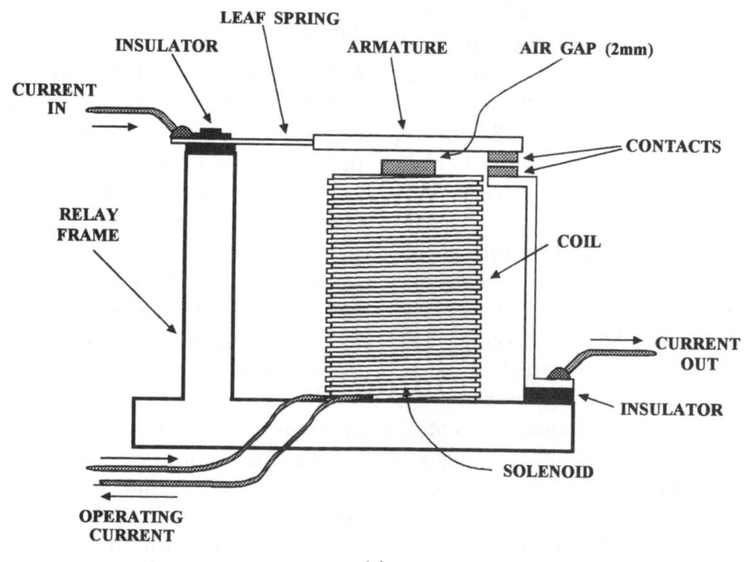
LEAF SPRING
INSULATOR
ARMATURE
AIR GAP (2mm)
CURRENT
IN
CONTACTS
RELAY
FRAME
COIL
CURRENT
OUT
INSULATOR
SOLENOID
OPERATING
CURRENT

(c)

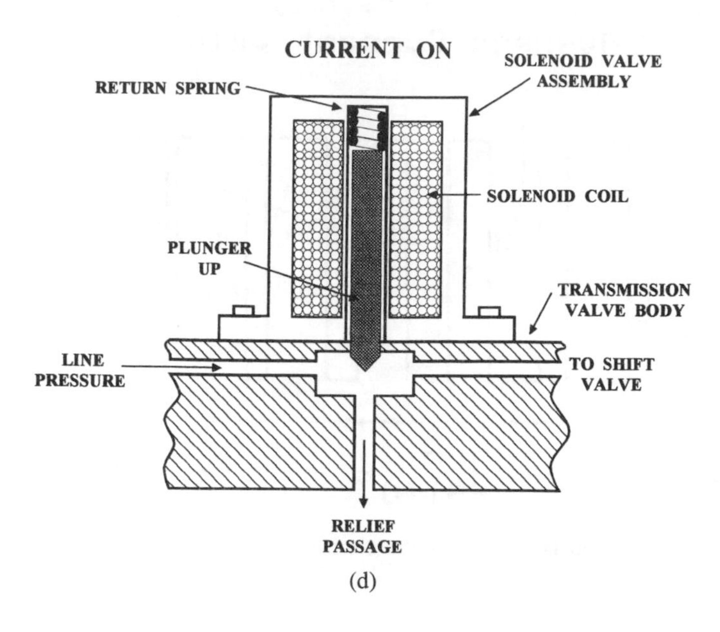

(d)

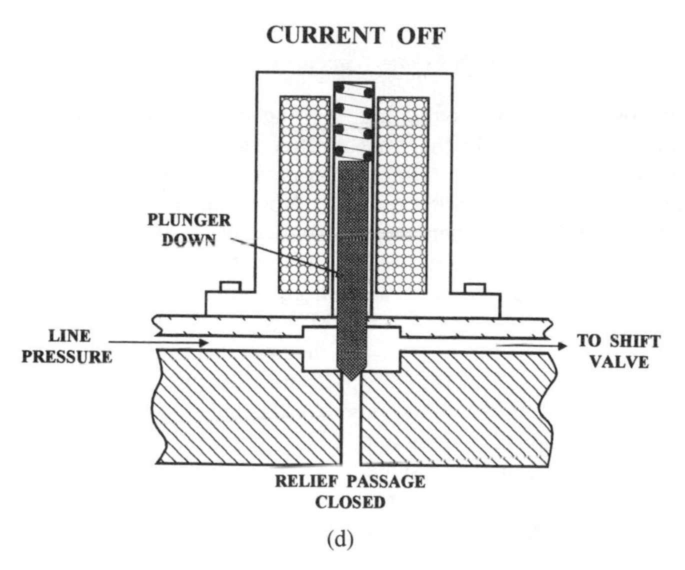

(d)

Figure 2.13 Applications of electromagnetism: (a) simple solenoid; (b) increasing the field with a core; (c) construction of a simple relay; (d) operation of a plunger solenoid valve used in automatic transmissions

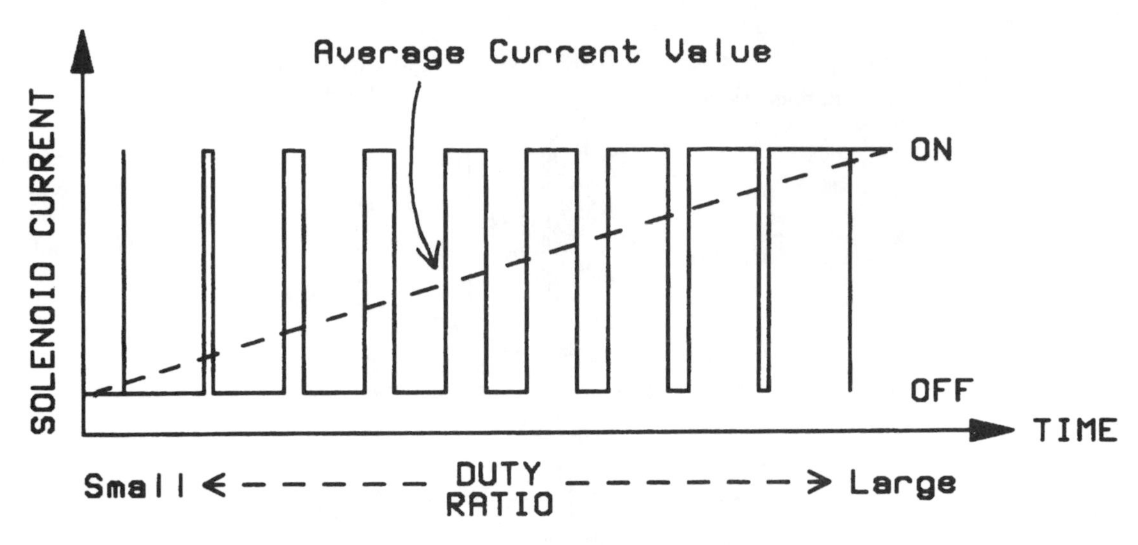

Figure 2.14 Use of pulse width modulation to vary the average current through a solenoid

for use in undemanding control circuits (switching for lights, motors, heaters, etc.). Typically the time taken for the contacts to close (the *operation time*) is about 20 ms and the time needed for the contacts to open (the *release time*) is about 10 ms. Where very fast switching or very high reliability is required, *power semiconductor* devices are used.

The plunger solenoid

Although plunger solenoids come in all shapes and sizes, the internal design is usually much the same. They are used in a wide variety of automotive applications ranging from the engagement of the pinion gear on pre-engaged starter motors through to remote boot release actuators. A common application is in the control of fluid flow in fuel injection systems, automatic transmissions and anti-lock braking systems.

To illustrate the operation of the solenoid a simple automatic transmission fluid pressure control solenoid is shown in Figure 2.13(d). The solenoid comprises a coil of fine copper wire wound around a thin-walled cylindrical former. The fluid control needle (the plunger) is a smooth sliding fit in the cylinder and, with no current applied to the coil, is held firmly against the pressure relief port by the action of the return spring. When current is applied to the solenoid, the plunger is pulled into the cylinder against spring pressure and fluid is allowed to exit into the relief passage, reducing the fluid pressure within the system.

Note that in an application such as this, *variable control* of fluid pressure is possible by continually *pulsing* the solenoid ON and OFF in rapid alternation; a technique known as *pulse width modulation* (PWM) or *duty cycle control*. Figure 2.14 illustrates the principle of PWM. A high-speed electronic switch alternately turns the solenoid current ON and OFF. As the ratio of the ON time (T_{on}) to the OFF time (T_{off}) is varied, the effective current supplied to the solenoid is also varied. Precise control of the solenoid position (to within a few per cent of its total travel) is possible. Such PWM control is common in fluid pressure and engine idle air control (IAC) valve applications.

Motor action

The motion resulting from the forces of magnetic fields is called *motor action*. An elementary example is the repulsion between the north poles of two bar magnets. More obscurely, motor action can occur between the magnetic field associated with the current in a wire and a separate, external, magnetic field. To obtain maximum force the conductor must lie in a direction *perpendicular* to the magnetic field, as shown in Figure 2.15.

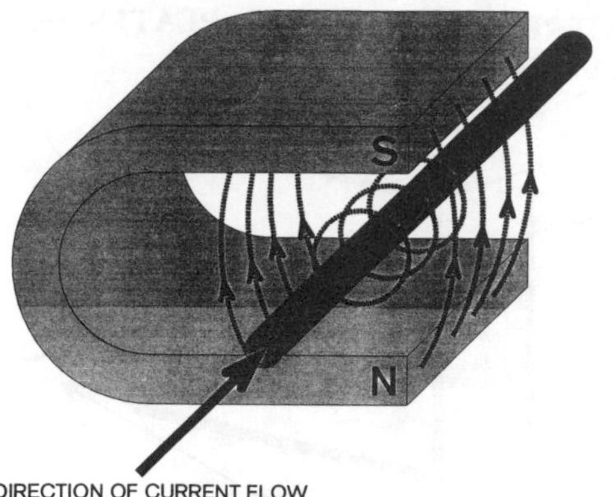

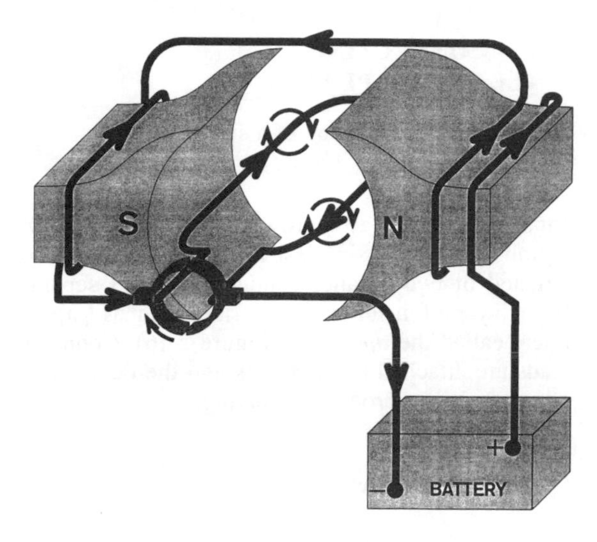

Figure 2.15 Principle of motor action

The force on the wire is proportional to,

(i) the length of wire, L, in the magnetic field.
(ii) the current, I, carried by the wire.
(iii) the strength, B, of the magnetic field.

$$\text{i.e. } Force = B \times I \times L \quad (2.19)$$

If a loop of wire is placed in the magnetic field, as in Figure 2.15, opposite sides of the loop will have opposite current directions; thus their associated magnetic fields will be opposite and the forces on each side of the loop will be in opposite directions. One side of the loop will be forced upward and the other downward. The result is a rotational force on the loop called a *torque*. The production of a torque is the basis of all electric motors and it will be illustrated further in the next chapter.

Electromagnetic induction

After Oersted had discovered the magnetic effects associated with electrical current, the British scientist, Michael Faraday, began to look for the converse; producing a current from a magnetic field. He finally succeeded in 1831. Faraday found that by *moving* a magnet through a coil of wire a voltage could be measured across the ends of the wire, Faraday called this an *induced emf.* If the ends of the coil were connected, to make a closed circuit, Faraday found that the induced emf would cause an *induced current* in the circuit. This important discovery subsequently enabled the construction of electrical machines such as power-station generators and automobile alternators.

The generation of induced current

Faraday soon realized that induced current was only produced by a change in the magnetic field producing it. He found that it only matters that there is *relative movement* between the coil and the magnet; it does not matter which item *actually* moves.

The induced current can be thought of as the result of motor action between the external magnetic field and the magnetic field of the free electrons in the wire. When there is no external magnetic field the free electrons move at random, with no particular direction, and they have no net field. When an external field is moved past the wire, motor action between the two sets of magnetic fields causes the free electrons to be 'pushed' along the wire and a current is observed.

For a length of wire moved through a magnetic field it is found that the induced emf increases as:

(i) the speed of movement of the wire, v, is increased.
(ii) the length of wire, L, in the field is increased.
(iii) the strength of the field, B, is increased.

This can be expressed mathematically by an equation very similar to equation 2.19, to give the induced emf, E_{ind},

$$E_{ind} = B \times L \times v \quad (2.20)$$

In order to make an electrical generator, a loop of wire can be placed in a magnetic field and rotated. The induced emf is then available at the ends of the loop. Note that this is simply the converse of the motor of Figure 2.15.

2.9 INDUCTANCE AND CAPACITANCE

Inductance and inductive reactance

Coils belong to a class of electrical components called *inductors*. When there is current in an inductor a magnetic field is created. If the current in the inductor rises, the magnetic field expands; if the current falls, the magnetic field contracts.

When the current through the coil changes, it causes the magnetic field to change; the changing magnetic field then induces a voltage that opposes the original change in current. Inductors thus permit direct currents to flow, but oppose alternating (or varying dc) currents.

Inductance (symbol L) is a measure of a coil's ability to generate an opposing induced voltage, as a result of a change in its current. The unit of inductance is the henry (H). The opposition presented to a current, by an inductor, is called *inductive reactance* (symbol X_L) and is measured in ohms. Inductive reactance increases as:

(i) the frequency, f, of the ac increases.
(ii) the inductance, L, is increased.

Energy storage in an inductor

The use of an inductor to store energy is very important in automobile electronics; it is the basis of most spark ignition systems.

An inductor stores energy in the magnetic field established by the current. The stored energy, E, can be calculated from the coil's inductance and the current,

$$E = LI^2/2 \qquad (2.21)$$

When L is in henrys and I is in amps then E is given in joules.

If the current through a coil is suddenly interrupted, the magnetic field around the coil rapidly collapses. All of the magnetic energy, E, is then quickly liberated in the windings of the coil and so a high voltage is generated. In petrol engined vehicles the ignition coil's inductance and operating current are carefully chosen so that voltages in the region of 10 000–40 000 V are generated to produce sparks at the spark plugs and so ignite the air-fuel mixture.

Capacitance

A *capacitor* is an electronic component that stores a small amount of electrical charge. A measure of a capacitor's ability to store charge is its *capacitance* (symbol C). In its simplest form, a capacitor is constructed of two parallel conducting *plates* separated by a layer of insulating material, such as paper or mica, called the *dielectric* (Figure 2.16). Connecting leads are attached to the plates and the device is encapsulated in a protective coating.

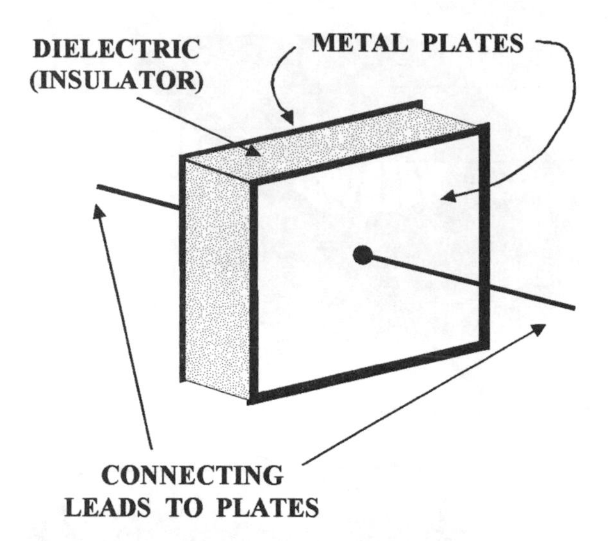

Figure 2.16 Construction of a simple capacitor

Charging and discharging a capacitor

When a voltage is applied to a capacitor a charge accumulates on the plates, resulting in a rising potential difference across the plates. When the capacitor voltage is equal to the applied voltage, no more charge flows onto the plates (i.e. current into the capacitor ceases) and the capacitor remains *charged*, whether or not the applied voltage remains connected. The capacitor *discharges* when a current path is provided between the plates. It acts as a momentary voltage source, causing a discharge current, until the voltage falls to zero. Figure 2.17 shows a simple circuit which illustrates capacitor action. When the switch connects the capacitor to the battery a charge current exists, large at first but falling as the voltage across the capacitor becomes equal to the battery voltage. The bulb thus glows brightly at first, becoming dimmer and eventually going out when the capacitor is fully charged.

When the switch is moved to the 'discharge' position, a large discharge current is initially present and the bulb glows brightly for a moment, fading as the charge flows off the plates.

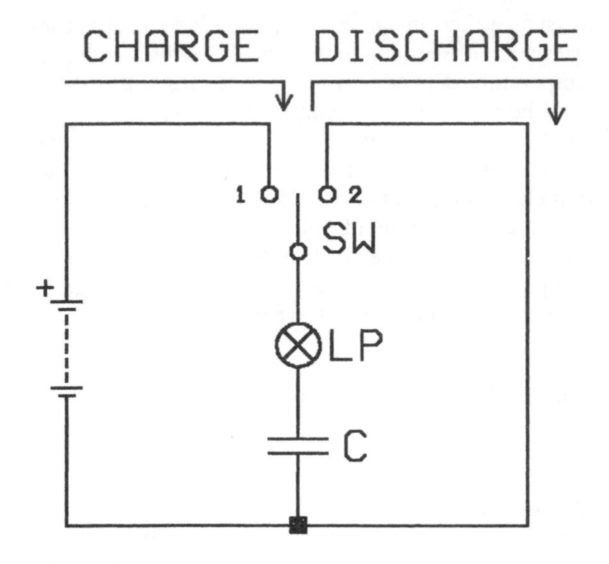

Figure 2.17 Charging and discharging a capacitor. The capacitor charges when the switch is in position '1', and discharges when in position '2'

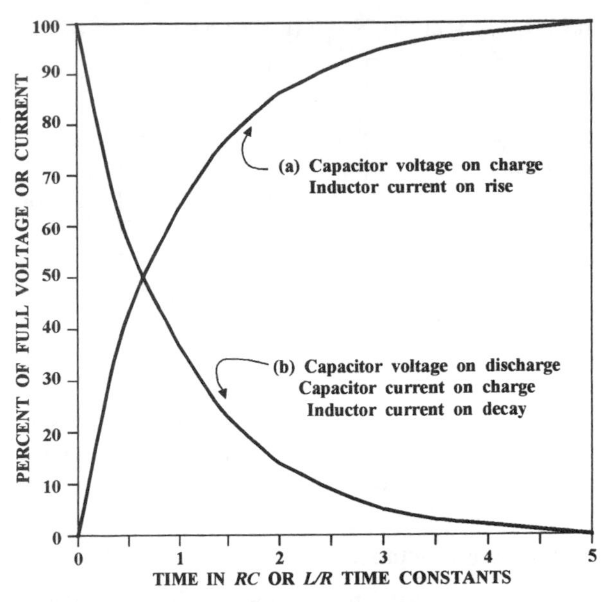

Figure 2.18 Exponential curves associated with capacitor–resistor and inductor–resistor circuits

Charge and energy storage

Capacitance is measured in units of *farads* (symbol F). One farad is an enormous amount of capacitance and so most capacitors have much smaller values. Values of from a few nanofarads (nF or 10^{-9}F) to a few microfarads (μF or 10^{-6}F) are common.

The amount of charge, Q, stored by a capacitor, C, when it is charged to a voltage, V, is given by,

$$Q = C \times V \qquad (2.22)$$

Capacitance can thus be defined as Q/V, and a one farad capacitor will store one coulomb of charge when one volt is applied across its plates.

The fact that a capacitor can be discharged to make a bulb glow shows that it is storing electrical energy and it can be shown that the amount of energy, W, stored by a capacitor is,

$$W = (Q \times V)/2 \qquad (2.23)$$

Although the amount of energy stored by a capacitor is very tiny indeed compared with that of a battery it can still be sufficient for use in applications such as capacitor-discharge ignition (CDI) systems.

Capacitor–resistor time constant

It is clear that when the capacitor in Figure 2.17 was connected to the battery, the current in the circuit was not constant because the bulb brightness changed. Looking in more detail, it is found that the charge current and the capacitor voltage follow curves like those drawn in Figure 2.18. These curves show that:

(i) The charging current, I_c, has a maximum when the battery is first connected, falling more and more slowly as the capacitor becomes charged, until it is zero.

(ii) The capacitor voltage, V_c, rises very rapidly at first, then more and more slowly approaches the battery voltage as the capacitor is fully charged.

Curves of this shape are called *exponential* curves and are characteristic of all capacitor–resistor (CR) circuits. The time in seconds, t, taken for the voltage or current to reach 63% of its final value (i.e. for the capacitor to reach 63% charge or discharge) is called the *time constant* for the CR circuit and is calculated as,

$$t = C \times R \qquad (2.24)$$

To completely charge or discharge a capacitor takes about five time constants (5 t). For example if C = 100 μF and R = 100 kΩ then the time to reach 63% charge or discharge is,

$$t = 100 \times 10^{-6} \times 100 \times 10^{3}$$
$$= 10 \text{ seconds}$$

And therefore the time required to completely charge or discharge is about 50 seconds.

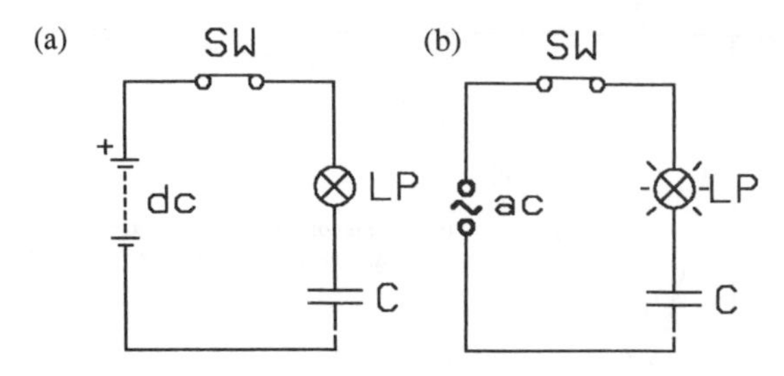

Figure 2.19 Capacitive reactance: (a) the capacitor blocks dc; (b) the capacitor passes ac

CR circuits are frequently used in electronics to provide a simple timing function, for example in wiper delay or courtesy-light delay applications.

Capacitive reactance

Figure 2.19 is used to illustrate another property of capacitors, that of *capacitive reactance*. In Figure 2.19(a) the bulb lights briefly as the switch is closed because there is a charging current, however after a short time the capacitor is fully charged. No more dc flows and the bulb goes out. The capacitor is thus said to *block direct current*. In Figure 2.19(b) the bulb is continuously illuminated and so it appears that the capacitor *passes alternating current*. In fact no current passes between the capacitor plates, but the electrons are able to flow on and off the plates in sympathy with the alternating voltage of the supply. This provides an alternating current through the bulb.

The opposition presented by a capacitor to a current is called its *capacitive reactance* (symbol X_C). Capacitive reactance depends on the size of the capacitance and the frequency of the ac in the following ways:

(i) capacitive reactance is infinite for dc;
(ii) capacitive reactance reduces as the ac frequency rises;
(iii) capacitive reactance reduces as capacitance is increased.

In electronic circuits, capacitors are often used to 'block' the unwanted dc component of a signal or to filter-out an unwanted ac component (by shorting it to ground).

2.10 SEMICONDUCTORS

Semiconductors are materials that have a higher electrical resistance than conductors, but a much lower resistance than insulators. A particularly useful characteristic of semiconductors is the strong light and temperature-dependency of their electrical properties.

Although there are many different types of semiconductor the most important one for the electronics industry is *silicon*. Silicon is easy to work with, extremely abundant and very cheap (glass and common sand are compounds of silicon), it also has electrical properties that make it ideal for the manufacture of components such as diodes, transistors, integrated circuits (ICs) and various types of sensors.

Conduction in semiconductors

Semiconductors differ from conductors in that at low temperatures there are very few free electrons available to conduct current and so electrical resistance is high. As a semiconductor's temperature is raised, ionization occurs (electrons break free from atoms), leading to an increase in the number of free electrons available for conduction (hence the resistance falls) and the formation of an equal number of positive ions.

If a voltage is applied across a sample of semiconductor, the negatively-charged free electrons will tend to drift from the negative contact toward the positive contact, outnumbering the fixed positive ions at that end. Meanwhile at the negative contact there will be few electrons left, but many positive fixed ions. Thus, whilst it is only the electrons that actually move, electron drift to the positive contact will appear to be balanced by a drift of positive charge toward the negative contact. In order to account for this apparent positive charge flow, electronics engineers invented the concept of a positively-charged particle called a *hole*. The hole has a charge of equal magnitude, but opposite polarity, to that of the electron and appears to drift through the semiconductor, just as the electron does.

Intrinsic and extrinsic semiconductors

Semiconductor material that is very pure is known as *intrinsic semiconductor*. The number of free electrons is exactly balanced by the number of holes and the resistance of the material is relatively high.

Extrinsic semiconductor has a lower resistance and is produced by taking intrinsic semiconductor and adding a tiny amount of *impurity* to give it an over-abundance of either holes or electrons. This process of adding impurities to intrinsic semiconductor (called *doping*), is the basis for the manufacture of most *active devices*, such as diodes and transistors.

N-type and p-type semiconductors

A semiconductor that has been doped to give it an excess of free electrons is called an *n-type* semiconductor (the *n*, here, stands for *n*egative charge). An n-type semiconductor still has many positively-charged holes available, but they are vastly outnumbered by the free electrons and so conduction in n-type material is largely due to electron flow, with holes acting as *minority carriers*.

A semiconductor that has an excess of holes is called a *p-type* semiconductor (*p* stands for *p*ositive charge) and is produced by adding a dopant that increases the number of holes in the semiconductor. In p-type semiconductor material conduction is largely due to hole flow, with the electrons acting as minority carriers.

Semiconductor devices

Semiconductor materials are used to make some of the most interesting and useful electronic devices. The range of components now available is enormous, and it would be quite impossible to look at them all. Those that are of importance to the automotive industry are described below.

Thermistors

One of the simplest applications of semiconductor materials is the thermistor. In contrast to normal resistors, which maintain a constant resistance almost independently of temperature, *thermistors* (*therm*al res*istors*) exhibit a resistance which is strongly temperature dependent. Since thermistors are cheap and durable, they are used for almost every temperature sensing requirement on the automobile (engine coolant, intake air, exhaust gas, etc.). These thermistors are generally of the negative temperature coefficient (ntc) type, which means that their resistance decreases with increasing temperature. A typical ntc thermistor response in shown in Figure 2.20, the resistance is about 30 kΩ at 20°C and falls steeply to about 180 Ω at 100°C. The resistance value is sensed by an electronic control module (ECU) and interpreted as a temperature for further processing or display (e.g. on a temperature gauge). Since the temperature–resistance response is very non-linear (i.e. not a straight line), the change in thermistor resistance usually needs to be linearized by the ECU.

Thermistors are made from the powdered oxides of metals such as nickel, cobalt and manganese, formed into small pellets and fitted into a threaded holder complete with connector. Figure 2.21 shows a typical

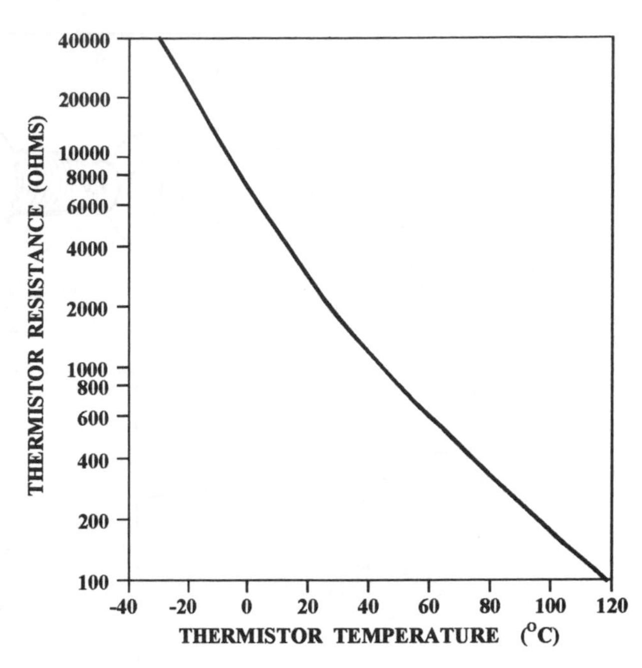

Figure 2.20 Temperature response of an ntc thermistor

application, that of an engine intake air temperature sensor system.

Another (less common) type of thermistor, the positive temperature coefficient (ptc) variant, is made from barium titanate and has a resistance that increases with increasing temperature. A typical application for ptc thermistors is in the control of electrical heaters, where a large current needs to flow at low temperatures and then be progressively reduced as temperature rises. The protection of stalled electrical motors from excessive currents (e.g. in a jammed power window) can also be achieved with ptc devices. Resistive heating in a series-connected ptc thermistor causes a sharp resistance rise, limiting the motor's stall current.

The Hall effect

In 1879 the scientist E H Hall noticed that a small voltage could be detected between the sides of a conductor when it was carrying current in a magnetic field. In metal conductors this voltage is tiny, but in semiconductors much larger voltages can be generated. The semiconductor material indium arsenide (InAs) is particularly effective.

Figure 2.22 illustrates the principle of the Hall effect. Electrons flowing through the Hall 'chip' are deflected by the strong magnetic field and so tend to be bunched-up at one side, giving an excess negative

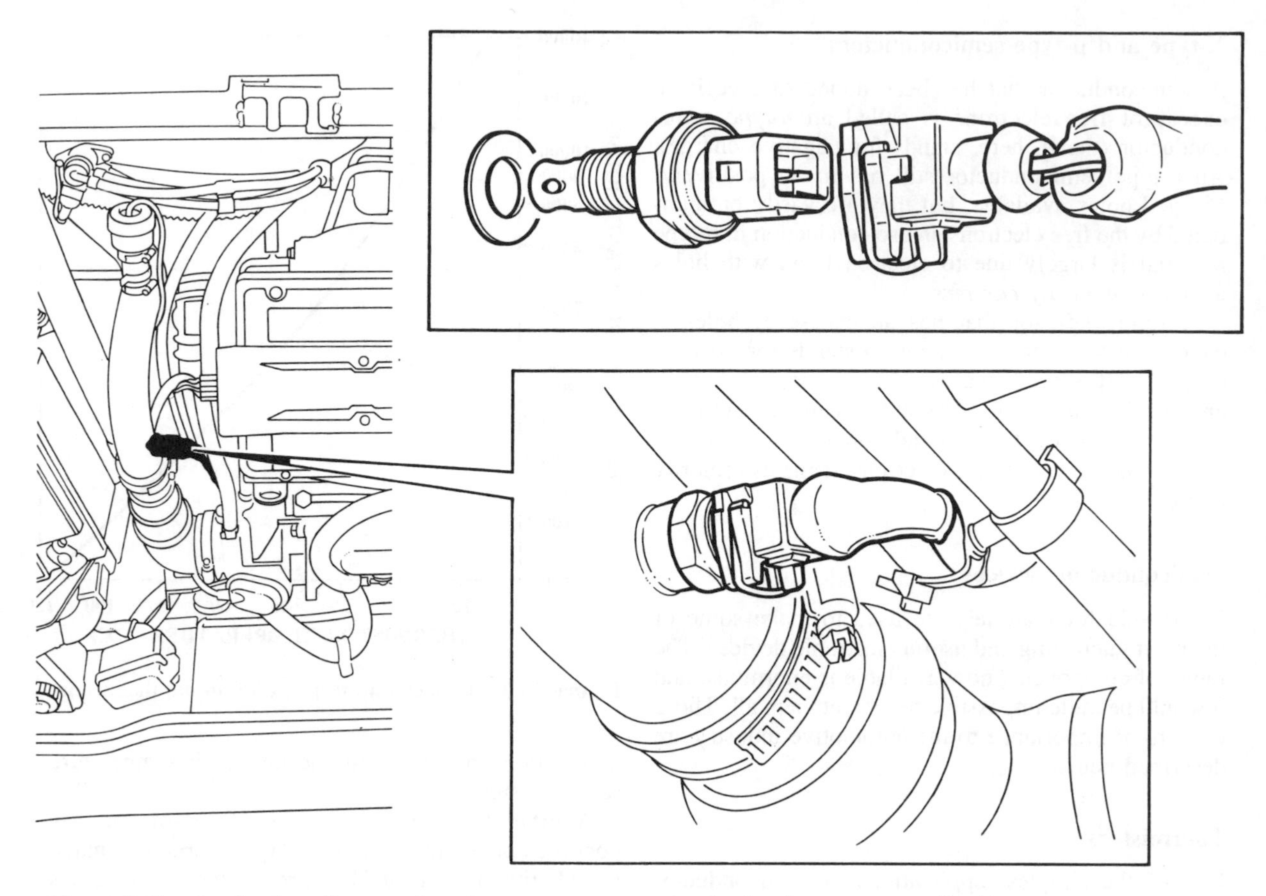

Figure 2.21 Application of an ntc thermistor to measure engine intake air temperature (*Saab*)

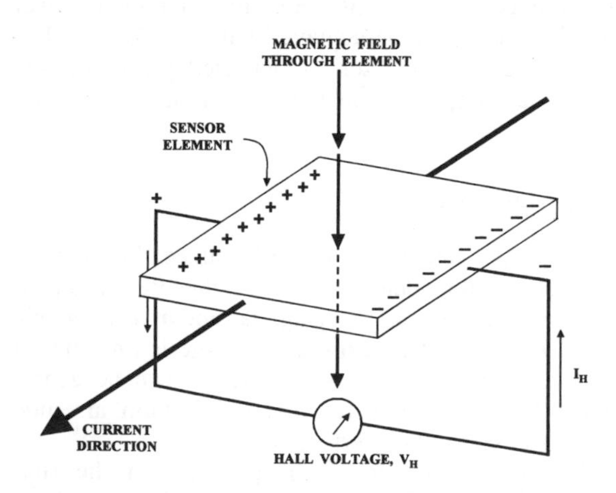

Figure 2.22 Principle of the Hall effect. The Hall voltage, V_H, is measured across the edges of an InAs sensor element which carries a current

charge in that region. At the other side there is a deficiency of electrons, leading to an excess positive charge. This difference in potential (the Hall voltage, V_H) increases and decreases, accordingly, in proportion to the magnetic field strength. When there is no magnetic field, V_H is zero. In general, with a modest magnetic field and a current of about 100 mA, a Hall voltage of about 50 mV would be produced.

Hall sensors are cheap and rugged, so the ability to detect the presence or absence of magnetism by using the Hall effect is widely used in automotive electronics as a means of measuring shaft rotation. Figure 2.23 illustrates a typical application; using a Hall sensor in place of contact-breaker points as a trigger for an electronic ignition system. The trigger system comprises a rotor made of soft iron, with four vanes located at the outside edge at 90° intervals, and a pick-up fixed to the distributor base-plate. The pick-up is a sensor composed of a Hall device and permanent magnet for detecting the presence or absence of the vanes.

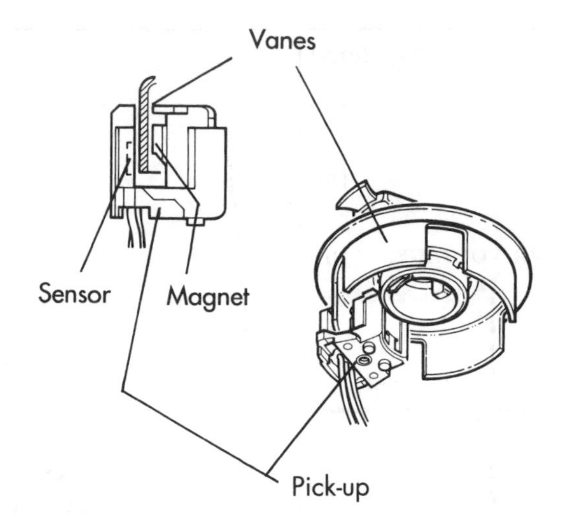

Figure 2.23 Hall-effect pick-up used in a distributor

The rotor is fixed to the distributor shaft so that as it turns the vanes pass between the Hall element and the magnet. When there is no vane between the sensor and the magnet, flux from the magnet passes through the Hall device and an output voltage is generated. When there is a vane between the Hall element and the magnet, the flux from the magnet is screened by the vane and does not reach the sensor. The output from the sensor is thus a series of voltage pulses corresponding to 90° of crankshaft rotation. These pulses can be processed and amplified for use by the ignition system ECU.

The pn-junction

The real usefulness of semiconductors is in having a *junction* between layers of p-type and n-type material. The simplest junction device is the *diode*, made by putting a piece of p-type material in contact with a piece of n-type material. This produces a component that only allows current to pass through in one direction (the *forward* direction). If an attempt is made to drive current through in the *reverse* direction, the diode will block it. The diode is, in effect, a one way valve for electric current.

Figure 2.24 illustrates the circuit symbol and electrical characteristics of the silicon diode. The diode symbol can be considered an arrow-head, pointing in the direction in which conventional current can pass. When a forward-bias voltage of more than about 0.7 V is applied across the diode the forward current rises very steeply and in a non-ohmic fashion (i.e. the diode does not obey Ohm's Law). When a reverse-bias voltage is applied to the diode it exhibits a very high resistance and only a minute leakage current will flow. If a very high reverse voltage is applied, the diode may suddenly fail (or break down) and pass a high reverse current, destroying itself in the process.

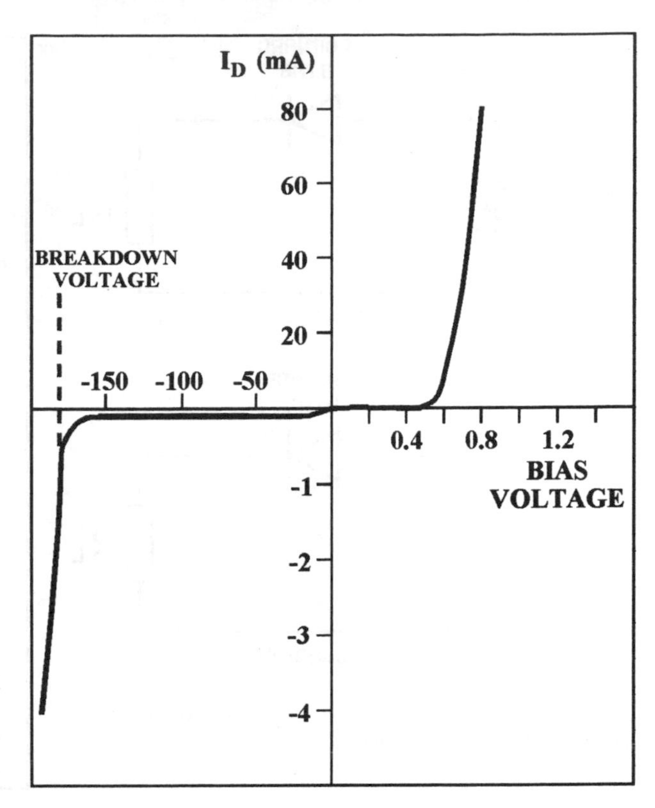

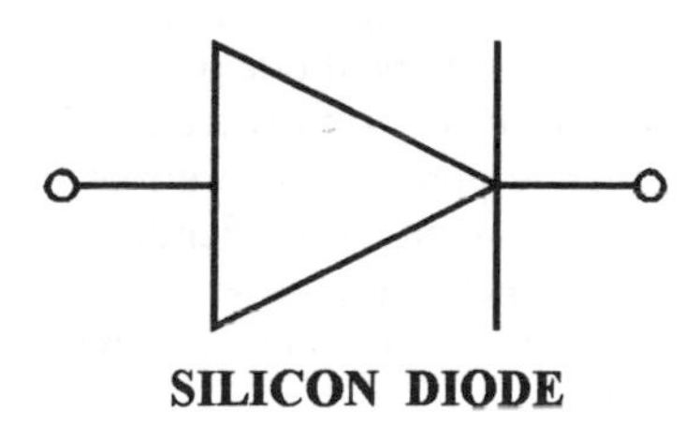

Figure 2.24 Electrical characteristics and circuit symbol of the silicon diode

Types of diode

Although all diodes work as electrical one-way valves they can be made to have additional, useful, properties.

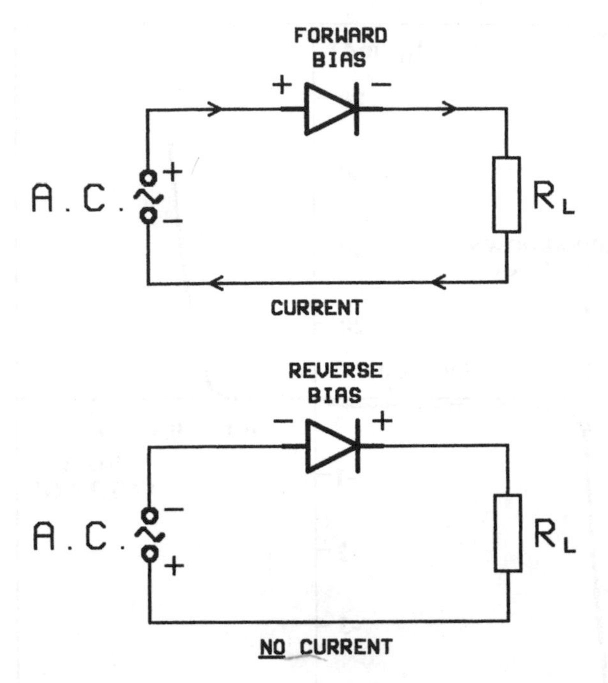

Figure 2.25 Using a diode to rectify. The diode only conducts when forward biased

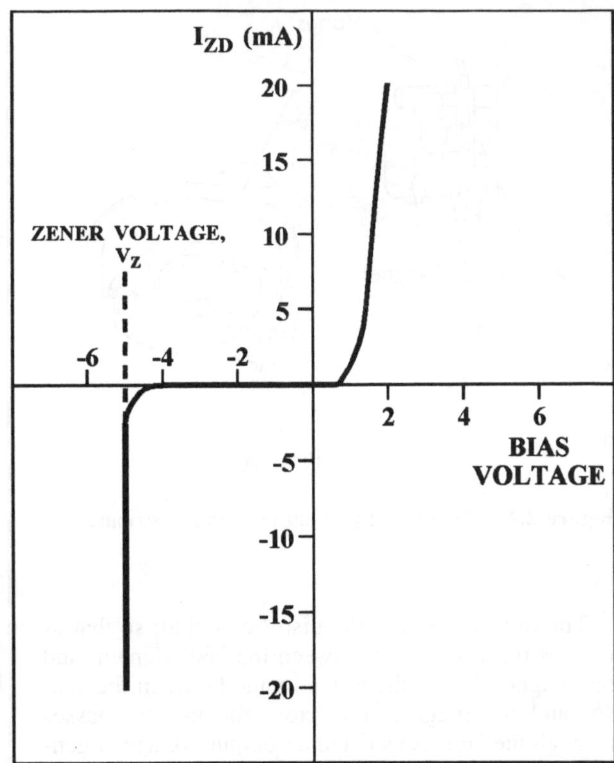

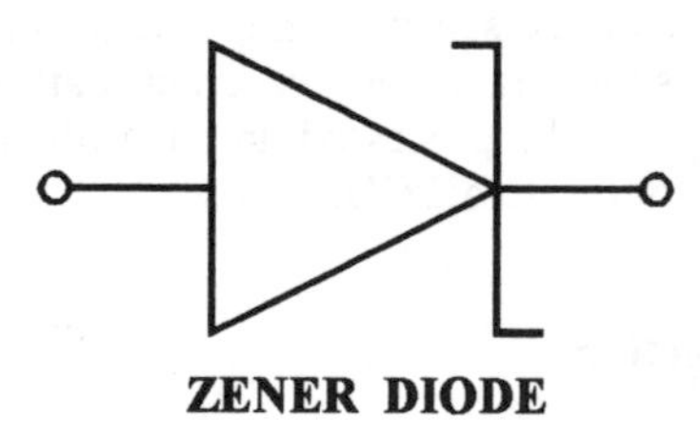

Figure 2.26 Circuit symbol and electrical characteristics for a 5 V Zener diode. The diode undergoes reverse breakdown at precisely 5 V

Power rectifier diodes

These diodes are used for the conversion of ac into dc, a process known as *rectification*. Figure 2.25 shows a simple rectifier circuit; the diode passes current only on the positive part of the sine-wave and blocks current on the negative portion. The current in the load is thus in one direction only. Rectifier diodes are found in all alternators, where they have to pass large currents (tens of amps) and so tend to be physically quite large components.

Zener diodes

When the reverse voltage applied to a *Zener diode* reaches a specified value the diode suddenly starts to conduct in a non-destructive fashion. Figure 2.26 shows the symbol for the Zener diode and also the current–voltage characteristics for a 5 V device.

Zener diodes have many applications, but in automobile electronics they are normally used in voltage regulating and stabilizing circuits. These are simply circuits that maintain a constant output voltage even during fluctuations in input voltage or load conditions. Figure 2.27 illustrates an example of a voltage regulator that uses a Zener diode. When the input voltage to this circuit changes, there are corresponding changes in the current through the series resistor, R_S. But the voltage at the output to the load circuit is

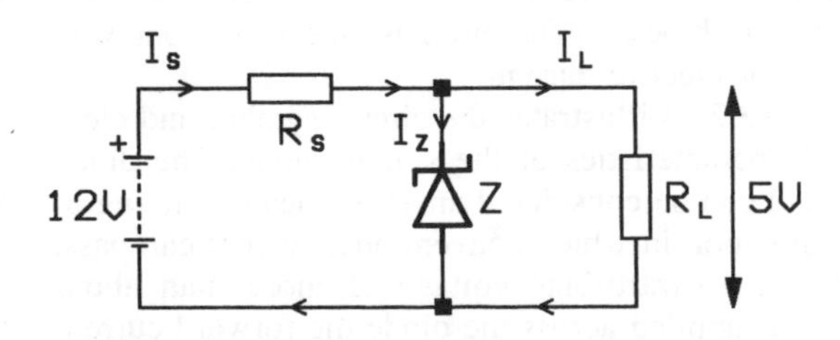

Figure 2.27 Voltage regulation using the Zener diode

held at a constant 5 V by the Zener diode. All excess current passes through the Zener diode. What must be noted, however, is that the input voltage must always be a few volts above the desired output voltage; the circuit cannot regulate if the input voltage is close to, or below, the desired output voltage.

Photodiodes

A *photodiode* is used to detect the presence or absence of light. The diode is connected in reverse bias so that when light falls on it extra free electrons are released in the silicon and the reverse current (normally tiny) rises sharply. A greater light intensity results in a higher reverse current. In automotive applications the photodiode usually forms part of an optical sensor system to detect, say, the rotation of a disc or shaft.

Light emitting diodes (LEDs)

A *light emitting diode* (LED) emits light when current is passed through it in the forward direction. Silicon is not a good light emitter and so most LEDs are made of other semiconductors such as gallium phosphide (GaP) or gallium arsenide (GaAs), operated at a forward voltage of about 1.6 V to produce a forward current of about 20 mA.

LEDs are available in several colours such as red (and infra-red), green, yellow and white, depending on the semiconductor material used. The LED chip usually comes encapsulated in a clear plastic lens which directs and magnifies the small amount of light coming from the semiconductor.

The advantages offered by the LED over the ordinary filament bulb are that it consumes very little power, does not get hot and goes on working almost forever. Unfortunately the light output is small and so LEDs are confined to applications such as sensors, dashboard displays and warning lights.

The thyristor

The *thyristor*, sometimes known as a *silicon controlled rectifier* (SCR), is a four-layer sandwich of p- and n-type silicon as shown in Figure 2.28. It functions as a very fast and efficient solid-state switch that never wears out. The symbol for the thyristor is like that for the diode, but with the addition of a third terminal called the gate. When forward biased, the thyristor does not conduct until a positive voltage pulse is applied to the gate. Conduction is then maintained, irrespective of the gate voltage, until the supply current falls to zero.

The circuit illustrated in Figure 2.28 demonstrates thyristor action. When the power switch, S_1, is closed, the thyristor does not conduct and the bulb stays off. If switch S_2 is now closed, the thyristor switches on ('fires') and conducts sufficient current for the bulb to light brightly. The thyristor continues to conduct, even if S_2 is opened, and the bulb will only go out if S_1 is opened.

Thyristors are good for switching large currents with a small gate pulse; a 50 mA gate pulse will typically switch a current of 10 A.

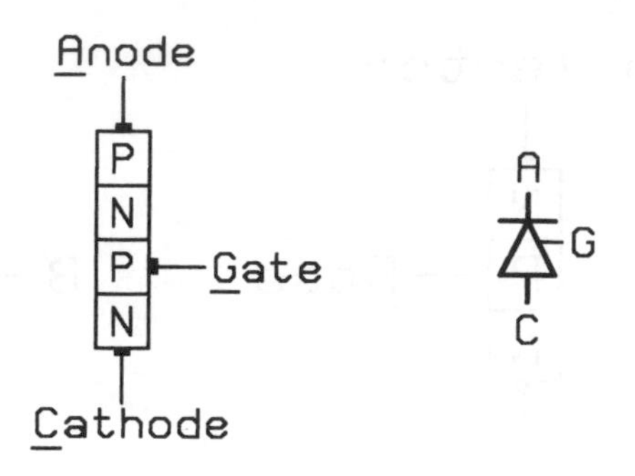

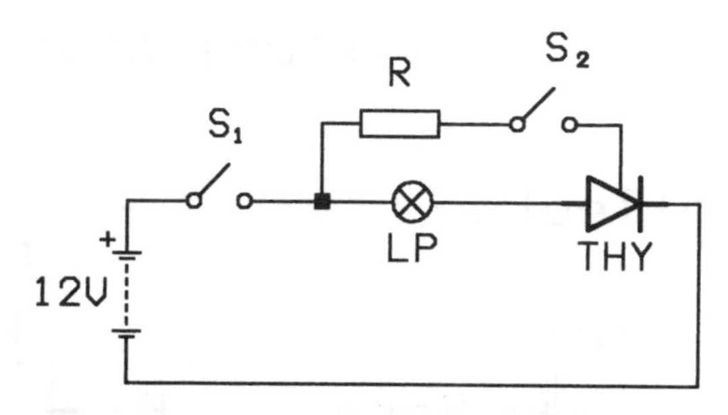

Figure 2.28 Construction, circuit symbol and circuit application of the thyristor

2.11 THE TRANSISTOR

The *transistor* is possibly the most important electronic component of all. It has the ability to *amplify*, in other words to produce an output signal with more power in it than the input signal. Transistors may also be used as switches (rather like solid-state relays); calculators and computers contain many hundreds of thousands of transistors used in this way. Most electronic circuits use large numbers of transistors since they are cheap, fast, extremely reliable and can be made microscopically small.

Since the first transistor was invented, in 1948, progress has been very rapid and numerous different types of transistor are now available. The two types

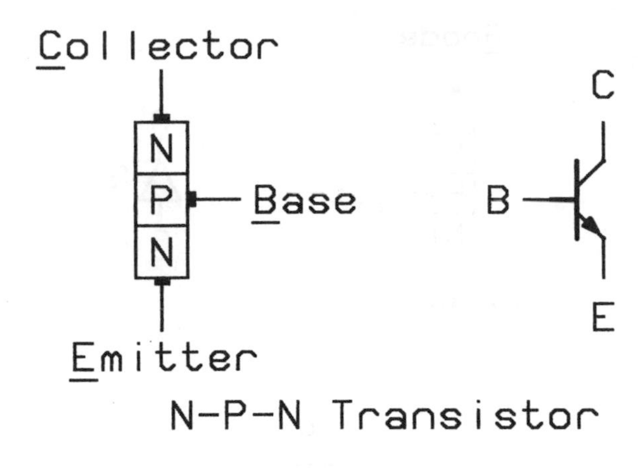

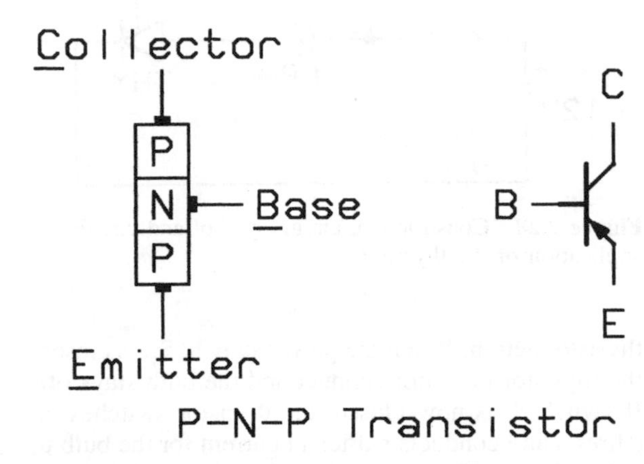

Figure 2.29 Construction and circuit symbols of npn and pnp transistors

of transistor most commonly encountered in automobile electronics are the *bipolar junction transistor* (BJT) and the *metal-oxide-semiconductor field effect transistor* (MOSFET).

The bipolar junction transistor (BJT)

The *bipolar junction transistor* (BJT) is a three-terminal device fabricated from two pn-junctions made in the same crystal of silicon. BJTs come in two varieties, *npn* and *pnp*, illustrated in Figure 2.29 along with their circuit symbols. The three terminals of the transistor are known as the *base* (B), *collector* (C) and *emitter* (E). Transistors owe their usefulness to the fact that a tiny change in current into the base terminal causes a much larger change in the current flowing between the collector and emitter terminals. How the BJT actually does this is extremely complicated and involves a considerable amount of mathematics, indeed, many professional electronics engineers could not claim to fully understand how it works. A much simplified description is given here.

The BJT as a current amplifier

The circuits of Figure 2.30 are used to illustrate the application of the npn BJT as a current amplifier (the pnp device works in just the same way, but all the polarities would have to be reversed). In Figure 2.30(a) the potentiometer (VR) is set so that the voltage between base (B) and emitter (E) is just a little less than 0.7 V. This voltage is very significant since silicon transistors are 'switched off' if the base-to-emitter voltage (V_{BE})is less than about 0.7 V. In this example, the *base current* (I_B) is zero and so there can be no current through the transistor from collector (C) to emitter (E).

In Figure 2.30(b) the potentiometer has been adjusted so that V_{BE} is now a little greater than 0.7 V. A small base current will flow into the transistor, switching it on and causing a much bigger current (called *collector current*, I_C) to flow from the collector to the emitter. The exact value of the base current will depend upon the type of transistor and the value of the base resistor, *R*, but a value of 0.1 mA can be assumed. The collector current, I_C, is typically 10 to 1 000 times greater than I_B, again depending upon the type of transistor, but may be assumed that I_C is about 100 times I_B, i.e. about 10 mA, and sufficient to dimly light the bulb.

If I_B is viewed as the input signal to the transistor and I_C as the output signal, then it is clear that the transistor is acting as a *current amplifier*. The ratio of I_B to I_C is a measure of the transistor's *current gain* (symbol β – 'beta') and a very important transistor parameter. Mathematically,

$$\beta = I_C/I_B \quad (2.25)$$

For the transistor in Figure 2.30(b), $I_C = 10$ mA and $I_B = 0.1$ mA and so,

$$\beta = 10/0.1 = 100$$

Note that the current coming out of the transistor through the emitter terminal (I_E) must be the sum of the currents going into the transistor (I_C and I_B), put mathematically,

$$I_E = I_B + I_C \quad (2.26)$$

And so in this example,

$$I_E = 0.1\text{ mA} + 10\text{ mA} = 10.1\text{ mA}$$

In Figure 2.30(c) the potentiometer has been adjusted to the 'maximum voltage' position and about 500 μA

(a)

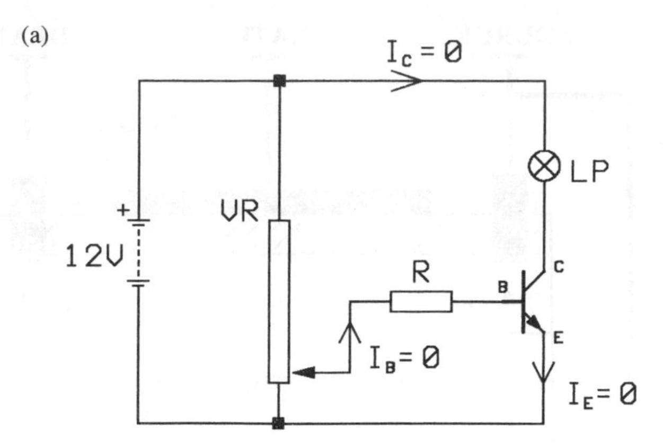

(b)

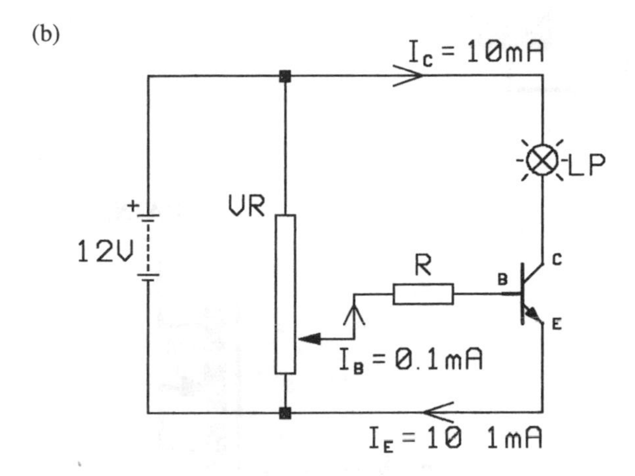

(c)

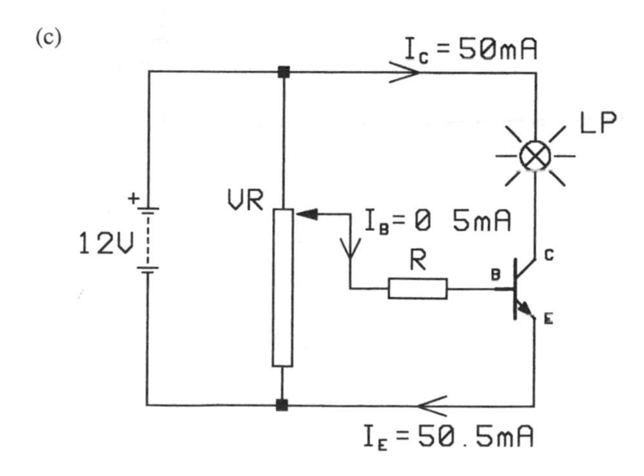

Figure 2.30 Operation of the npn transistor as a current amplifier; (a) V_{BE} is less than 0.7 V so the transistor does not conduct; (b) V_{BE} is a greater than 0.7 V and so the transistor conducts and the lamp glows; (c) the transistor is fully turned on and the lamp glows brightly

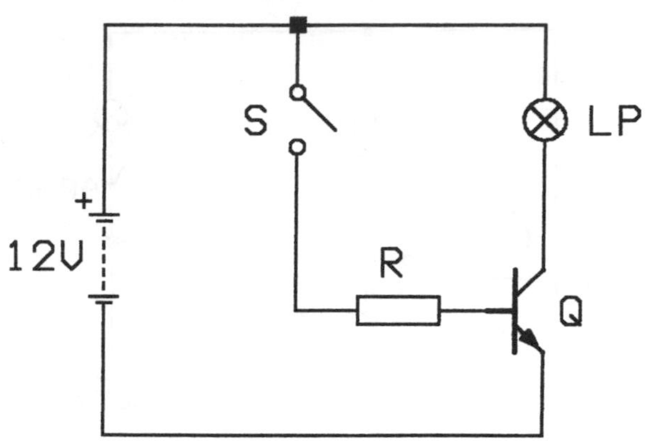

Figure 2.31 The transistor switch

of base current flows, limited by the base resistor, R. R prevents excessive base current (which might destroy the transistor) and should always be present in a circuit. I_C is again $\beta \times I_B$ and so $I_C = 50$ mA. This current causes the lamp to glow brightly and so verifies the calculations.

Note that, although in the above examples a potentiometer has been used to provide a varying input current, other sources of current could also be used. Pressure, temperature and mechanical position sensors are all devices that may need their output signals amplifying by transistors.

The BJT as a switch

The circuit drawn in Figure 2.31 is called a transistor switch. From the preceding example it should be easy to understand.

When the mechanical switch is open, there is no base current and so no collector current. The lamp is off.

When the switch is closed, the base-emitter voltage, V_{BE}, rises to about 0.7 V and the transistor turns on, lighting the lamp. Bearing in mind that $I_C = \beta I_B$, the base resistor, R, needs to be carefully selected to ensure the correct value of I_B, and so the correct I_C.

In comparison with electromechanical relays, transistor switches have many advantages; they are small and light, enable very rapid switching (in a fraction of a microsecond) and many circuits can be switched with a single control signal.

The Darlington pair

A single BJT might typically provide current gain of up to about 700 times its input. When more gain is

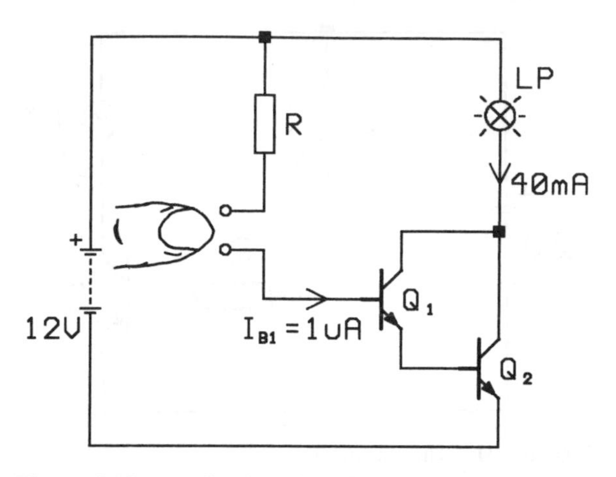

Figure 2.32 Application of the Darlington pair in a 'touch switch'

required, the *Darlington pair* circuit can be used. Figure 2.32 illustrates the application of this circuit in a 'touch switch'. Each transistor has a current gain, β, of 200 and so the cascaded gain of Q_1 and Q_2 is $\beta \times \beta$ (i.e. 200×200), which is 4 000.

When there is no connection between the switch contacts, no current will flow in the circuit and the bulb stays off. But if a finger-end is touched across the contacts, a tiny leakage current of about 1 μA will flow and the current gain of the Darlington pair is then sufficient to light the bulb.

In automotive applications the Darlington pair is often used to provide large amounts of current gain for driving inductive loads such as relays, ignition coils, fuel injectors and solenoid valves.

The MOSFET

Unlike the BJT, which is a current amplifier requiring a current input to control a current output, the MOSFET is an amplifier that uses an input *voltage* to control an output *current*. The input current to a MOSFET is usually negligible, which is useful since this makes it a very low-power device.

Principles of MOSFET operation

One of the attractions of the MOSFET is that, in contrast to the BJT, the principles of its operation are quite easily understood. Figure 2.33 shows an *n-channel* MOSFET together with its circuit symbol (*p-channel* devices are also used – just interchange the n- and p-regions). The basic npn silicon structure is similar to that of the BJT; the differences arise from the way in which it is connected to the outside

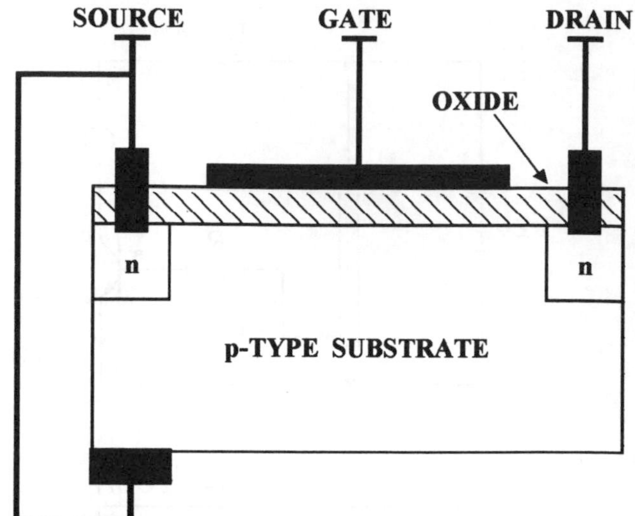

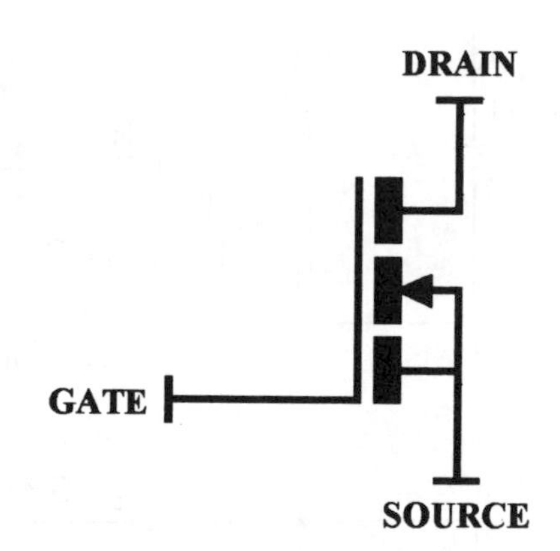

Figure 2.33 Construction and circuit symbol of the n-channel MOSFET

world. The MOSFET has three metal contacts and an additional controlling terminal called a *gate*. The gate is insulated from the npn structure by a thin layer of silicon oxide (i.e. glass) and so acts rather like a capacitor plate. As long as the voltage between the body and the gate is not positive, the device behaves like two 'back-to-back' pn-junction diodes and so no current passes between the n-region contacts. The device is 'OFF'.

Although the p-region has many more holes than electrons (that is what makes it p-type) there are still *some* free electrons, and if the gate is made positive with respect to the body some of these electrons will be attracted toward it, as in Figure 2.34(a). The gate-insulator-body sandwich structure thus forms a capacitor which accumulates charge. As the gate is made

(a)

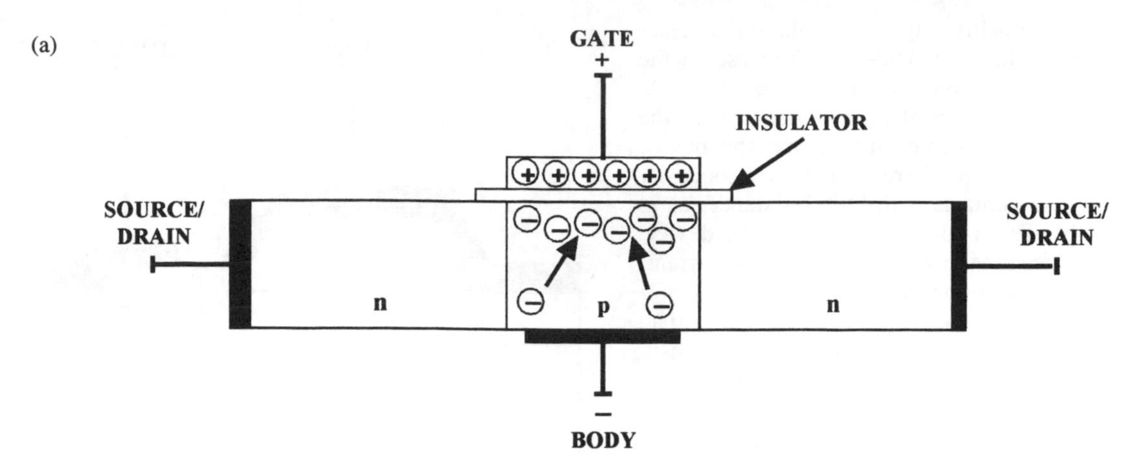

(b)

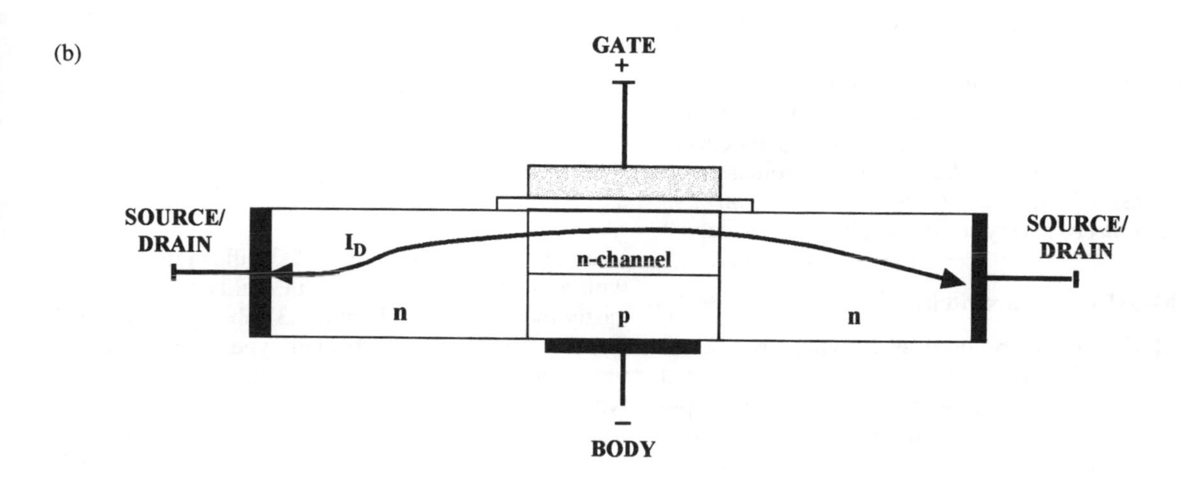

(c)

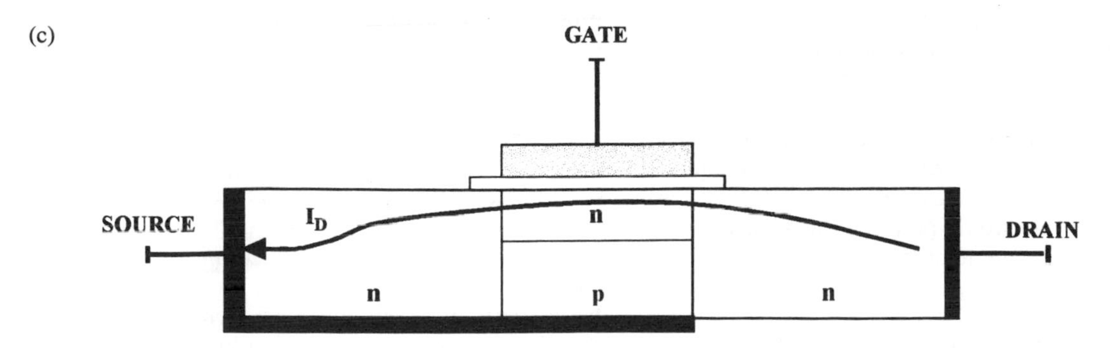

Figure 2.34 Operation of the n-channel MOSFET: (a) application of a positive voltage to the gate terminal draws free electrons to the silicon surface; (b) an n-type channel is formed and the transistor passes current between the end contacts; (c) connection of the 'body' terminal to one end terminal to give 'source' and 'drain' terminals

more and more positive, the accumulation of electrons just under the gate oxide will increase to the point where the electrons outnumber the holes. At this stage a small portion of the p-region (called the *channel*), inverts to become n-type and the device structure is now n–n–n. Current can now pass easily between the end contacts with little resistance, as in Figure 2.34(b). This is the 'ON' state for the device. Channel resistance (and hence the channel current) may be controlled by the gate voltage.

In general most MOSFETs have three, rather than four, leads. This is because the body terminal is usually connected directly to one of the n-regions, as in Figure 2.34(c). This particular n-region is then called the *source* connection and the other n-region is called the *drain* connection. To turn the device 'ON', the gate is made positive with respect to the source and conventional current (*drain current*, I_D) flows from drain to source.

MOSFETs manufactured for use in amplifier circuits are designed so that I_D is controlled only by the gate voltage, V_G, and is almost completely independent of variations in the drain-to-source voltage, V_{DS}. The MOSFET then behaves as a *voltage-controlled current source*, a very useful amplifying device.

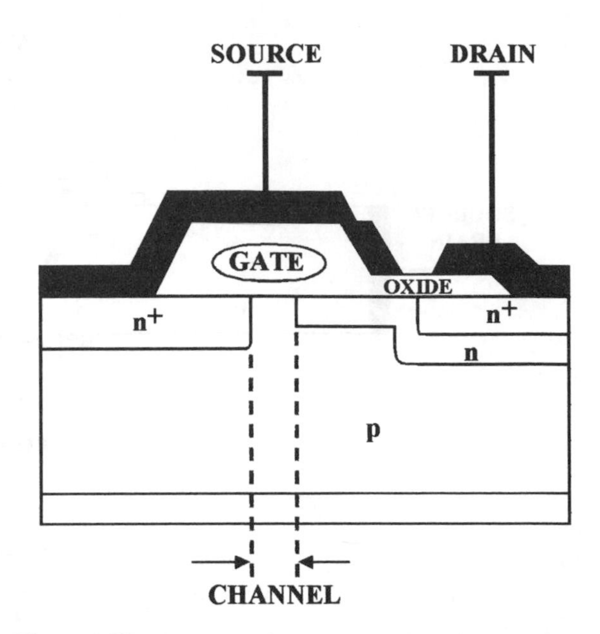

Figure 2.35 Structure of a 'power MOSFET' used for controlling large currents

The MOSFET as a switch

The MOSFET can be used as a switch in just the same way as the BJT, the only difference being that a voltage, rather than a current, is used as the input signal. For this reason MOSFET switches are more efficient than the BJT type, consuming about one-tenth the power and so generating a lot less heat. This makes them ideal for applications where large numbers of switching transistors are needed in a small space, for example in microcomputers.

'Power MOSFETs' with complex structures (Figure 2.35) have also been developed. They are used for controlling large currents and are slowly replacing BJTs and relays for power switching in automotive applications. Some devices (termed 'SmartFETs') even have built-in electronic protection against excessive current or excessive temperature.

2.12 INTEGRATED CIRCUITS (ICs)

Although all of the devices discussed so far are available as discrete (i.e. separate) components, the greatest use of them is in *integrated circuits* (ICs). An integrated circuit contains a range of circuit devices (diodes, transistors, resistors, small capacitors) as well as wiring tracks, on one single substrate (usually a small tablet of silicon). Figure 2.36 illustrates the idea with a simple IC structure – in reality most ICs are vastly more complex. Figure 2.37 shows an enlarged image of a very sophisticated type of IC called a microcontroller. This particular microcontroller is a type 68HC05 manufactured by Motorola and used in many engine management ECUs – the total area of the chip is only a few square millimetres and yet it contains many thousands of transistors.

Since 1959, when the first integrated circuit was produced, the number and variety of ICs has grown enormously. The low cost, excellent reliability and small size of ICs means that they are now at the heart of almost every electronic control unit (ECU) used in automobiles. ICs are available for nearly every electronic circuit function; where a particular function is not catered for, an IC manufacturer will usually design a chip to order. These 'custom chips' (more properly known as *application-specific integrated circuits*, or ASICs) are widely used by vehicle manufacturers, some of whom now have their own 'in-house' ASIC design and production facilities. With a simple chip costing just pence, the financial incentive to use ICs is very strong.

Integrated circuit design and manufacture

The low cost of ICs owes much to the sophisticated techniques that have been developed for their design

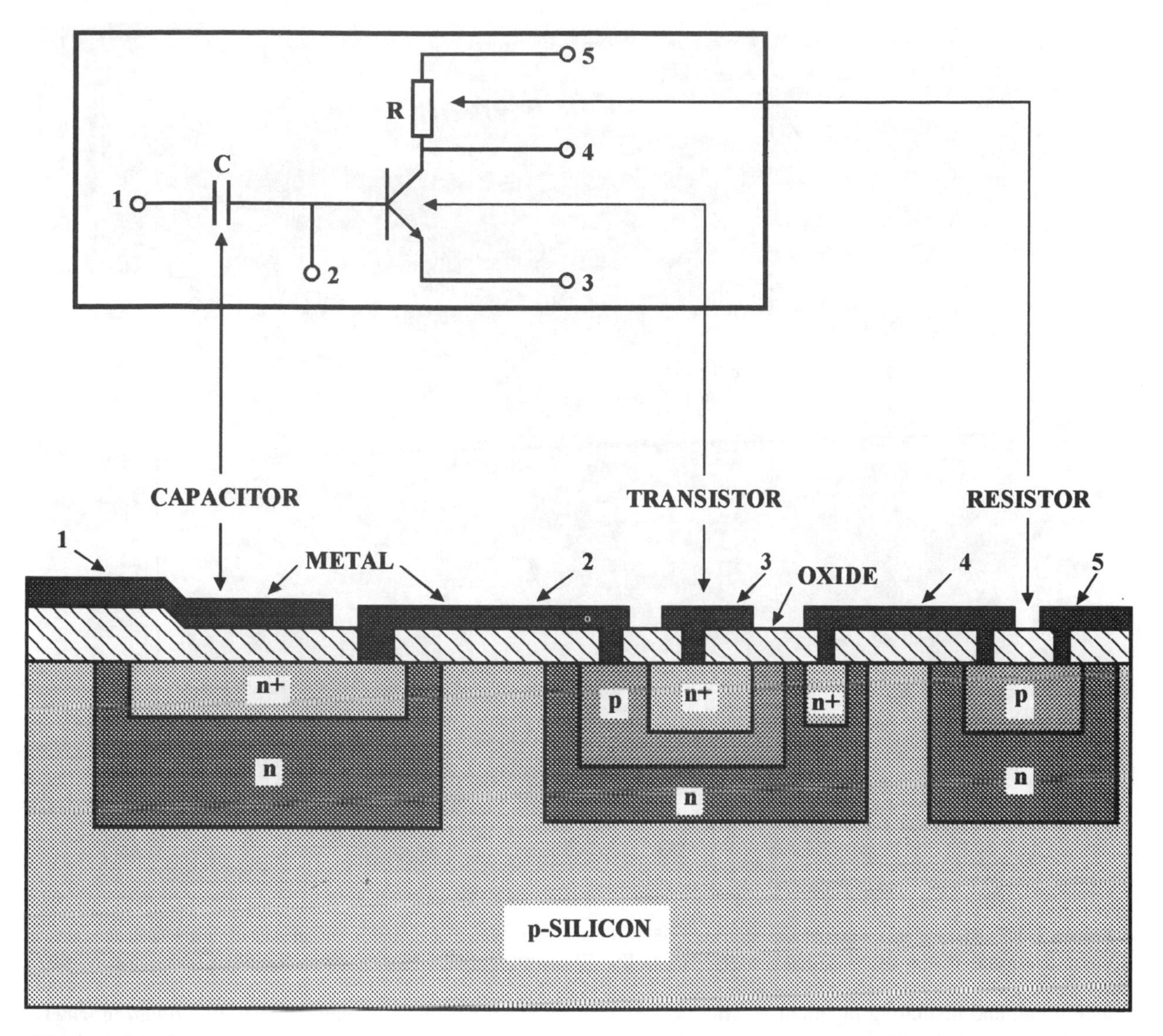

Figure 2.36 Section through a simple integrated circuit and relationship with its circuit schematic diagram

and manufacture. Modern ICs are invariably designed with the aid of computer programs (*computer-aided design*, or CAD programs). These programs enable the designer to 'draw' the required circuit on the computer display screen, simulate its electronic operation, and then place all of the components onto the silicon in their theoretically correct positions. Once the computer software has verified the operation of the IC, all of the design data can be committed to a floppy disc and sent to the chip factory for manufacture.

Figure 2.38 shows the steps in the manufacture of a typical IC. The raw material for production is a bar of very pure monocrystalline silicon, generally about 15 mm in diameter. The bar is sawn into discs, called *wafers*, of about 0.7 mm thickness which are then lapped and polished to give a perfectly flat surface. A large number of identical ICs are then produced on each wafer.

The IC production process is achieved through a technique called *photolithography*; the circuit is literally 'photographed' onto the silicon using ultraviolet light and a tiny photographic mask. This is performed by coating the silicon with photosensitive lacquer, exposing, developing the image and then etching away unwanted material using acid. P- and n-regions can then be created using a selective doping

Figure 2.37 Magnified image of the Motorola 68HC05 microcontroller, a complex integrated circuit which contains many thousands of transistors. *Facing page*—key to the different functional blocks within the 68HC05 microcontroller

technique, and insulating layers of silicon dioxide are produced by oxidizing the silicon in an oxygen-rich atmosphere. A typical IC might require from 10 to 100 such separate photolithographic stages, each precisely aligned with the previous, and all performed in conditions of extreme cleanliness.

When all of the electronic components have been fabricated, metal connecting tracks are laid down using aluminium vapour (a process termed *metallization*).

Depending on the complexity of the circuit, the finished wafer might contain in the region of fifty separate, but identical, ICs. A computer-controlled tester is then used to make a functional check on each IC, and the faulty ones are marked with a spot of dye. The wafer is subsequently separated into chips using a diamond saw. Good chips are mounted into either plastic or ceramic *packages* and connected to the connecting pins by means of very fine *bonding wires*. A final functional test takes place after the package has been sealed with an air-tight lid.

Due to the use of photolithographic techniques, the cost of chip production is very dependent on volume. It is generally not economic to produce chips in quantities of less than about 5 000. Most chips are therefore made in batches of 10 000–100 000 or more.

Hybrid integrated circuits

The integrated circuits discussed so far have consisted of a number of circuit elements (transistors, resistors, diodes, etc.) integrated onto a common silicon substrate. This type of IC is called a *monolithic IC*. Figure 2.39 shows the structure of a different type of IC, called a *hybrid IC*. It consists of discrete compo-

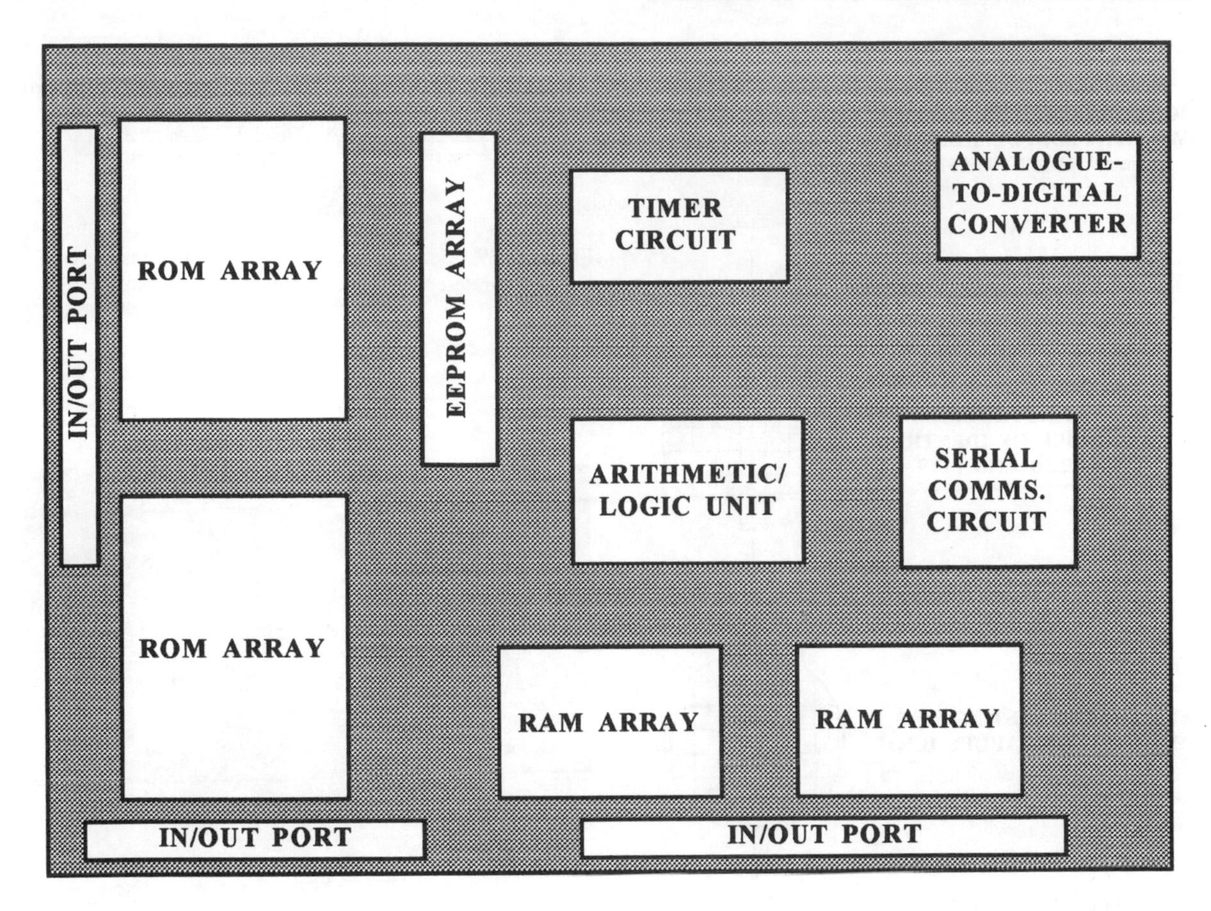

Figure 2.37 (*cont.*)

nents, such as diodes, transistors, resistors and capacitors, as well as monolithic ICs, all mounted on a common glass or ceramic substrate. Although hybrid ICs make up only a small fraction of the total number of ICs in production, they are often found in automotive electronic systems. The reason for this is that the automotive design cycle is now very short and the development time required for a hybrid IC is often shorter than that for a monolithic IC. Additionally, the hybrid IC is generally able to handle higher voltages and larger currents than its monolithic counterpart. Figure 2.40 shows a typical hybrid IC application, an alternator voltage regulator. Figure 2.41 depicts a more complex hybrid IC, integrating two monolithic ICs onto a common ceramic substrate.

Analogue and digital signals

When a varying voltage or current is used to convey information from one place to another it is called a *signal*. A signal may indicate the temperature of the engine coolant, or the quantity of fuel in the tank, for example. In electronics we deal with two categories of signal;

(i) analogue signals;
(ii) digital signals.

Analogue signals are voltages or currents that vary over time in a smooth and continuous fashion, and can take any value within a specified range, as in Figure 2.42(a). The information in an analog signal is contained in the instantaneous value of its voltage or current. For example, if a coolant temperature of 10°C is indicated by a 1 V signal, then 50°C might be indicated by a 5 V signal and 100°C by a 10 V signal. Analog circuits are used to process analog signals and they use transistors as *amplifying elements.*

Digital signals, on the other hand, are signals that can have only two possible voltage levels; an 'ON' voltage (usually the supply voltage), and an 'OFF' voltage (usually 0 V), as in Figure 2.42(b). The information in a digital signal is conveyed via the timing

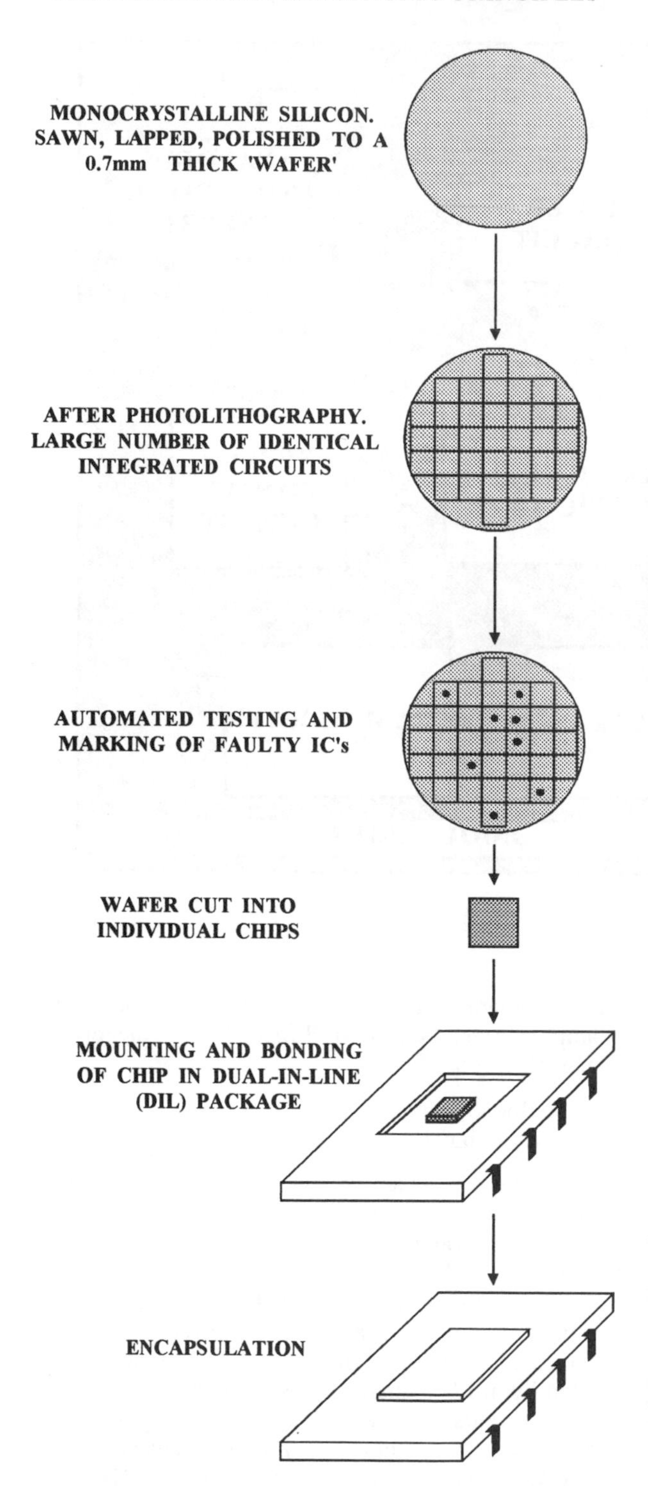

Figure 2.38 Typical steps in the manufacture of an integrated circuit

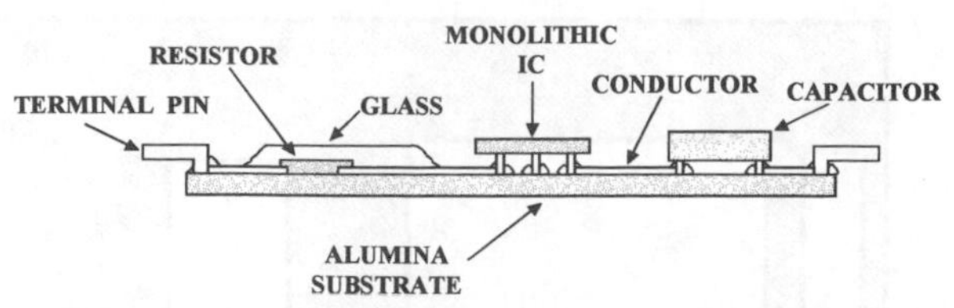

Figure 2.39 Section through a hybrid IC consisting of many separate components mounted on a ceramic substrate

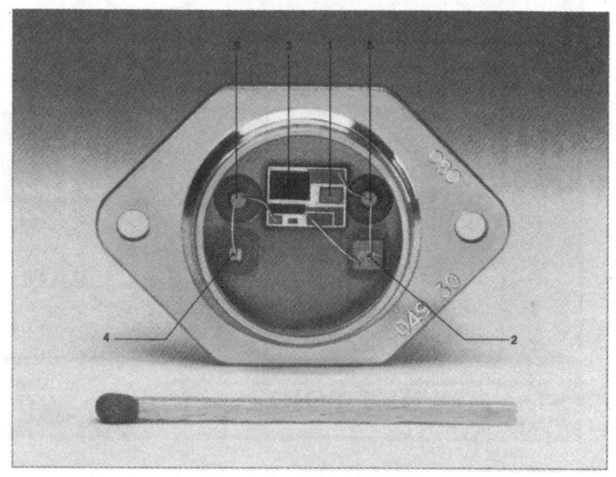

Figure 2.40 Hybrid IC voltage regulator used in alternators. [1] Control IC; [2] Darlington output circuit; [3] dropping resistors; [4] diode; [5] terminals (*Bosch*)

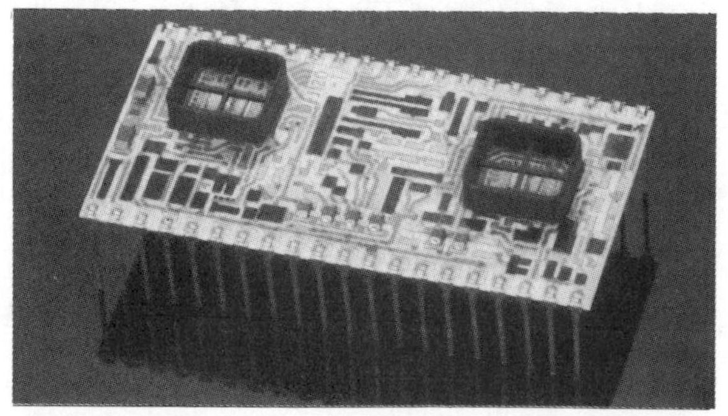

Figure 2.41 Hybrid IC integrating two monolithic ICs on one substrate

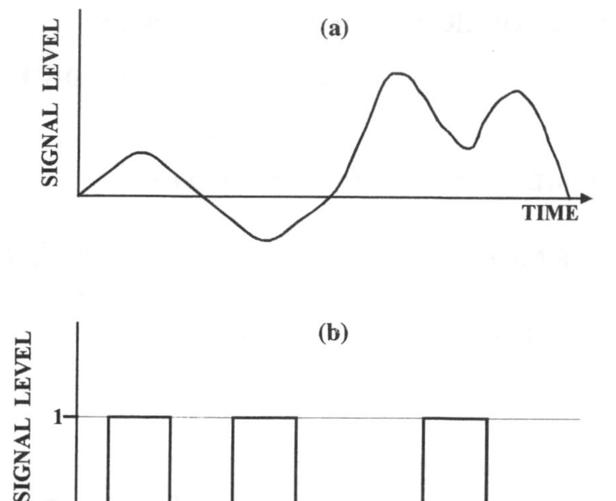

Figure 2.42 Analogue and digital signals: (a) the analogue signal can take any value within a specified range; (b) the digital signal can take one of only two specified voltages

of the digital pulses, or as a coded *binary number*, indicated by a particular sequence of 'ON' and 'OFF' pulses (binary numbers are discussed in more detail in Chapter Four). Digital signals are processed by digital circuits, using transistors simply as *switching elements.*

Integrated circuits may be of the analogue or the digital type and may use either BJTs, MOSFETS or a combination of both. Some ICs, known as *mixed-signal* ICs, contain both digital and analogue circuitry on the same chip. A classification of the types of IC available, in terms of their technology and the category of signal that they process, is given in Figure 2.43. Most of the circuits within automobile electronic control units (ECUs) are digital ICs, a rather smaller number of analogue ICs are also used.

Analog integrated circuits

One of the most versatile analog ICs is the *operational amplifier* (usually shortened to 'op-amp'). An op-amp is a very high performance amplifier, built up from about 20 to 30 transistors, and giving a voltage gain in the region 500 000 to 1 000 000.

Most op-amps come in an 8-pin dual-in-line (DIL) IC package and the manufacturer's data sheet is needed to identify the function of each pin. Figure 2.44 shows a typical op-amp package, together with its terminal connections.

A signal entering terminal 3 (the *non-inverting input*) emerges from terminal 6 (the *output*) unchanged in shape, but with voltage gain. The same signal entering terminal 2 (the *inverting input*) also emerges from the output with gain, but in an inverted (upside-down) state. The op-amp thus functions as a *differential amplifier*; it amplifies the difference in voltage between the inverting and non-inverting inputs. If the *same signal* is applied to both inputs, then there will be *no signal* from the output terminal.

Depending on the connections made to the inputs, and the way in which external components such as resistors, capacitors and diodes are connected, the op-amp can be made to perform a great variety of analog functions. One of the most common op-amp functions used in automobile ECUs is the *comparator.*

The comparator

A comparator makes a comparison between the voltages applied to its two inputs and provides a binary output. In an automotive application, a comparator is frequently used to compare the input from a sensor (a thermistor, potentiometer, etc.) with a fixed reference value.

Typically, when the non-inverting terminal voltage is greater than the inverting terminal voltage, the op-amp's output is approximately equal to the supply voltage. Conversely, when the inverting terminal voltage is greater than the non-inverting terminal voltage, the output is approximately 0 V. Figure 2.45 shows a typical comparator circuit called a *Schmitt trigger.* The voltage from a throttle position potentiometer is applied to the inverting terminal of the Schmitt trigger and compared with a fixed reference voltage of 0.5 V (set by R_1 and R_2). A third resistor (R_3) is used to provide *hysteresis*; this makes the reference voltage alter slightly when the output is 'low' and so makes the comparator more tolerant of a 'noisy' input signal.

When the output from the throttle sensor is below 0.5 V the comparator output is 'high', indicating a closed throttle. When the throttle is opened, the comparator input voltage rises above 0.5 V and so its output goes 'low'. The engine ECU microprocessor can use this signal, together with others relating to vehicle speed, to decide whether or not to implement engine idle speed control.

Digital integrated circuits

Digital circuits categorize input signals as either 'high' (near to the supply voltage) or 'low' (near to 0 V). Engineers often term these the 'H' and 'L', or '1' and '0' *logic states.* Minor variations in the '1' or '0' voltages are ignored by digital circuits, something that makes digital signals particularly resistant to electrical noise.

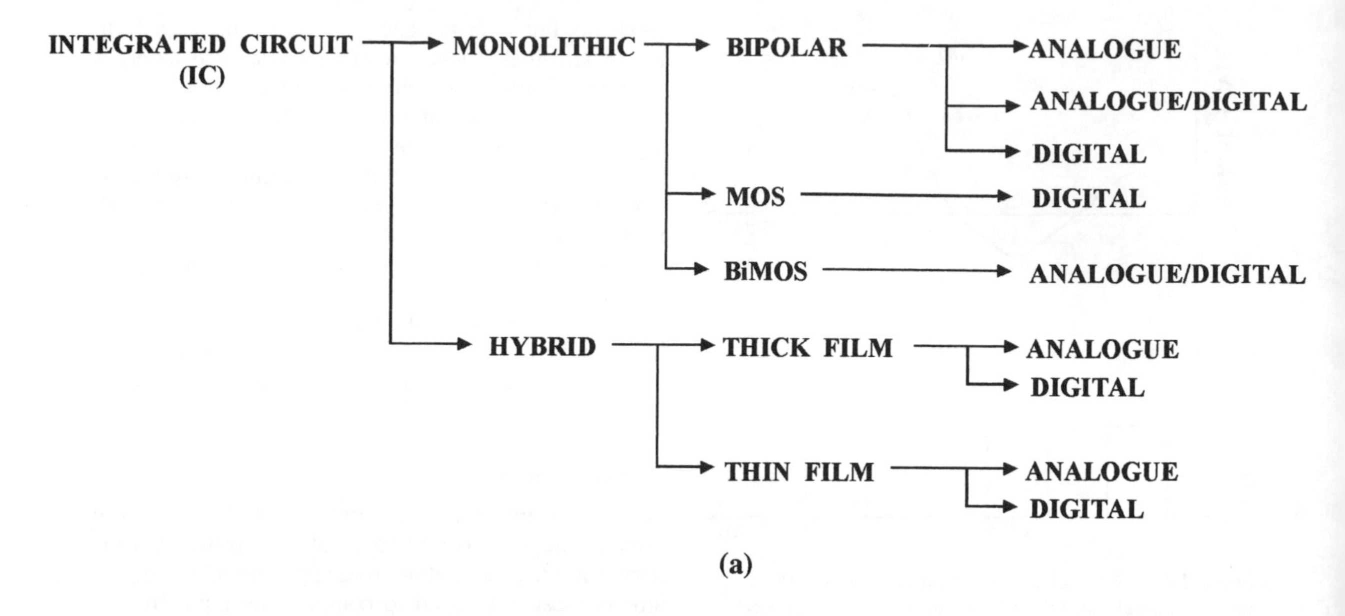

SSI (Small Scale Integration) —— **Up to 100 gates per chip**

MSI (Medium Scale Integration) —— **Up to 1000 gates per chip**

LSI (Large Scale Integration) —— **Up to 100 000 gates per chip**

VLSI (Very Large Scale Integration) —— **More than 100 000 gates per chip**

(b)

Figure 2.43 Classification of integrated circuits: (a) by fabrication technology; (b) by number of digital gates on a chip

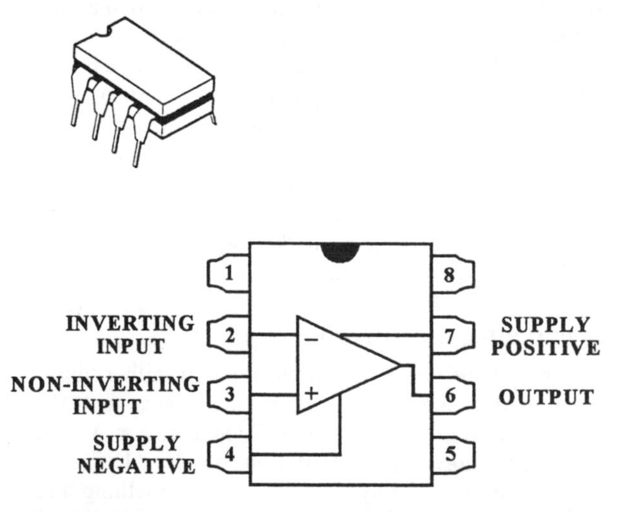

Figure 2.44 Typical 8-pin dual-in-line (DIL) op-amp package and terminal connections

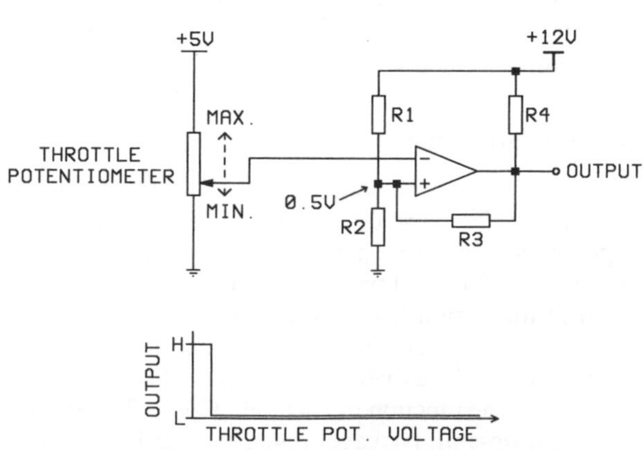

Figure 2.45 Schmitt trigger comparator circuit used to detect when a throttle potentiometer is at the 'idle' position (indicated by a 'HIGH' output voltage)

Circuits that *output* a particular logic state in response to a particular set of *input* logic states are known as *logic circuits* or *logic gates*.

Digital logic ICs come in two main varieties; TTL (*transistor-transistor logic*) which consists of many interconnected BJTs, and CMOS (*complementary metal-oxide-semiconductor*) which consists of interconnected pairs of p- and n-MOSFETS.

TTL chips can switch very quickly between logic states and are often used where high-speed processing of digital signals is required. The drawback is that they consume quite a lot of power and demand an very accurately regulated 5 V supply. CMOS ICs are slower than TTL, but offer the benefits of much lower power consumption, greater noise tolerance and a wide operating voltage range of from 3 V to 15 V. These are important advantages in automotive applications and so automobile circuit designers tend to use CMOS ICs wherever possible.

Logic gates

The basic logic gates are the *AND*, *OR*, *NOT*, *NAND*, *NOR* and *EXCLUSIVE OR* circuits. In an ECU, many logic gates are connected together to provide a particular logic function. In general, the cheapest way to do this is to use an ASIC, which allows the chip designer to pick from a 'library' of logic gates and lay out the circuit exactly as required.

For logic gates with two inputs, called A and B, the operation of each gate type can be summarized as follows;

(i) *AND*: outputs a 1 only if A *AND* B are 1.
(ii) *OR*: outputs a 1 only if A *OR* B is 1.
(iii) *NOT* (1-input): outputs *NOT* A (i.e. inverts the logic level).
(iv) *NAND*: outputs 0 only if A *AND* B are 1.
(iv) *NOR*: outputs 0 only if A *OR* B is 1.
(v) *EXCLUSIVE OR*: outputs a 1 when A and B are *different*.

Figure 2.46 shows circuit symbols and 'truth tables' for each basic 2-input logic gate. The truth table indicates the gate's output, Q, for all combinations of A and B inputs. Note that the symbols shown are to the American standard; a British standard also exists, but it is rarely encountered in automobile circuit diagrams.

2.13 ELECTRONIC CONSTRUCTION TECHNIQUES

When the design of an electronic circuit has been completed the next step is to turn it into a usable, mass-produced, product. The best technique for the manufacture of circuits in quantity is to use a *printed circuit board* (PCB). PCB technology has now advanced to the stage where compact and reliable products can be rapidly manufactured at low cost. A PCB starts out as a rigid sheet of fire-resistant epoxy-bonded fibreglass, about 1–2 mm thick. Both sides of the board are covered with 0.05 mm thickness copper film.

To make the PCB, most of the copper must be etched away to leave a dense pattern of interconnecting tracks. This track pattern is the circuit's 'wiring'. The business of producing a dense and efficiently routed track layout is very involved and is performed with the aid of a computer.

Once the track routeing has been defined, a photographic process, involving a light-sensitive 'etch-resist' and a developer, is used to transfer the track pattern onto the copper. Unwanted material is etched away with ferric chloride, leaving copper tracks which are then electroplated to protect them from corrosion.

Depending on the complexity of the final circuit, several track layers can be sandwiched together to make a 'multilayer PCB', four or six layer PCB's are often used in ECUs. In order to mount the components and provide an electrical pathway ('via') between tracks on different layers of the board, it is necessary to drill a large number of small holes (about 0.8 mm diameter) through the PCB. These holes are then internally copper-plated to provide connectivity and ensure a good solder joint. Finally, a computer-controlled machine is used to 'populate' the board with electronic components before all the joints are simultaneously soldered in a solder-wave bath. Automated test equipment then checks the entire circuit for functionality prior to shipment to the vehicle assembly plant. Figure 2.47 illustrates a typical engine ECU circuit board.

Surface mount devices

In conventional *through-hole mounting*, as described above, the component leads pass through holes in the PCB and are soldered into place underneath. An alternative approach is to use leadless components that solder directly to the surface of the PCB tracks, with no requirement for a hole. Such components are termed *surface mount devices* (SMDs) and because they are much smaller and lighter than conventional components they are preferred by ECU manufacturers.

In practice, most ECUs use a combination of conventional and SMD technology, with most of the heavy and bulky components being leaded.

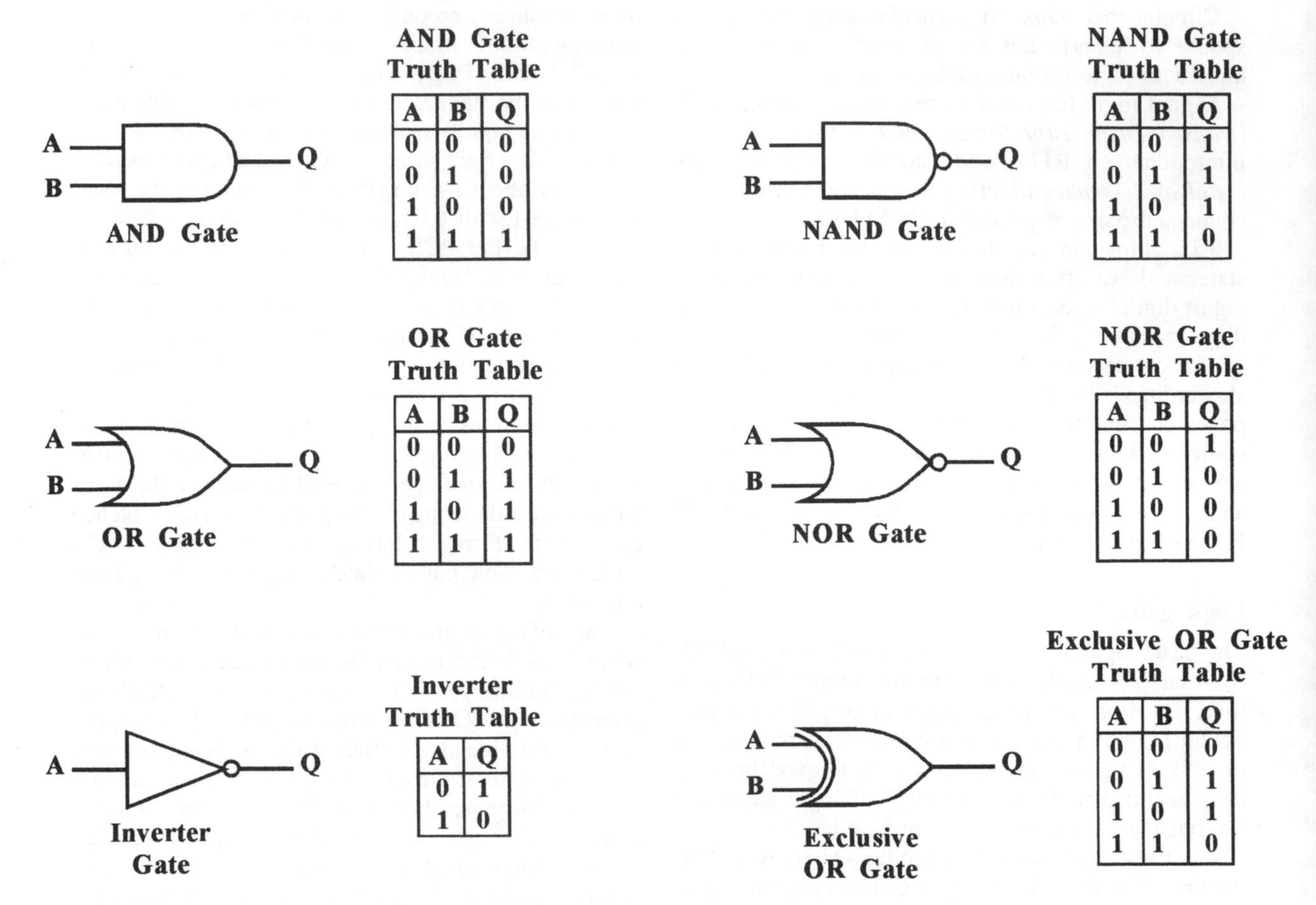

AND Gate Truth Table

A	B	Q
0	0	0
0	1	0
1	0	0
1	1	1

NAND Gate Truth Table

A	B	Q
0	0	1
0	1	1
1	0	1
1	1	0

OR Gate Truth Table

A	B	Q
0	0	0
0	1	1
1	0	1
1	1	1

NOR Gate Truth Table

A	B	Q
0	0	1
0	1	0
1	0	0
1	1	0

Inverter Truth Table

A	Q
0	1
1	0

Exclusive OR Gate Truth Table

A	B	Q
0	0	0
0	1	1
1	0	1
1	1	0

Figure 2.46 Circuit symbols and associated truth tables for commonly encountered logic gates

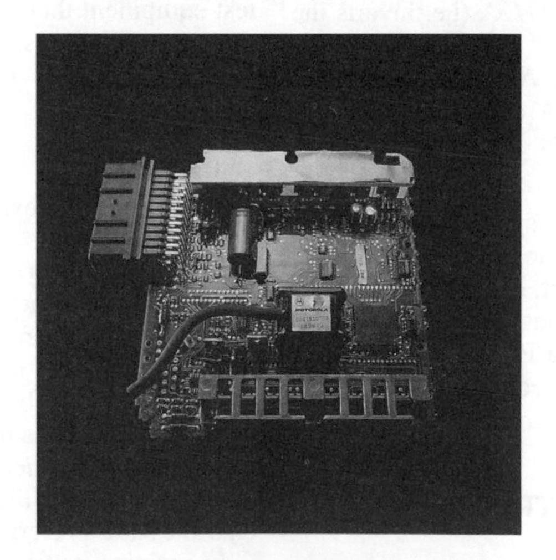

Figure 2.47 Printed-circuit board used in Rover's Modular Engine Management System ECU

Repair of electronic control units

Since the first electronic control units were introduced in the late 1960s great strides have been made in manufacturing technology. Although the automotive environment is far harsher than that in which electronic equipment generally has to operate, ECUs are specially designed to resist it. A typical ECU is designed to operate from −40°C to +125°C and in conditions of high humidity and vibration. As a result, ECUs tend to be highly reliable and, in general, most electronic system failures can be attributed to either maltreatment or to faults in peripheral components such as connectors, sensors or actuators. Repair of ECUs is almost impossible without specialist knowledge, expertise and the ready availability of replacement electronic components. In the majority of cases it is therefore best not contemplated.

3

The automobile electrical system

3.1 INTRODUCTION

The electrical systems of vehicles manufactured up to the mid-1970s were comparatively simple, consisting of just a few circuits for lighting, wiper and heater motors, and a points-type ignition system. The 1980s saw remarkable advances in electronics technology which brought about a corresponding growth in electrical system functionality and a rapid increase in the number and complexity of electrical modules incorporated into vehicles. However, everything has to be interconnected and that means that vehicle electrical wiring systems have become alarmingly complex, with more and more connectors, terminals, relays and control units being required.

In order to allow the automobile technician to grasp the operation of the vehicle electrical system it is normal to decompose it into a number of smaller sub-systems. If an electrical fault arises, diagnosis can then be confined to the particular sub-system affected.

Vehicle sub-systems

Although there are no formal definitions for vehicle sub-systems, the following will be applicable to most automobiles:

Battery and charging system
The battery is a device cable of converting chemical energy into electrical energy, and vice versa. It acts as a store of energy to operate electrical equipment when the electrical load exceeds generator output and when the engine is off. A major task of the battery is to provide the power necessary to crank the engine and get it started. Once the engine is started the charging system (comprising an ac generator, rectifier circuit and voltage regulator) maintains the battery in an optimally charged state and supplies electrical power at a regulated dc voltage to the whole vehicle.

Engine starting system
The engine starting system comprises a powerful (1–2 kW) dc starter motor which cranks the engine when the ignition key is turned. The starter motor is invariably the greatest electrical load on any vehicle and so the whole starting system must be designed to handle very large currents (up to 500 A in some cases).

Lighting system
Legal requirements for vehicle lighting vary from country to country, but in general all vehicles must be equipped with lights that allow the vehicle to be seen by other drivers and its movements anticipated (via brake, turn signal and reversing lights). In order to ensure safe night-driving, most countries enforce minimum requirements for headlight brightness and aim.

Engine management system
The engine management system controls the basic fuelling and ignition parameters for the engine. It generally consists of a microcomputer-based electronic control unit (ECU) to which various sensors and actuators are connected. The ECU's microcomputer is programmed to ensure that the engine is always operating at a condition for minimal exhaust-gas emissions and maximum power and economy. The engine ECU is generally the most powerful control system on the vehicle and may well have overriding control of heater/air-conditioner systems, automatic transmission control systems and so on.

Chassis control systems
This area is generally taken to include anti-lock braking systems (ABS), traction control systems (TCS), four-wheel steering systems (4WS) and electronically controlled suspension systems. They provide enhanced safety and comfort by assisting the driver in controlling the vehicle, especially when driving in adverse or unpredictable conditions.

Body electrical systems
These systems are largely concerned with passenger comfort and safety. They include a broad range of electrical and electronic equipment such as the heater/air-conditioner, instrument pack, washer-wipers, power windows, sunroof, mirrors, seats, etc. As customer expectations have risen, the available range of body electronic systems has expanded enormously in recent years and now includes features such as the supplementary restraint system (SRS or 'air-bag').

In-car entertainment (ICE)
Most vehicles are supplied with a radio-cassette unit ready fitted and many also have provision for a CD player. The ICE system will include a radio aerial, power amplifier and a number of loudspeakers.

3.2 VEHICLE ELECTRICAL DISTRIBUTION SYSTEM

The increased application of electrical and electronic systems in automobiles has resulted in the need for complex electrical distribution systems. A mid-priced, medium-size car of the 1990s typically contains more than 1.5 km of wiring and more than 2 000 terminals, connectors and relays. The weight of such an electrical distribution system exceeds 30 kg.

The major component of the electrical distribution system is the wiring harness. This consists of bundles of cables that connect all of the electrical parts in a vehicle. It has two primary functions; (i) to act as a power distribution network, and (ii) to act as an information distribution network, connecting sensors and actuators with electronic control units.

As may be imagined, the wiring harness presents vehicle manufacturers with many problems; it is very expensive to produce (often only the engine and transmission cost more) and it cannot be accurately specified until late in the car design process (when the exact location of all components has been defined). Moreover, since the majority of vehicle breakdowns are caused by electrical failure, vehicle reliability is critically dependent upon good wiring harness design and installation.

In general, the wiring harness is divided into a main harness that runs the length of the vehicle (connecting the battery to the charging system, vehicle interior, lighting and accessory circuits) and various sub-harnesses (door wiring sub-harnesses, tailgate wiring sub-harness, roof wiring sub-harness). In order to aid vehicle assembly and servicing it is normal for the sub-harnesses to connect to the main harness via connector blocks.

Cable types

Electrical cables for automotive applications consist of several strands of annealed copper wire bunched together and encased in an insulating covering (generally PVC – polyvinyl chloride) of 0.2–0.4 mm thickness. Each strand of copper is typically about 0.32 mm diameter, and the number of strands determines the size, and hence current-carrying capacity, of the cable.

Table 3.1 Sizes and current ratings of wires used in automobile wiring harnesses

Wire size	Current rating (amps)	Resistance per metre (Ω)	Application
7/0.3	4.0	0.032	Side/tail lamps,
9/0.3	5.5	0.027	signals, music
14/0.3	9.0	0.017	system
28/0.3	17.5	0.009	Lights, heater, motors
44/0.3	25.0	0.006	
65/0.3	35.0	0.004	Alternator, main
84/0.3	42.0	0.003	supply to fuse-
97/0.3	50.0	0.002	box
120/0.3	60.0	0.002	
37/0.9	170.0	0.001 or less	Starter motor

Where great flexibility is required, such as in a taxi door sub-harness, very flexible cable is used, made from 0.18 mm annealed copper strands.

In applications where high temperatures are encountered (usually in the engine compartment) ordinary PVC insulation is not used, instead special plastics such a PTFE, PFA, FED or X-ray treated cross-linked PVC or polyethylene are employed.

Cables are generally specified in terms of the strand diameter and number of strands used; for example, a cable specified as 7/0.3 is made from seven strands, each of 0.3 mm diameter. With an insulation thickness of 0.35 mm, such a cable would have a finished diameter of about 1.6 mm and be suitable for carrying currents up to about 4 A.

In order to reduce costs, vehicle manufacturers use the thinnest possible cable for a given application, without incurring too much voltage drop. As a 'rule-of-thumb', a maximum voltage drop of about 5% (i.e. 0.6 V in a 12 V system) is permissible for general lighting and control circuits. In general, a current rating of about 8.5 A per square millimetre of cable cross-section is assumed. This may be derated to about 4 A per square millimetre for continuously loaded cables. Table 3.1 summarizes typical current ratings and resistances for commonly used automobile wire sizes.

Harness routeing

The bundle of cables which make up the wiring harness generally has a diameter in the range 10–30 mm. This bundle must be carefully routed around the vehicle, avoiding trim screws, mounting bolts and extremes of temperature (exhaust system, air-conditioner components). Clamps are used at regular intervals to secure the harness to the bodyshell and guard against stressing of the wires. Where the harness passes through a metal panel it is normal to fit a rubber grommet, thereby protecting the harness against chafing and preventing the ingress of moisture and dirt.

Where many cables go to the same location (for example, the rear light clusters) ribbon cable is often used. Ribbon cable is simply a number of conductors laid side-by-side to make a wide, flat, harness which is easy to conceal underneath carpeting and along flat body panels.

Cable colour coding

The complexity of modern vehicle wiring harnesses means that some form of cable colour coding is essential for fault diagnosis and repair purposes. The colour codes used vary from one manufacturer to another (and sometimes between different models from the same manufacturer!) and so it is absolutely vital to consult the vehicle repair manual before commencing any electrical work.

Most manufacturers code cables by giving each a base colour and then adding a contrasting tracer. These colours can then be indicated by a letter code on the vehicle wiring diagram. For example, a cable indicated 'BO' on a Rover wiring diagram would be black with an orange tracer. Note that in order to cut costs, the 'tracer' may be a thin painted line added *only* where the cable enters a connector. Some manufacturers (for example Citroën) use just a few base colours and then code the wires by adding small coloured sleeves at each end.

Connectors

Connectors join cables together, both physically and electrically. They have a great impact on vehicle reliability and so good design is very important.

Up until the late 1950s, most automobile electrical connections were achieved by crimping a small eyelet onto a cable and then securing it to the electrical component with a bolt. Such bolt-and-eyelet terminations are still used in certain applications (typically starters and alternators), but for light-current uses (up to about 17.5 A) the familiar 'Lucar' type spade terminal was introduced in the 1960s. During the mid-1970s, however, more electronics came into use and this increased the demand for connectors of greater reliability. Block connectors, in which a large number of terminal pins are held in one housing, were developed; locking tabs and water seals were also introduced. In recent years there has been a great proliferation in the types of connector available, but most incorporate the following features:

(i) An anti-backout lock. This locks the terminal pins into the connector housing, preventing the pins from sliding out of the housing as the connector is pushed together.
(ii) A snap lock. This is a locking tab that produces an audible click as the connector is pushed together – indicating to the technician that the connection is pushed fully home and is locked.
(iii) Provision for waterproofing. A rubber seal may be fitted so that the terminal pins are protected from the corrosive effects of water-splash.
(iv) A keyway. This ensures that the connector is fitted in one orientation only.

Since electrical system reliability increases as the number of connections decreases, keeping the number of connections to a minimum is a major design goal. In the early 1990s manufacturers introduced bolt-tightened multipole connectors. These combine what were four or more small connectors into one large connector block of 80-plus pins, cinched home by a bolt through its middle.

In general, the connector terminal pins may be circular or flat, and can be gold-plated where low-level signals are to be carried. The terminals are usually crimped onto the cable and then inserted into the connector-block housing. Polyamide 66 plastic, reinforced with glass fibres, is normally used for the housing. It is very strong, light and fire-resistant. Because of the proliferation of connector types the degree of interchangeability between vehicles is minimal. When a connector is faulty or damaged it must be replaced with an identical, original equipment, part.

A very recent trend in connector technology is to replace the connector parts with terminated wiring in which the cables and connector form an integral, moulded, unit. This increases reliability still further and makes for faster vehicle assembly.

Fusing

A fuse is a vital safety component in all electrical circuits. It is essentially a 'weak link' which melts and isolates the circuit if the current exceeds a predetermined limit. Fuses are necessary because a car battery is capable of providing a very large current (hundreds of amps) and in the event of an accidental short-circuit such a current could generate sufficient heat to melt plastic insulation or even start a fire.

Vehicle designers generally equip a car with one main fusebox (often incorporating relays and electronic modules) and other, smaller, supplementary fuseboxes serving optional equipment (ABS, electric windows, seats, etc.). Current from the battery reaches the main fusebox via the main supply cable and is then distributed to each electrical circuit via a fuse of appropriate rating.

Three types of automobile fuse are in common use:

(i) moulded plastic housing with flat blades;
(ii) tubular glass cartridge;
(iii) ceramic-bodied fuses.

A fuse always has a current value marked on its body. In the case of flat-blade and ceramic type fuses this value is an indication of the maximum continuous current that it can carry. For glass cartridge fuses it is the current at which the fuse will 'blow' (typically twice the maximum continuous current). As a rule-of-thumb, a fuse should have a rating at least 50% higher than the nominal continuous current in the circuit, this is for two reasons:

(i) A fuse will eventually blow due to thermal fatigue if continuously run near its rated current.
(ii) Some margin must be incorporated to allow for current rise due to variations in supply voltage or load resistance.

The electrical system designer will have determined the correct fuse rating for each circuit, so always follow the manufacturer's recommendation when replacing blown fuses. A circuit which repeatedly blows fuses has a fault which requires rectification. Most such faults are due to intermittent short-circuits caused by loose terminations or damaged insulation.

It is very important to appreciate that fuses are designed only to protect electrical wiring. They offer little protection to electronic circuits. If an electronic module is wrongly connected an excessive current may flow; but in the small time it takes for the fuse to heat up and melt, permanent damage may be done to the circuitry. Where it is necessary to protect an electronic system against improper connection, special circuitry (employing zener diodes and current-limiting devices) is incorporated within the electronic module.

Thermal circuit protection

Fuses are a very cheap and reliable means of protecting electrical circuits. In some instances, however, it is desirable to use other types of protection.

Thermal breakers are occasionally used in lighting and motor-drive circuits. A typical application is in an electrically operated window mechanism, where jamming would cause the motor to stall and overheat. A thermal circuit breaker (consisting of a pair of contacts and a bimetallic strip) is connected in series with the motor. If the current to the motor becomes excessive, the bimetallic strip heats up and bends, causing the contacts to open and thus isolating the circuit. When the strip cools down the contacts close again and so power is reapplied to the window mechanism.

3.3 MULTIPLEX WIRING SYSTEMS

As the electrical content of the typical car has grown, the wiring harness servicing the electrical equipment has also grown, both in size and complexity. Since the early 1980s automotive engineers have been working on other methods to switch power to the various circuits around the vehicle. Multiplexing is one such technique.

The main feature of a multiplex wiring system is that the same wiring is used repeatedly. Each piece of information (from switches, sensors and ECUs) is converted into a digital (ON–OFF) signal and transmitted serially throughout the system. Figure 3.1 uses a simple example to illustrate the technique. The four motors A,B,C and D are operated from four switches. Information on the position of each switch is sent by the sending unit, in a fixed sequence, along just one transmission line called a data line. The receiving unit decodes the digital data and operates relays or transistor switches to supply current to the appropriate motor. Since each burst of data is sent many times every second, the response of the system appears almost instantaneous as far as the driver is concerned, just as if the switches were wired directly to the motors. A practical multiplex system would incorporate many sending and receiving units, located at strategic points around the vehicle and all connected to the same data bus.

A practical multiplex system

The additional cost of multiplex systems, along with design conservatism, has meant that relatively few systems have reached the market. Multiplexing, so far, has largely been confined to just a part of the electrical system (for example, the door sub-harnesses). As an example of practical low-cost 'whole vehicle' multiplexing, the system designed by Lucas Rists Wiring Systems is described.

The Lucas Rists system uses what is known as a 'multi-master' multiplexing architecture and employs a simple two-wire data bus which carries information around the vehicle. Electronic modules called active junction boxes (AJBs) are attached to the data bus

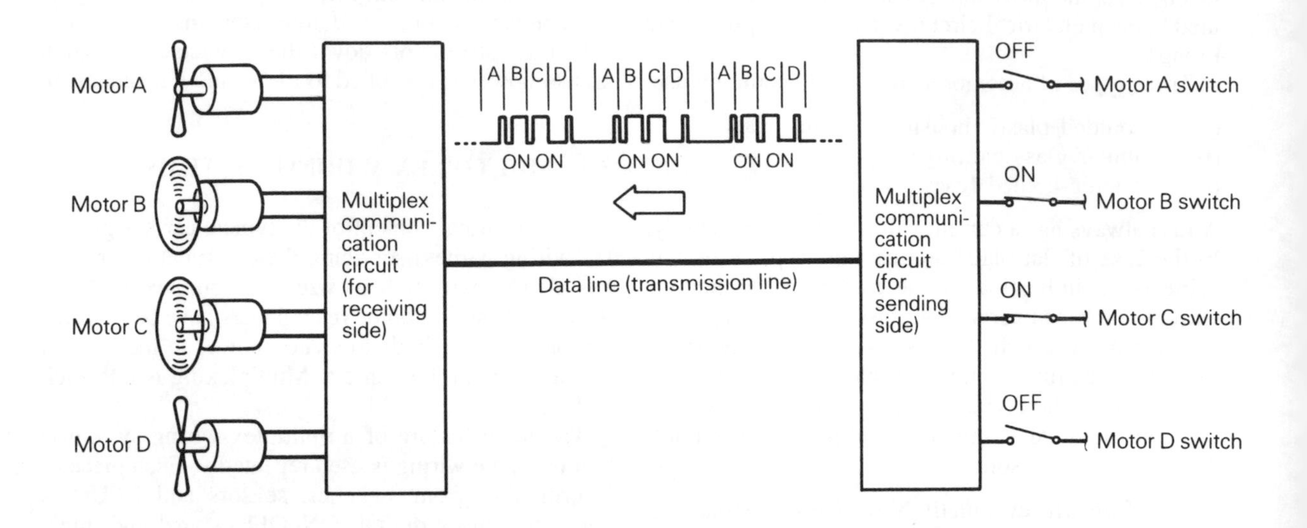

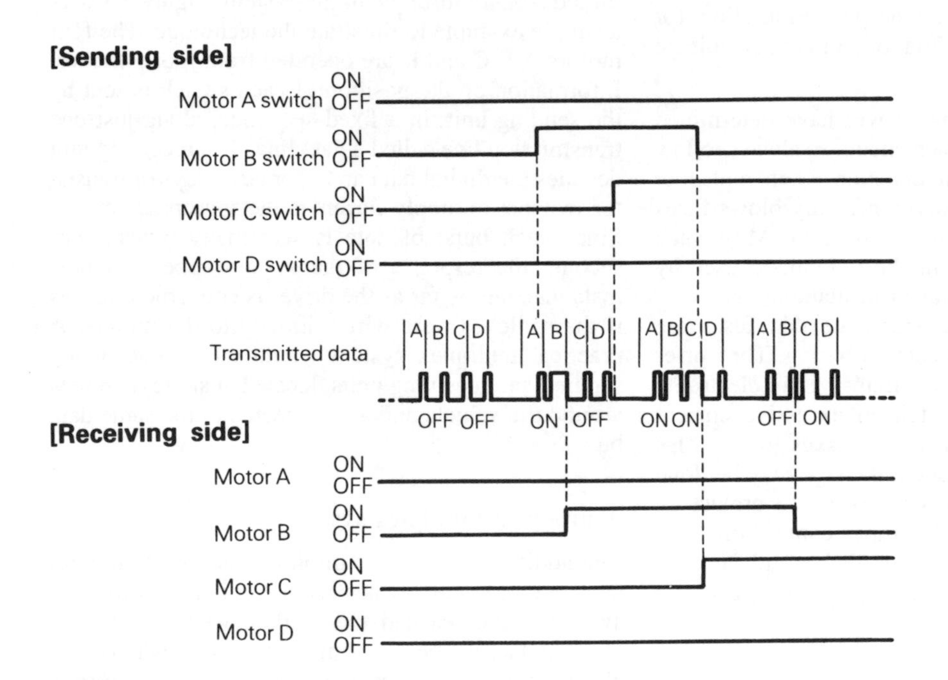

Figure 3.1 Use of multiplexing to transmit multiple signal information along a single wire by transmitting different signals at different times (*Mitsubishi*)

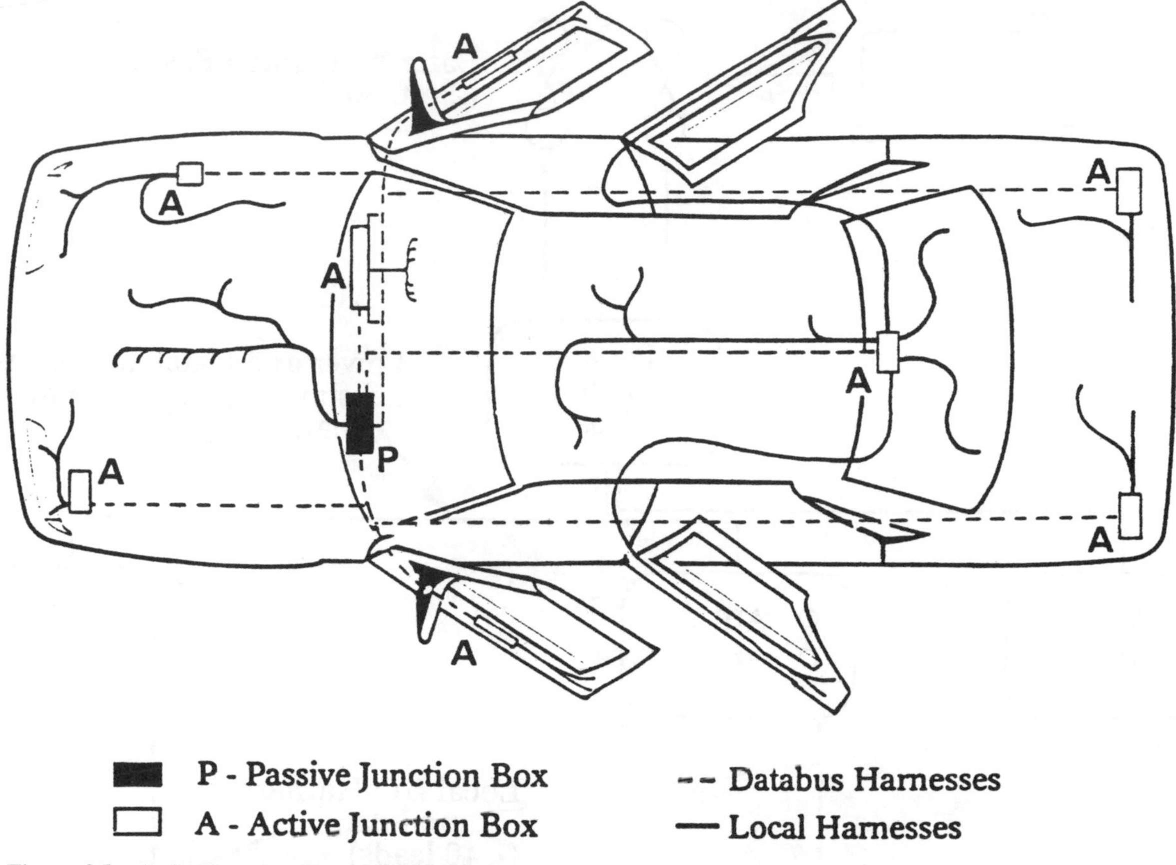

Figure 3.2 Active Junction Box (AJB) locations around the vehicle (*Lucas Rist's Wiring Systems*)

and each is capable of transmitting and receiving data to or from any of the others. Each AJB module is connected to a local wiring sub-harness which controls the actuators and monitors the switches/sensors in its own region of the vehicle (Figure 3.2). The system operates by having control data sensed at the AJB module nearest to the relevant switch or sensor. This information is then broadcast on the bus and received by all the other AJB modules, with those that need to act on this information doing so via their local sub-harness. The interconnectivity of the modules is shown in Figure 3.3.

From the battery, current is delivered to the passive junction box (PJB) via two fused cables. From here it is distributed to all of the AJBs around the vehicle. Each AJB contains a number of silicon power switches that control locally fused power feeds to the electrical loads in its vicinity.

The passive junction box has three main functions:

(i) It distributes fused power to the AJBs.
(ii) It acts as a central 'star-point' for the data bus.
(iii) It provides interconnection of the harnesses associated with (i) and (ii).

There are eight active junction boxes located around the vehicle and they each have four main functions:

(i) Transmission and reception of information via the data bus.
(ii) Fused power switching to local loads.
(iii) Acceptance of data from local switches and sensors.
(iv) Provision of an interconnection of the local sub-harness.

Advantages of multiplexing

Multiplex wiring systems offer a number of advantages over a conventional 'point-to-point' wiring harness, in particular:

(i) Multiplex wiring harnesses are smaller and simpler, saving on cost and weight.

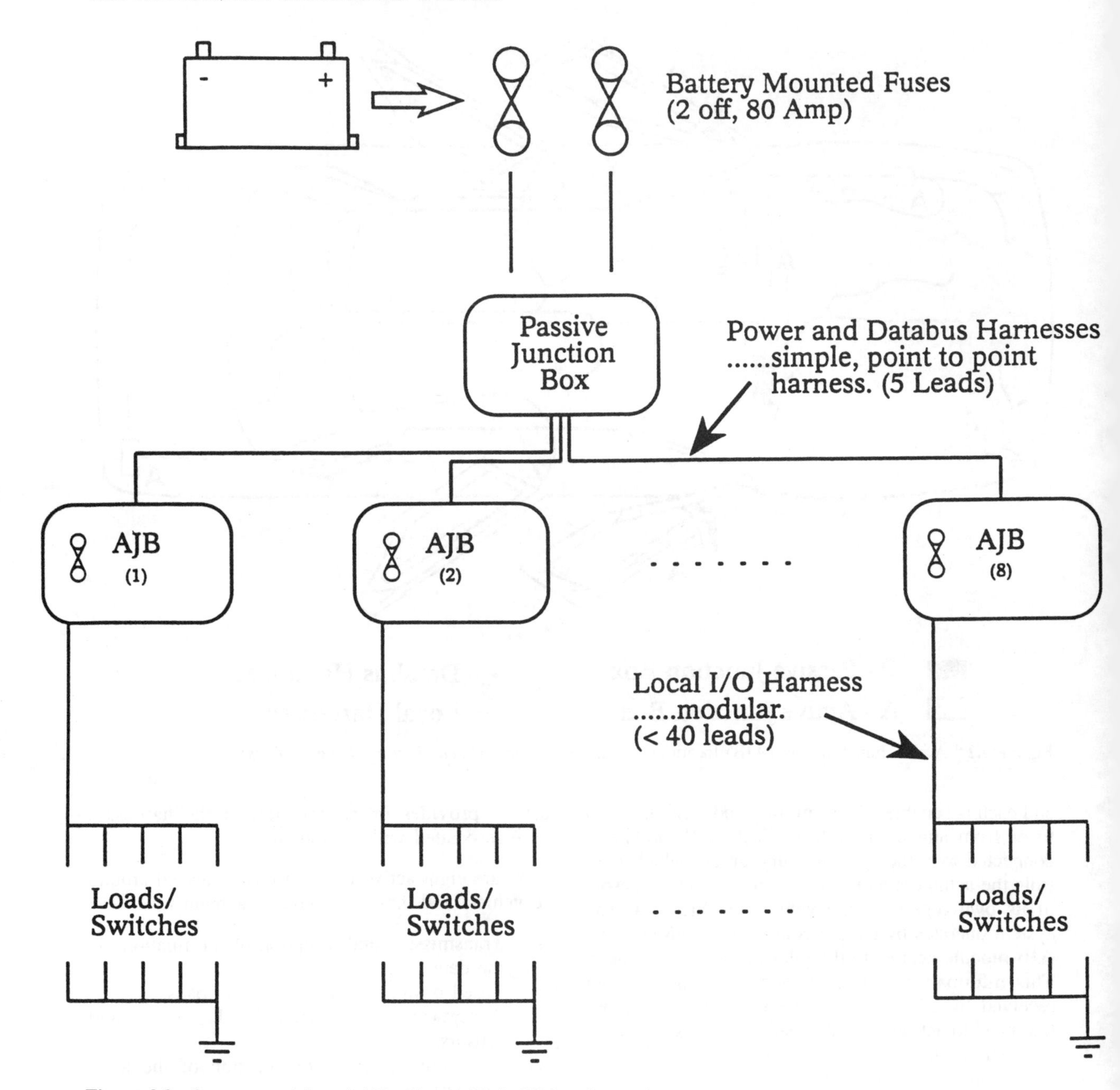

Figure 3.3 Interconnectivity of AJBs (*Lucas Rist's Wiring Systems*)

(ii) Automated manufacture of the harness is made easier.
(iii) Much faster harness installation on the vehicle assembly track is possible.
(iv) Reliability is improved due to a reduction in the number of wires and connectors.
(v) Self-diagnosis features can be built into the AJB electronics to help the technician trace faults.

Controller area network (CAN)

Multiplex wiring systems of the type described above are categorized by the Society of Automotive Engineers (SAE) as Class A systems. Class A systems are 'low-speed buses', capable of transferring several thousand pieces of data per second and suitable for use in the control of body electronic systems such as windows, sunroof, lights, etc. When data must be exchanged at a much faster rate (as is the case with engine and transmission management, ABS and traction control electronics) Class B or Class C bus systems must be used. During the 1980s many vehicle manufacturers investigated such systems and a number of competing protocols were proposed, notably ABUS (Volkswagen-Audi), VAN (consortium of French manufacturers) and J1850 (US manufacturers), however CAN (Bosch) has become the most widely accepted standard for Europe.

In 1987 Bosch defined the Controller Area Network (CAN) as a Class C system for intracar (within the vehicle) communications and in 1988 Intel Corporation produced the first network controller chip, the 82526 CAN IC, making CAN a practical proposition for vehicle manufacturers.

CAN is widely predicted to become the standard bus system for Europe. It was proposed as an ISO standard (TC22/SC3/WG1) in 1990 and in 1991 Daimler Benz installed a CAN network in their new S-series Mercedes-Benz cars.

Basic concepts of CAN

CAN is a high speed serial data bus that can transfer up to 1 million bits of data per second. All engine and chassis control ECUs can be connected to the common CAN bus, and the direct mode of communication between the components means that it is possible to reduce the overall number of sensors and actuators by 'sharing' their information between the various ECUs. Ideally, it allows vehicle manufacturers to design vehicle electronic systems using ECUs from different suppliers, all working together in a cooperative manner with a minimum of interconnecting wiring.

With CAN, only one sensor is required per measurement variable to supply all systems with this signal. For safety and reliability, however, CAN can use two sensors for double testing of the same measurement parameter.

CAN has the following basic properties:

(i) It is a multi-master system, meaning that each ECU has the authority to temporarily control the action of all the other ECUs.
(ii) When the bus is not transferring data, any ECU may start to transmit. If two or more ECUs start to transmit at the same time, the ECU with the most important data gains bus access.
(iii) An ECU can request any other ECU to send data.
(iv) The system is able to detect and signal that data transmission errors have occurred. If data is destroyed by errors during transmission, then it will be automatically retransmitted.
(v) The system can distinguish between temporary errors and permanent failures of ECUs. Defective ECUs can be automatically switched off.

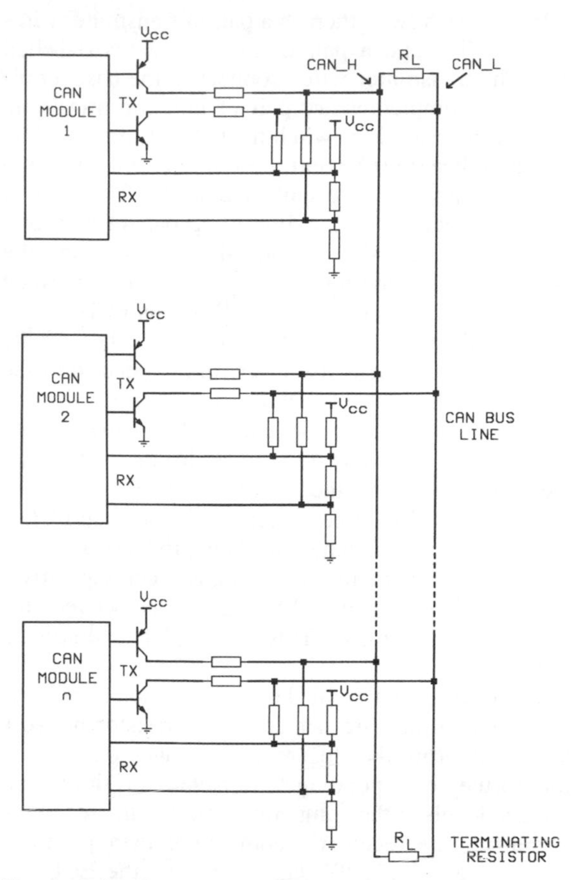

Figure 3.4 Connection of CAN modules onto the CAN databus

CAN bus operation

CAN uses a common bus line of two parallel wires which can be screened or unscreened, depending upon the application. The individual wires are designated as CAN_L and CAN_H. The names of the corresponding connector pins on ECUs are also denoted by CAN_L and CAN_H respectively.

The total length of the bus wires can be up to 40 m, and up to 30 ECUs may be connected to the bus line using short stub wires which can be up to 30 cm in length (Figure 3.4). The bus line is terminated at each end with a load resistor, denoted R_L, which suppresses electrical reflections.

Within each ECU there is a pair of transmitter wires (labelled Tx) and a pair of receiver wires (labelled Rx). The transmitter wires connect to the base terminals of a complementary pair of npn/pnp transistors which can be used to switch the CAN_L line to a low voltage (close to 0 V) and the CAN_H line to a high voltage, denoted V_{CC} (usually about 5 V). The receiver wires connect to the bus line via a pair of resistors.

CAN transfers data around the bus by using the principle of generating a voltage difference between the CAN_L and CAN_H wires. When an ECU wishes to transmit a logic '1' data bit, it switches off its transistor pair. The resistor network connected to the bus causes both the CAN_L and CAN_H wires to assume approximately the same voltage level (about 2.5 V), and so there is no voltage difference between them. This is called the 'recessive state'.

If a logic '0' is to be placed on the bus, the ECU's transistor pair is switched on. This produces a current through the terminating resistors, and consequently a differential voltage of 2–3 V is generated between the two wires of the bus. This is called the 'dominant state'.

To receive data from the bus, the 'dominant' and 'recessive' states are detected by a resistor network that transforms the differential voltages on the bus line to the corresponding 'recessive' and 'dominant' voltage levels at the comparator input of the receiving circuitry. The receiver's comparator then produces logic '1' and logic '0' data for use by the ECU.

3.4 BATTERIES

Batteries are electrochemical devices for storing electrical energy in a chemical form. Active materials in the battery react chemically to produce a flow of direct current whenever motors, lights or other current consuming loads are connected across the terminals.

In automotive applications the battery performs four functions:

(i) It supplies current to accessories when the engine is not running.
(ii) It supplies energy to the starter motor and ignition system when the engine is started.
(iii) It intermittently supplies current for the lights, heater and other accessories when the electrical demands of these devices exceed the output of the generator.
(iv) It acts as a voltage stabilizer for the electrical system.

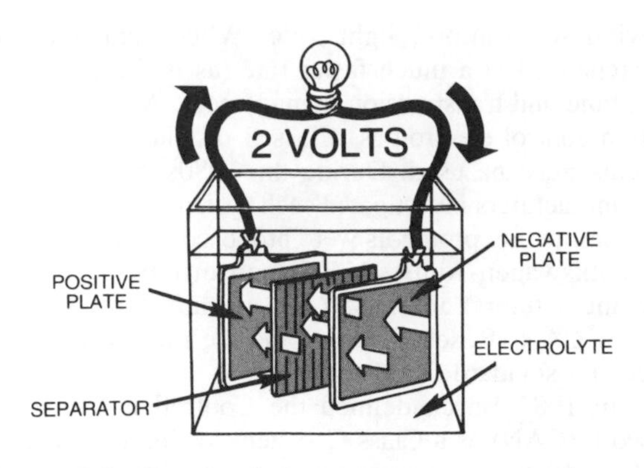

Figure 3.5 The chemical reaction generates a voltage of about 2 V

The basic cell

A basic battery cell is formed when two plates of dissimilar material are placed in a solution called an electrolyte. In automobile batteries the plates are made of lead (chemical symbol Pb) and lead dioxide (PbO_2), and the electrolyte is a solution of sulphuric acid and water (H_2SO_4 and H_2O).

Due to the chemical reaction that occurs between the electrolyte and the dissimilar plates, a voltage of about two volts exists between the two plates. If the cell is connected to a 2 V bulb, current flows from one plate through the electrolyte to the other plate, and then through the bulb to complete the circuit (Figure 3.5).

Since virtually all car electrical systems are designed to operate at a voltage of about 12 V, a car battery is comprised of six separate 2-volt cells, connected in series.

Battery construction

Grids

The basic component of the battery is the grid. Grids are flat and ridged mesh-like structures, designed to hold the soft plate material which is pasted onto them. Traditionally, grids were made of lead, with a small amount of antimony added as a hardening agent to give strength. Unfortunately, antimony reacts chemically within the battery leading to the consumption of a small amount of water – hence the requirement to periodically 'top-up' old-style batteries. In the early 1970s grids using a lead-calcium alloy were introduced. These do not react and so water usage is greatly reduced, resulting in the concept of a 'low-maintenance' or 'no-maintenance' car battery that needs no 'topping-up'. Most original equipment car batteries are now of the lead-calcium type construction.

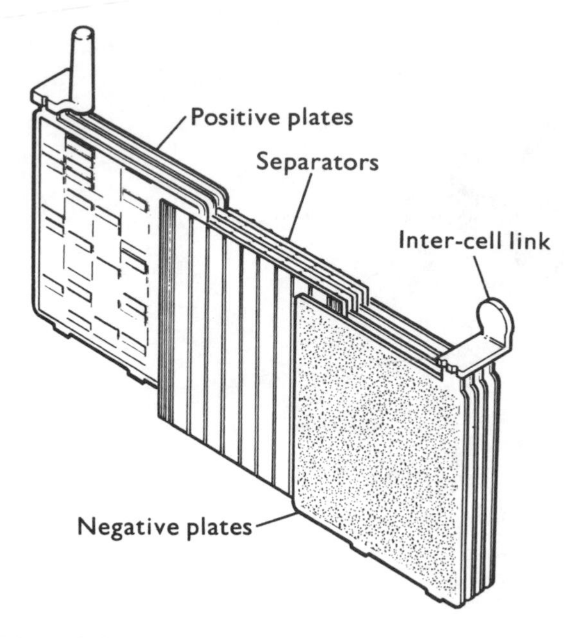

Figure 3.6 Plate group (*Lucas*)

Plates
Once the grids have been pasted with active materials they are called plates. Negative plates consist of grey coloured sponge lead (Pb) and positive plates are made from dark-brown lead dioxide (PbO_2). Groups of plates of the same type are made by attaching a cast strap to their upper edge. Plate groups of opposite polarity can then be interlaced so that positive and negative plates alternate, as shown in Figure 3.6. In order to prevent the plates from touching (and producing a short-circuit) they are placed into porous envelope separators, made from an insulator such as paper or PVC. The final assembly of plates, separators and two connecting straps is called a cell element. An element always provides around two volts, but its current capacity increases as the number of plates is increased.

Case
Battery cases are usually manufactured from polypropylene, a strong, light and durable plastic. The case is partitioned into six cell enclosures (for a 12 V battery) and a cell element is located in each enclosure. The cells are then series-connected by using a through-the-partition (TTP) inter-cell link to join the negative plates of one element to the positive plates of its neighbour. The positive plates at one end of the battery, and the negative plates at the other are then taken to external terminal posts. These are the output connections of the battery.

Table 3.2 State of charge as a function of specific gravity reading for a healthy battery

Specific gravity	State of charge
1.270	100%
1.230	75%
1.220	65%
1.200	50%
1.170	25%
1.090	0%

Electrolyte
The electrolyte is a solution of about 36% sulphuric acid (H_2SO_4) and 64% pure water (H_2O). It has a specific gravity (or SG) of 1.270, meaning that for equal volumes it weighs 1.270 times as much as pure water. The specific gravity of the electrolyte will vary in accordance with its chemical composition, and hence can be used to determine the state of charge of the battery. Table 3.2 shows how specific gravity can be related to state of charge; as the battery becomes discharged, its electrolyte becomes more and more like water. Specific gravity (and therefore state of charge) is usually measured with a hand-held device called a hydrometer.

A hydrometer is simply a glass bulb containing a calibrated float. Electrolyte is drawn out of the battery, into the bulb, and the SG then read off the float scale. A good battery will have a dense electrolyte, causing the float to sit high, giving a SG reading of 1.250 or more. A flat battery will have a watery electrolyte and so the float will sit low, giving an SG reading of 1.170 or less. Certain batteries of the 'no-maintenance' type are entirely sealed and so have a hydrometer built into the plastic cover. Figure 3.7 shows the construction of the Delco 'Freedom' battery, which incorporates this useful feature.

Chemical reactions in the battery

The chemical reactions occurring during battery charging and discharging are shown diagrammatically in Figure 3.8.

During discharge, oxygen from the positive plate combines with hydrogen in the electrolyte to form water. Simultaneously, lead from the positive plate reacts with the sulphate ion (SO_4) to form lead sulphate ($PbSO_4$). A very similar reaction takes place at the negative plate, where lead combines with the sulphate ion (SO_4) to form lead sulphate ($PbSO_4$). Thus, as the battery is discharged, lead sulphate is produced on both plates and the electrolyte becomes

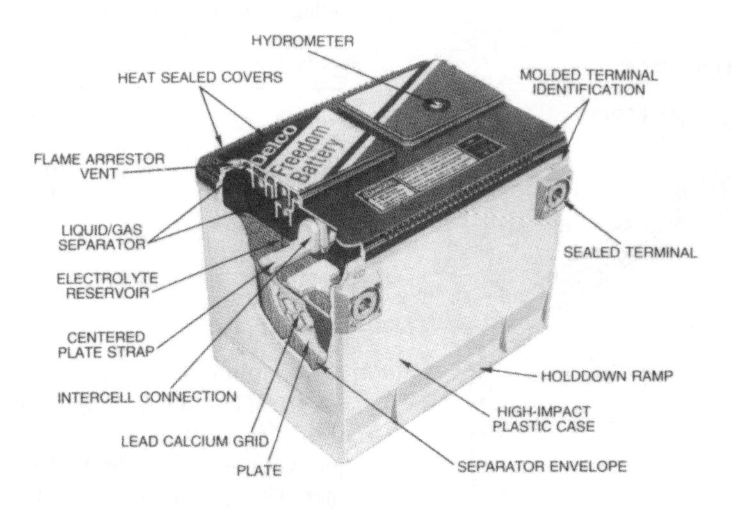

Figure 3.7 Construction of the Delco 'Freedom' battery with built-in hydrometer (*Delco*)

more dilute (lower specific gravity). Eventually dilution of electrolyte and a build-up of sulphate on the plates causes the reactions to stop – the battery is then discharged.

Sometimes a very high current discharge (as when cranking a cold engine) will produce localized plate sulphation and electrolyte dilution, reducing the battery output. Waiting for a few minutes (the *recovery time*) will allow the chemical concentrations to equalize and the battery will then provide an additional cranking period.

The chemical reactions occurring during charging are simply the reverse of those occurring during discharge. The lead sulphate on both plates separates into lead (Pb) and sulphate (SO_4). The sulphate leaves the plates and reacts with hydrogen (H_2) in the electrolyte to form sulphuric acid (H_2SO_4), raising the specific gravity. The negative plates return to their original lead form. Simultaneously, oxygen (O_2) from the electrolyte combines with lead at the positive plates to form lead oxide (PbO_2). Once a battery is fully charged the charging current should be switched off. Overcharging causes water in the electrolyte to break down and be released as gas (oxygen at the positive plate, hydrogen at the negative plate) and can also lead to overheating of the battery, resulting in warped and damaged plates.

It is interesting to note that during discharge the plate material expands slightly, whilst during charging it contracts back to its normal size. When the battery is in use this continuous expand-and-contract cycling eventually causes the plate material to loosen and drop off, resulting in a 'worn-out' battery that does not hold charge. This is one of the reasons why batteries need replacing after several years' service.

Battery charging

The battery is normally kept in a charged state by the vehicle's generator. When the generator is running, its output voltage must be carefully controlled in order that the battery is not overcharged and possibly damaged.

In order to charge the battery, the generator must force a charging current through the cells in a direction opposite to that of the discharge current. Obviously, for a charge current to flow, the generator's output voltage must be greater than the battery's terminal voltage, V_B. Another factor which must be taken into account is the opposition to the charging current I_G, presented by the internal resistance of the battery, R_i. Mathematically, the generator output voltage must be at least V_G for charging to occur, where,

$$V_G = V_B + I_G R_i \quad (3.1)$$

Because lead-acid batteries have a very low internal resistance (typically less than 0.005 Ω), $I_G R_i$ will only amount to about 0.1 V at typical charging currents. However, since a fully-charged battery will attain a voltage of about 2.4 V per cell (Figure 3.9) this

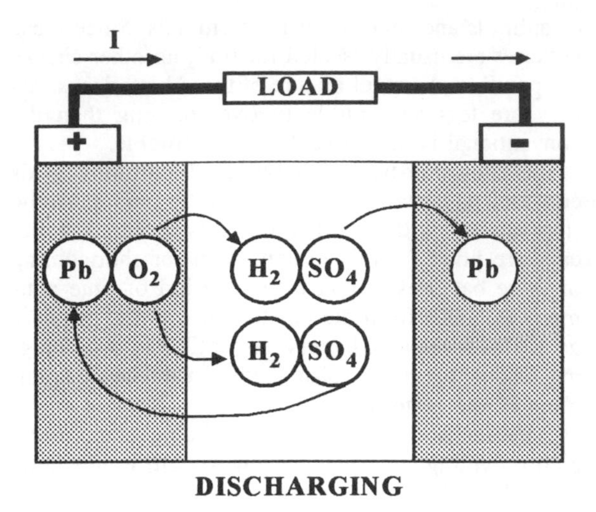

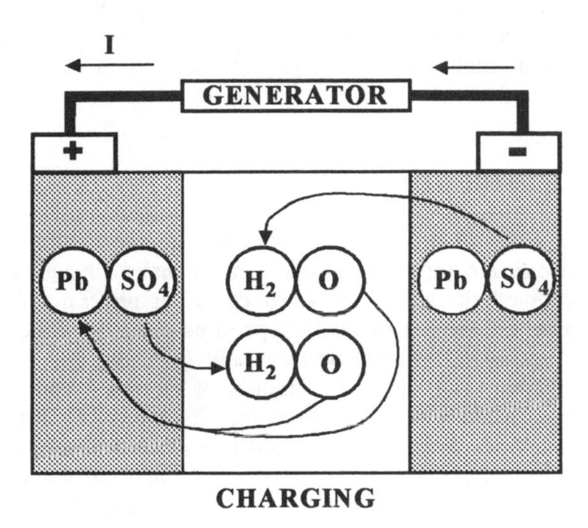

Figure 3.8 Chemical reactions occurring during battery charging and discharging

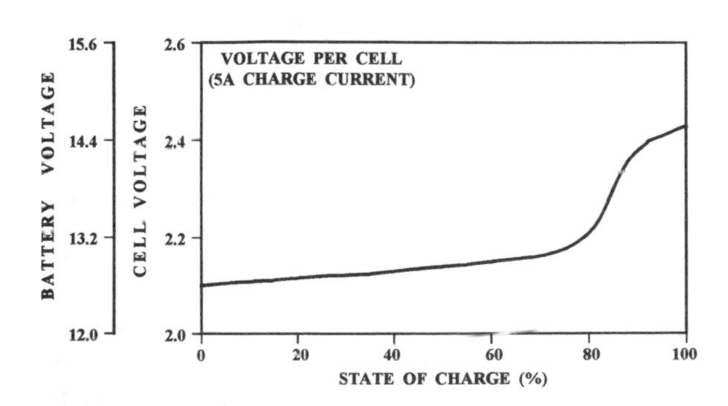

Figure 3.9 Charge curve for lead-acid battery

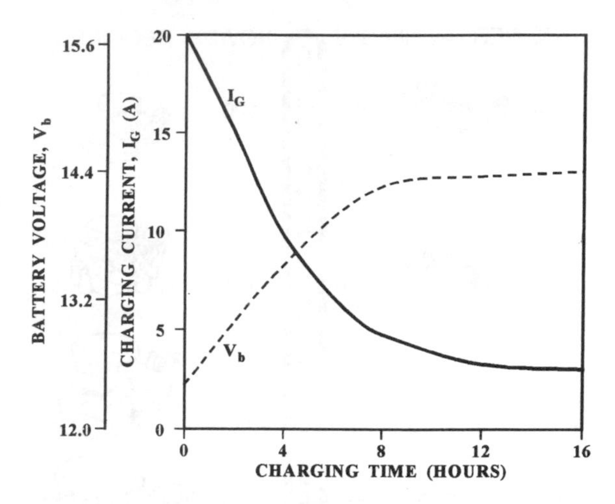

Figure 3.10 Charging current, I_G, at 14.5 V charging voltage

indicates a maximum V_B of about 14.4 V for a six-cell '12 V' battery, and so most generator voltage regulators are set to give an output of approximately 14.5 V.

The charging current, I_G, is dependent on R_i and the difference between the battery terminal voltage, V_B, and the generator output voltage, V_G. Mathematically,

$$I_G = (V_G - V_B)/R_i \tag{3.2}$$

Equation 3.2 indicates the characteristic shown in Figure 3.10 where I_G is large when V_B is low (i.e. a discharged battery) and falls as V_B rises, becoming very small when V_B is about 14.4 V (almost fully charged). This characteristic automatically helps protect the battery from overcharging and overheating.

Battery specification

Although all car batteries supply a nominal 12 V output, they are available with different storage capacities. When vehicle designers specify a battery, they need to be sure that it will perform adequately under all likely loads and temperatures. Cold-starting imposes a particularly heavy demand on the battery. This is for two reasons; first, at low temperatures the chemical reactions that generate current slow down and so battery output falls, secondly, a cold engine has higher oil and bearing resistances and so is more difficult to crank. This situation is illustrated in Figure 3.11, and it shows why choosing a battery of sufficient capacity is essential. For European markets, most manufacturers specify a battery capable of providing the required

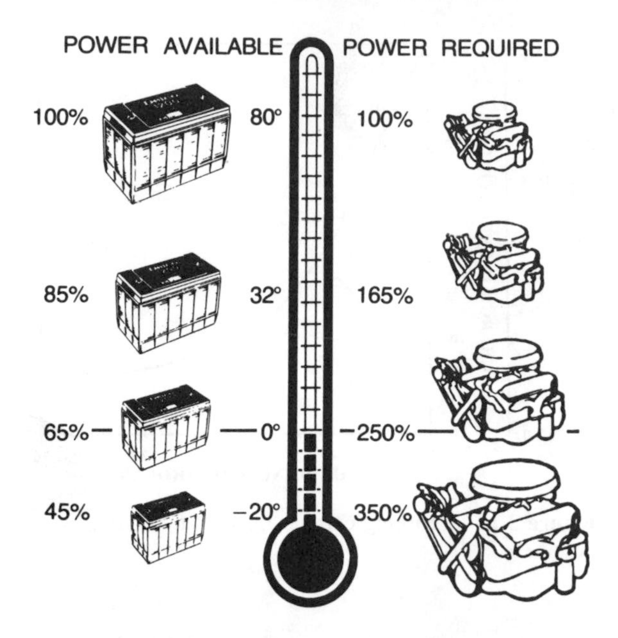

Figure 3.11 Power required by starter motor versus power available from battery at various temperatures (°F) (*Delco*)

cranking speed when 80% charged and at a temperature of around –7°C (20°F).

In Britain, the performance of a battery is indicated by two test parameters that are usually marked on the battery casing:

(i) The cold cranking amperage (CCA).
(ii) The reserve capacity.

The British Standard cold cranking amperage is an indication of the battery's ability to crank the engine at low temperatures. It is defined as the minimum current that the battery will deliver for one minute at a temperature of –18°C, without the cell voltage dropping below 1.4 V (i.e. a terminal voltage 8.4 V for a '12 V' battery). A typical small car would require a battery with a cold cranking amperage of 150–200 A.

The Reserve Capacity rating of a battery indicates how long the battery will keep the vehicle running in the event of a generator failure. It is defined as the length of time for which a battery operating at 25°C will deliver a current of 25 A before it falls to a cell voltage of 1.75 V (i.e. a terminal voltage of 10.5 V for a 12 V battery). A typical car battery has a reserve capacity in the region of 50–100 minutes.

Service requirements of batteries

Batteries constructed with lead-calcium grids require no maintenance other than a periodic check on the cleanliness and security of the terminals. Since these batteries are usually 'sealed for life', no other checks are possible. A useful feature of these batteries is that they are less susceptible to overcharging than the conventional type and seldom give trouble.

Batteries constructed using lead-antimony grids consume a certain amount of water and periodic 'topping-up' with distilled water is required. Filler caps are fitted for this reason. A major shortcoming of these batteries is that over a period of time antimony migrates from the positive grid to the negative grid. This makes the battery susceptible to overcharging, leading to positive plate oxidation and hence crumbling of the plate material.

Batteries are generally very reliable components and, providing that the charging system is operating correctly, will give at least 4 years of reliable service. Due to their relative cheapness (and the great amount of inconvenience that results when one loses charge) it is normally wise to simply replace them at the first sign of trouble.

3.5 THE GENERATOR

In order to provide power to operate the vehicle's electrical equipment and maintain the battery in a good state of charge an on-board electrical *generator* is required. Early cars used a type of dc generator called a *dynamo*. Dynamos were relatively heavy and inefficient, and so as the amount of electrical equipment fitted to vehicles grew, manufacturers began to develop a more powerful and efficient ac generator called an *alternator*. Alternators were first introduced into series production in the early 1960s, and by 1975 where standard equipment on almost every vehicle sold in the UK. Since their introduction, alternators have been improved enormously, in particular the power output and efficiency of modern alternators is far higher than that of early models. This development has come about for a variety of reasons. In particular the fitment of additional electrical equipment has meant that between 1960 and 1990 there was, approximately, a six-fold increase in the average car's electrical power consumption. Additionally, vehicles spend a much greater proportion of their operating time with the engine at idle, often with heavy electrical loads such as heaters and lights switched on. In 1950, for example, only about 10% of the average city journey was spent with the engine at idle, as against about 35% today. Modern alternators are therefore designed to provide a high output at low engine rpm, with a maximum output current in the range 30–100 A.

Operating principles

In the review of electrical fundamentals presented in Chapter 2 an explanation was given of how a voltage can be induced in a conductor when a magnetic field is moved across the conductor. Figure 3.12 shows how this principle is exploited in the alternator.

The magnetic field is provided by a rotating electromagnet called a rotor (or field coil) which is energized by a small *excitation current* supplied to its windings via a pair of carbon brushes that bear against slip-rings. The rotor revolves within a fixed coil of wire which is called the stator. The moving magnetic field of the rotor causes an induced current in the stator, and it is this current which is the alternator's output.

The output current is at a maximum when the rotor's magnetic poles are lined up with the stator winding (i.e. when the rate of magnetic flux cutting is greatest) and falls to zero when the magnetic poles are at 90° to the stator winding. Since the north and south poles of the rotor follow each other in alternation, the output from the stator is ac. Thus, as the rotor is spun, the output from the stator rises to a maximum at 90°, falls to zero at 180°, rises to a negative maximum at 270° and falls back to zero at 360°. This produces the familiar sinusoidal output voltage variation illustrated in Figure 3.13. Note that both output frequency *and* voltage increase as the speed of rotation of the rotor is increased. In order to make this varying ac output suitable for connection to the car's electrical system it must first be rectified to dc with a *diode pack*, and then regulated to the required system voltage by a *voltage regulator* circuit.

A practical alternator is, of course, constructed of more than just a single electromagnet turning within a single loop of wire. Figure 3.14 shows an exploded diagram of a modern alternator.

The stator

In order to obtain the greatest possible efficiency, most alternators use a stator assembly consisting of three entirely separate windings, mounted at 120° intervals around a laminated iron frame. Each winding contains many turns of copper wire and a completely separate ac output is induced in each. On account of the 120° spacing each output cycle has, accordingly, a 120° phase separation with respect to its neighbours.

In order to simplify the rectifier circuitry it is usual to connect the three stator windings together in one of the two ways illustrated in Figure 3.15, these are referred to as:

(a) the star connection;
(b) the delta connection.

In the star connection, one end of each stator winding is joined together to make a star point. An output phase is then available at the remaining terminal of each winding.

In the delta connection, the phase terminals are simply joined in pairs. Both types of connection provide the same output power, but the delta connection provides about 1.73 times the current output of the star, whereas the star provides about 1.73 times the voltage output of the delta. Because of the higher current output of the delta connection many manufacturers prefer it for their high-output alternators (those producing currents above about 50 A).

The rotor

The rotor is typically constructed of a soft iron former which is pressed over a steel shaft. Many turns of relatively fine copper wire are wrapped around the former and these comprise the coil of the electromagnet. Iron pole pieces are then fitted to each end of the former in order to direct the magnetic field outward.

The most common design for the poles is that of the 'claw' which uses two pole pieces, each having six interlocking fingers. The rotor coil is then wired to the two slip-rings and when supplied with an excitation current produces twelve alternate north and south poles, one for each finger of the claw.

The strength of the magnetic poles (and hence the induced stator current) depends on the level of excitation current supplied to the rotor winding. With a 12-pole rotor, six complete cycles of ac will be produced in each stator winding for every rotor revolution.

The rotor shaft is subject to considerable side loads from the drive belt and so is supported in substantial pre-lubricated ball bearings. A cooling fan is also fitted to the rotor shaft, and may be located either inside or outside the alternator housing. The rear of the housing encloses the brush assembly, the rectifier pack and the voltage regulator electronics.

Rectifier pack

The diodes in the alternator's rectifier pack are designed to perform three functions, namely:

(i) To rectify the generated ac current in order to provide a dc output suitable for use by the battery and electrical equipment.
(ii) To rectify a small proportion of the generated ac in order to provide the rotor excitation current.

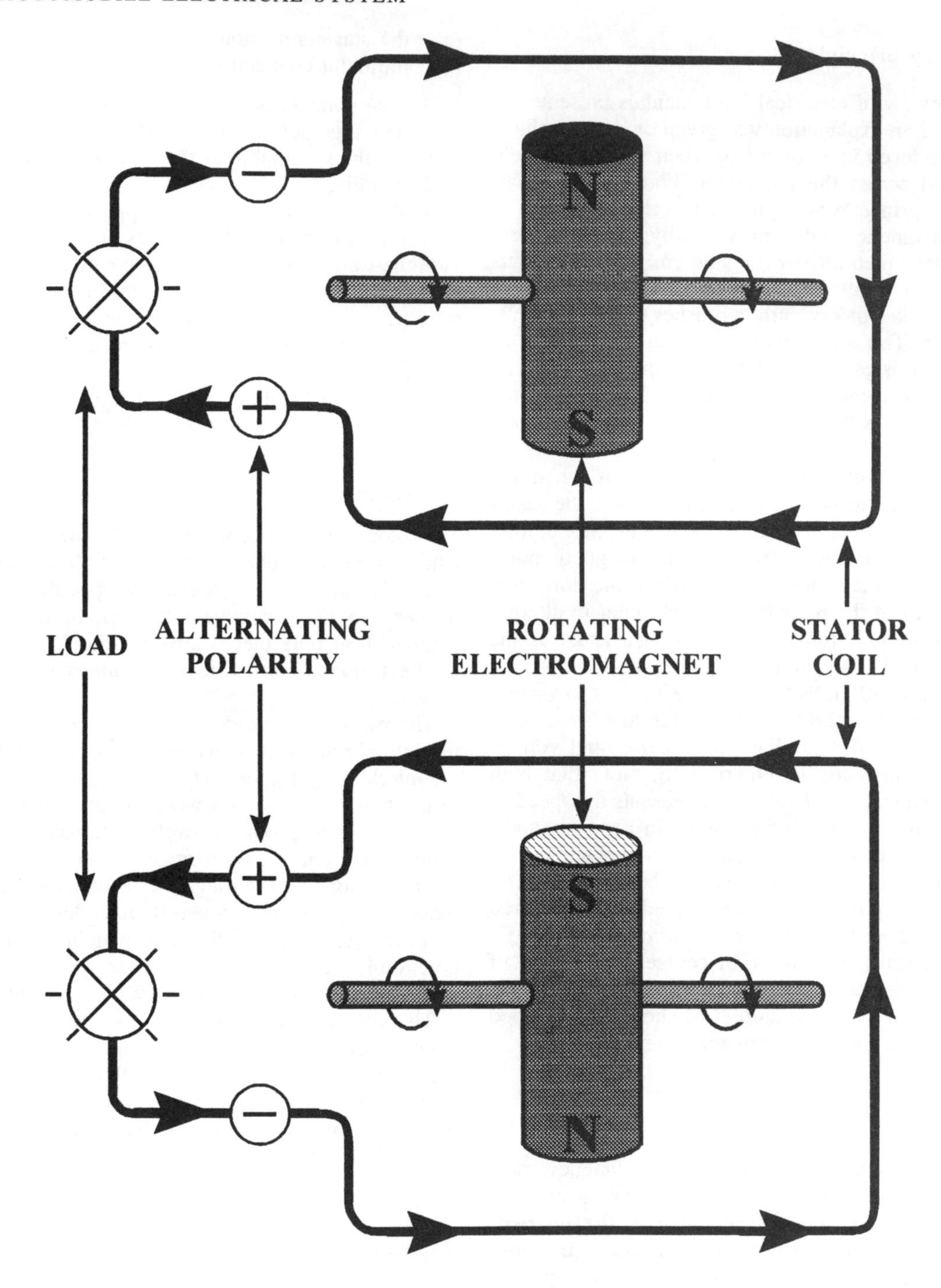

Figure 3.12 Principle of the alternator

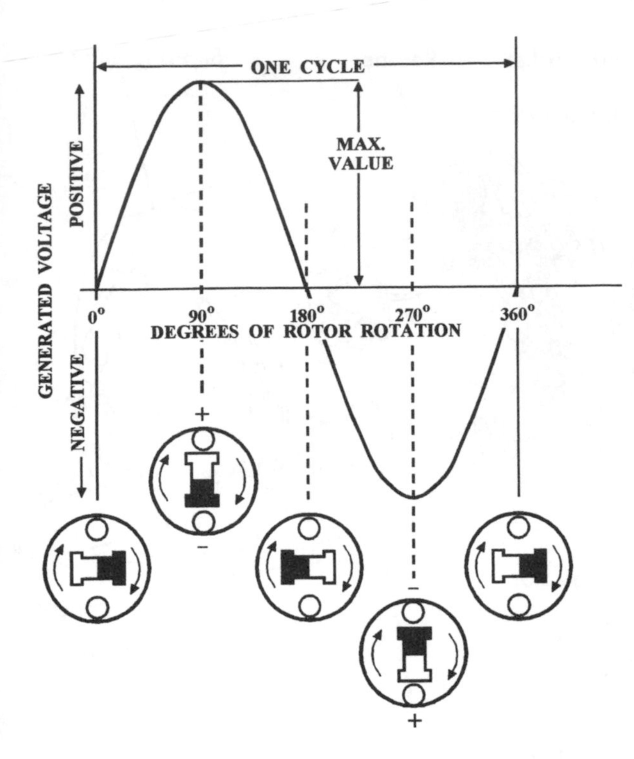

Figure 3.13 Alternator output voltage as a function of rotor position

(iii) To prevent the battery from discharging through the stator windings when its voltage exceeds that of the alternator (i.e. when the engine is stopped or running at a very low speed).

Figure 3.16 shows the type of rectification circuitry which is now almost universally used in alternators. The diodes D1–D6 are high-current silicon power diodes which provide full-wave rectification of the ac output from the stator windings. The positive half-cycles pass through the diodes connected to the battery positive (B+) terminal and the negative half-cycles pass through the diodes connected to the battery negative (B–) terminal. Thus, slightly 'rippled' dc is available at the B+ and B– connections. This ripple is largely removed by the battery.

The diodes D7–D9 rectify a small proportion of the stator output in order to provide excitation current for the rotor winding. The amount of current passing through the winding, and hence the alternator's output, is controlled by the voltage regulator.

Additional output from the star point

Alternators using star-connected stator windings are often fitted with an extra pair of rectification diodes wired to the star point. These diodes rectify high-frequency ('third harmonic') voltage fluctuations which arise at the star point due to the slight variations in the number of turns on each stator winding that occur during manufacture. Approximately a 15% increase in alternator output is available by using this additional energy.

Voltage regulator

Since alternator output rises with rotational speed an unregulated alternator designed to generate adequate output (say, 14 V) at engine idle would produce several hundred volts when operated with no load at maximum engine rpm. In order to maintain a suitable output voltage at all engine speeds and with varying electrical loads a voltage regulator circuit is therefore essential.

Voltage regulation operates on the principle of switching the excitation current on and off at a high frequency (hundreds of times per second). Control is achieved by adjustment of the duty-cycle ('on–off' time) of the excitation current, as illustrated in Figure 3.17. At low rotational speeds the excitation current may be switched on for 90% of the time and off for 10% of the time. This produces a strong magnetic field in the rotor and so maintains the required output voltage from the stator windings.

As the alternator speed increases, only a weak rotor field is needed to generate the desired output voltage and so the duty cycle changes to reduce the average excitation current. At high rpm the current may be on for only 10% of the time and off for 90% of the time.

Because the regulator is constantly monitoring output voltage it always provides the correct excitation current to generate the correct system voltage, irrespective of speed and load.

An additional feature provided by most voltage regulators is compensation for changes in ambient temperature. As the under-bonnet temperature rises, the regulator slightly reduces the alternator output voltage. This prevents overcharging and water loss from the battery. Conversely, during cold weather, the battery requires a slightly higher charging voltage and the regulator makes this adjustment.

Regulator operation

Hybrid IC voltage regulators are now universally used on alternators. Although the IC contains hundreds of components its operation may be understood from the much simplified circuit of Figure 3.18.

The voltage regulator samples the stator output

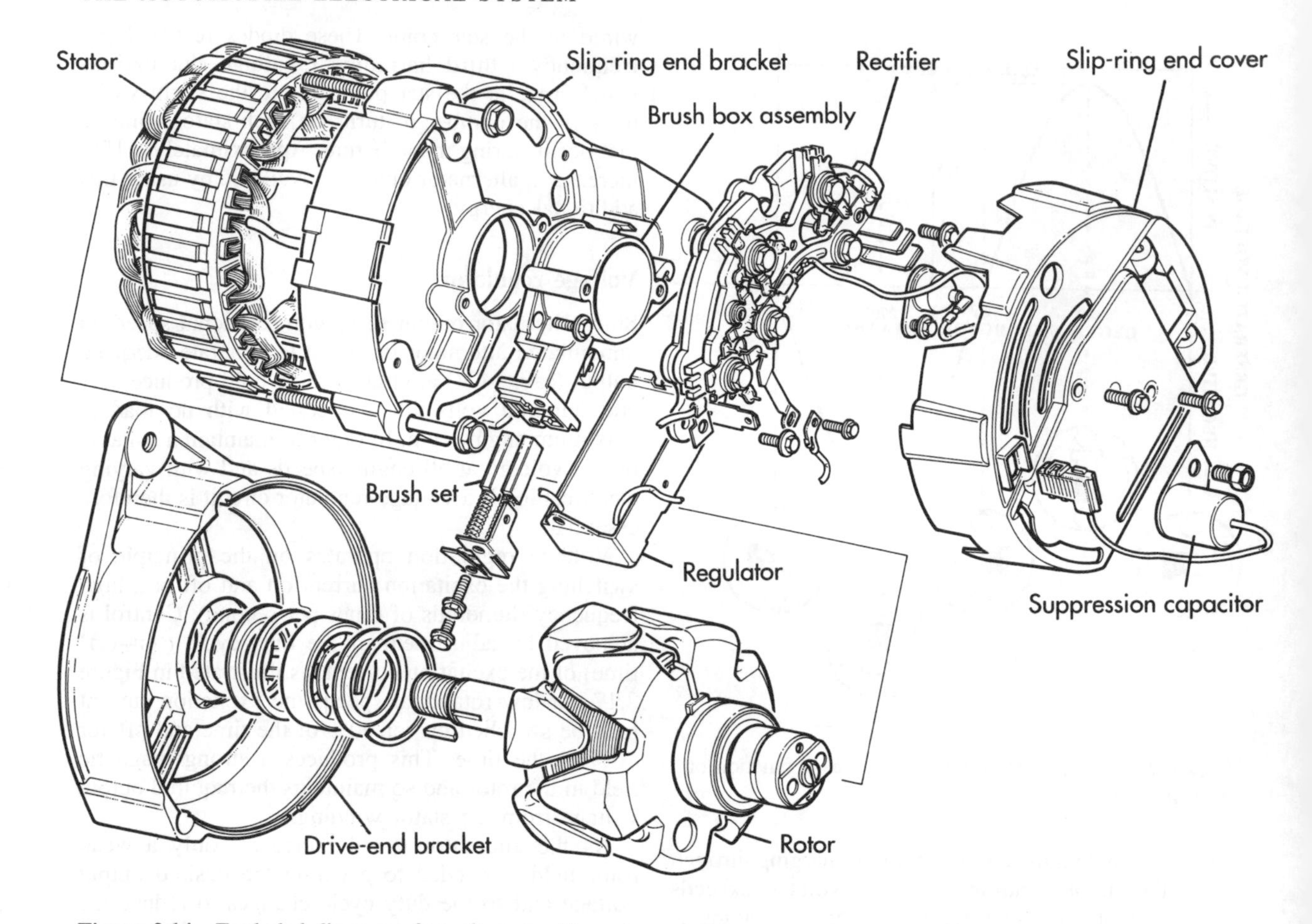

Figure 3.14 Exploded diagram of an alternator (*Lucas*)

voltage at the exciter diode terminal, D+. This voltage is divided-down by the potential-divider resistors R1 and R2 and applied to the diodes D11, D12 and ZD. ZD is a Zener diode, which conducts in the reverse direction only when its Zener voltage is exceeded. It is the value of the Zener voltage which controls the output voltage of the generator. Diodes D11 and D12 each provide a small forward voltage drop (about 0.7 V) which varies slightly with temperature, so providing ambient temperature compensation of the regulator voltage.

Voltage too low
If the fraction of alternator output voltage appearing at the Zener diode, ZD, is below the Zener voltage then ZD will not conduct and the control transistor Q1 will be OFF. With Q1 OFF, current from the D+ terminal reaches the base of Q2 via resistor R3. This switches the Darlington pair Q2 and Q3 ON. Excitation current then flows from the D+ terminal, through the rotor winding and to ground via Q3. This rise in excitation current causes the generator output voltage to rise.

Voltage too high
When the output voltage rises too high the Zener diode begins to conduct. Control transistor Q1 switches ON, so maintaining the base of Q2 at a voltage close to ground and thus holding Q2 and Q3 in the OFF state. With Q3 OFF, no excitation current can flow through the rotor winding and so the generator output falls. This sudden interruption in excitation current leads to a large self-induced voltage in the rotor winding, which is discharged through D10, the 'free-wheel' diode.

As soon as the output voltage falls below the set voltage, the regulator switches the excitation current back on again to increase it. Since this switching process occurs very rapidly (hundreds of times a second), the excitation current assumes an average value that maintains the correct output voltage.

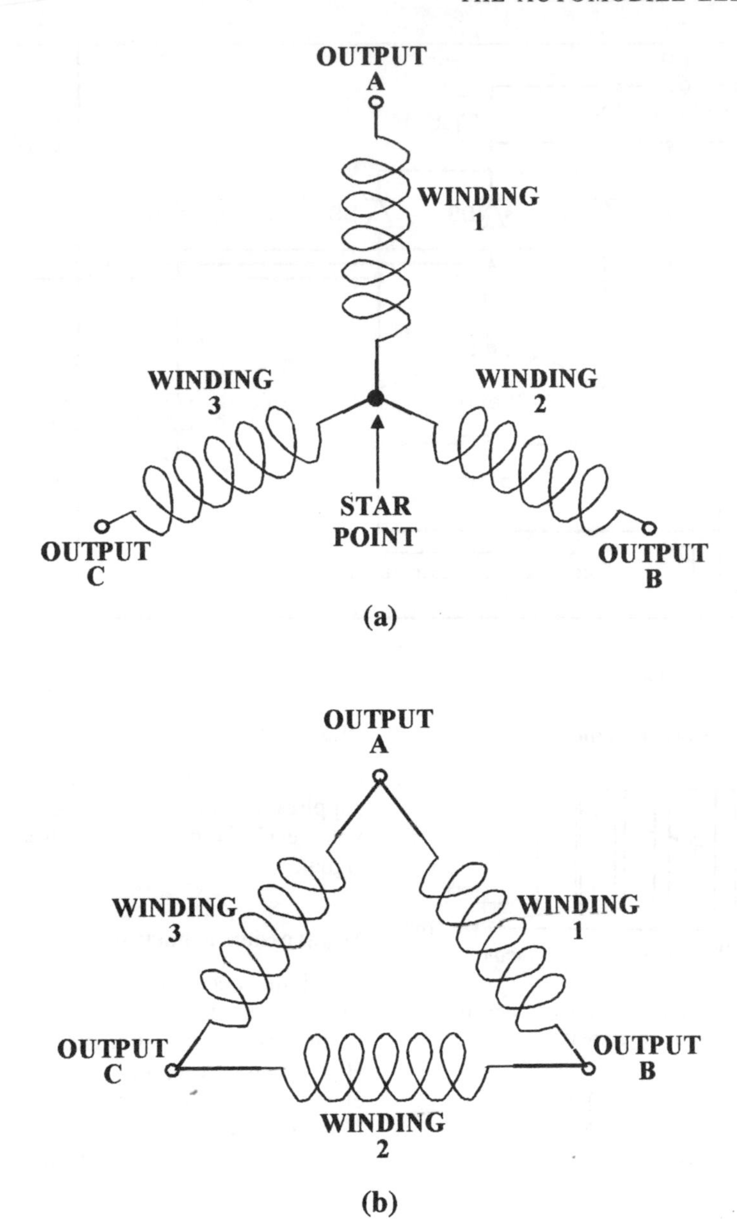

Figure 3.15 Interconnection of alternator stator windings: (a) star connection, (b) delta connection

Pre-excitation current

When the engine is started, the alternator is initially stationary and so no excitation current is available at the exciter diodes. Under this circumstance a pre-excitation current is supplied from the battery.

When the ignition key is operated a battery-sourced pre-excitation current of about 200 mA flows to the D+ terminal via the charge warning lamp. This produces sufficient magnetic field in the rotor to allow the generation process to begin. A resistor (R6 in Figure 3.16) is often wired in parallel with the warning lamp as a precaution against a blown bulb.

Once the engine is running, the voltage generated at the exciter diodes rises until it is similar to that of the battery and so the warning light goes out and the alternator becomes self-exciting.

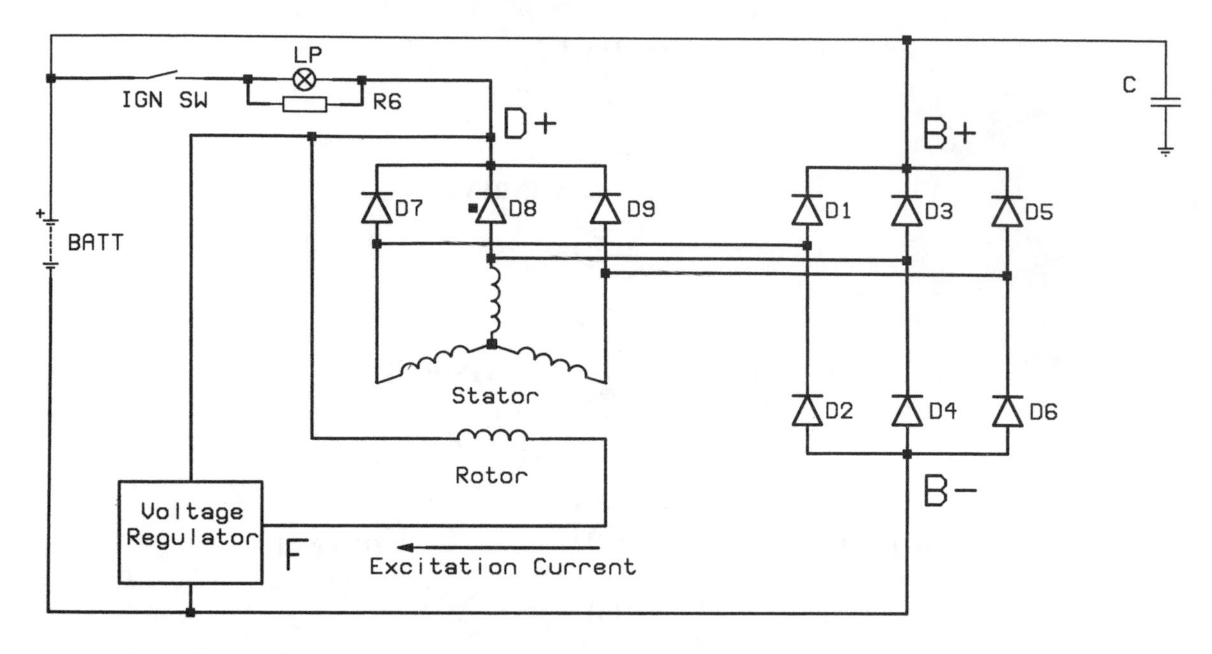

Figure 3.16 Alternator rectification circuitry

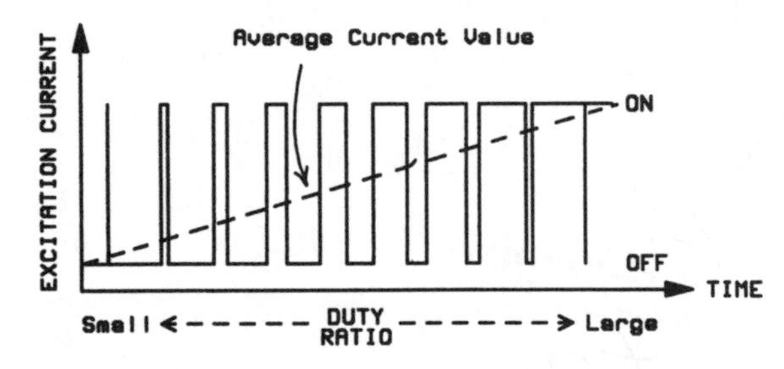

Figure 3.17 Control of alternator output achieved by duty-cycle variation of excitation current

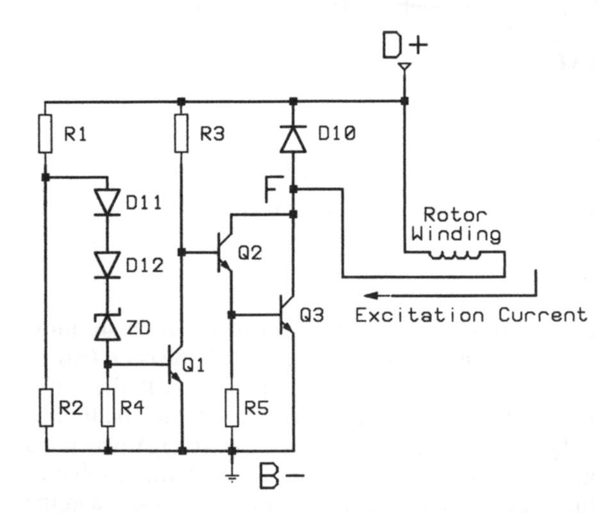

Figure 3.18 Simple voltage regulator circuit

Phase terminal

On some alternators a phase terminal is provided. This supplies a sinusoidal output at about half system voltage (7 V) and may be used as an engine speed signal.

Suppression capacitor

The inductance of the stator windings often leads to the generation of small currents at harmonics (multiples) of the alternator's ac output frequency. High-frequency harmonics may interfere with other electronic systems on the vehicle and so must be filtered to ground with a large value 'suppression' capacitor which is connected to the output terminal.

Drive pulley ratios

Most alternators produce a useful output in the range 1 000–15 000 rpm. Since engines operate over about 700–6 000 rpm, a drive pulley ratio of about 2:1 is generally used.

3.6 ELECTRIC MOTORS

The function of the electric motor is to convert electrical energy into mechanical energy, with the greatest possible efficiency. On an automobile, electric motors are used to start the engine and to drive various

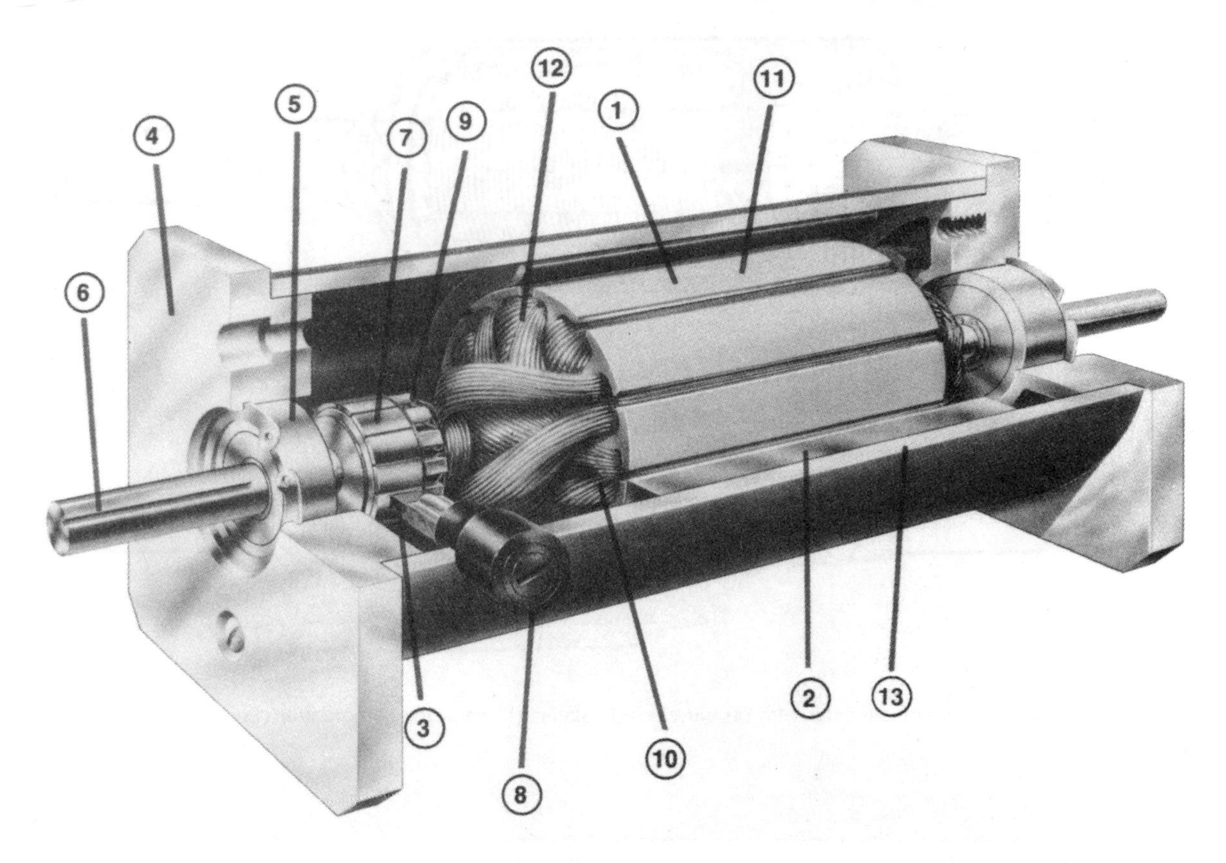

Figure 3.19 Construction of the dc permanent magnet motor. [1] Armature laminations; [2] ceramic, alnico or rare-earth field magnets; [3] brushes; [4] end caps; [5] armature shaft support bearings; [6] armature shaft; [7] commutator; [8] brush holders; [9] armature wire welded to commutator; [10] armature windings; [11] insulation between armature laminations; [12] insulation between armature windings; [13] motor enclosure

mechanisms (window and sunroof winders, wipers, seat adjusters, cooling fan and so on). As cars become more and more highly specified the number of motors used continues to increase. Some prestige vehicles now carry close to 100 motors. The majority of these are of the cheap and simple permanent magnet variety, but for some applications more sophisticated motors are needed. For example, where a precisely controlled angle of shaft rotation is required the stepper motor has been introduced, often controlled by a microprocessor.

Motor operation

The basic principles of dc motor operation were reviewed in Chapter Two.

All practical dc motors operate on the principle of interaction between two magnetic fields, one field is produced by the stator and the other is produced by current flowing in the rotor winding.

Figure 3.19 illustrates the construction of a simple dc motor. Here, the stator field is provided by powerful permanent magnets which generate a strong magnet field in which the rotor (usually called the armature) can rotate. Current is supplied to the armature winding by a type of rotary switch called a commutator. The commutator consists of spring-loaded carbon-based brushes which bear against copper contacts. The contact pads are soldered to the armature winding and are so designed that as the winding rotates the brushes switch the armature current in an alternating fashion. The commutator is thus a clever way of creating an alternating current in the armature winding, even though the supply to the brushes is dc.

When current is supplied to the commutator, a magnetic field is produced around the armature winding, interacting with the stator field to produce a turning force, or torque, on the loop. The action of the torque causes the loop to rotate, and so mechanical power is available.

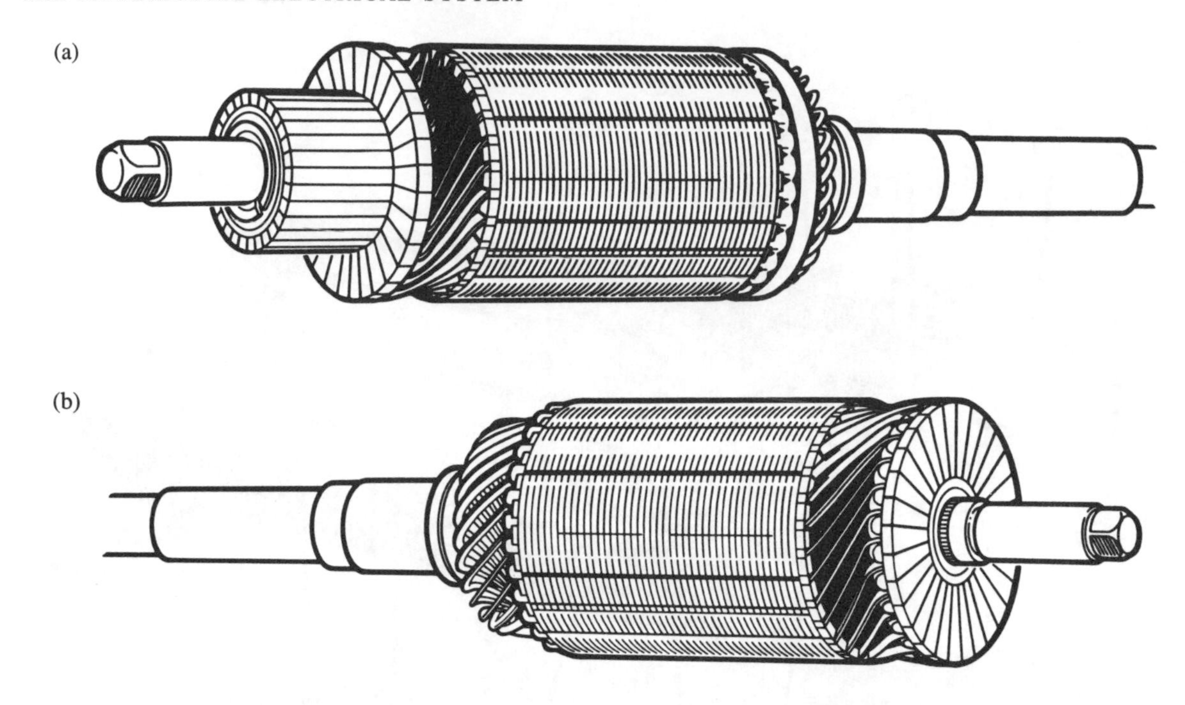

Figure 3.20 Typical commutator designs: (a) barrel commutator; (b) end-face commutator (*Lucas*)

Back electromotive force

From the description of motor operation given above, it may be noted that all the conditions needed for the induction of current are met in the motor. Indeed, a motor can be operated in 'reverse', and will function as a generator. The electromotive force, or emf, produced by moving an armature winding through the field of the stator, always acts to oppose the current that is powering the motor and so is known as a back electromotive force, or bemf.

Since the bemf is proportional to the speed at which the armature winding is moving through the stator flux, it follows that it must increase as the motor speed increases. This increase in bemf opposes the supply current and so results in a reduction in current drawn from the battery. As the speed of rotation rises, the bemf will eventually approach the battery voltage and the motor will have reached its maximum speed.

Since the motor torque is directly proportional to the supply current, the maximum torque is produced when the armature is stationary. Under this condition the supply current is at a maximum because there is no bemf at all. As the rotational speed of the armature increases, the bemf increases, the supply current falls and so the torque decreases.

Practical motors

The actual design of a practical motor depends upon its application and the design constraints (cost, weight and so on), but all dc motors have certain common constructional features.

The armature

In a practical motor the armature comprises a steel shaft onto which a commutator and a stack of soft iron laminations are located. Windings of relatively heavy gauge copper wire are soldered to the commutator bars and fed along slots in the laminations. The commutator itself consists of a number of copper segments insulated from each other by a material such as mica or plastic. The segments can be arranged around the shaft, in a 'barrel' fashion, or aligned radially around the shaft to give an 'end-face' commutator, as illustrated in Figure 3.20. The number of commutator segments is usually the same as that of armature slots.

The armature is supported within the stator by the shaft ends, which locate in bearings that are assembled into the motor end-covers. One end of the shaft is usually extended beyond the cover and acts as the motor output.

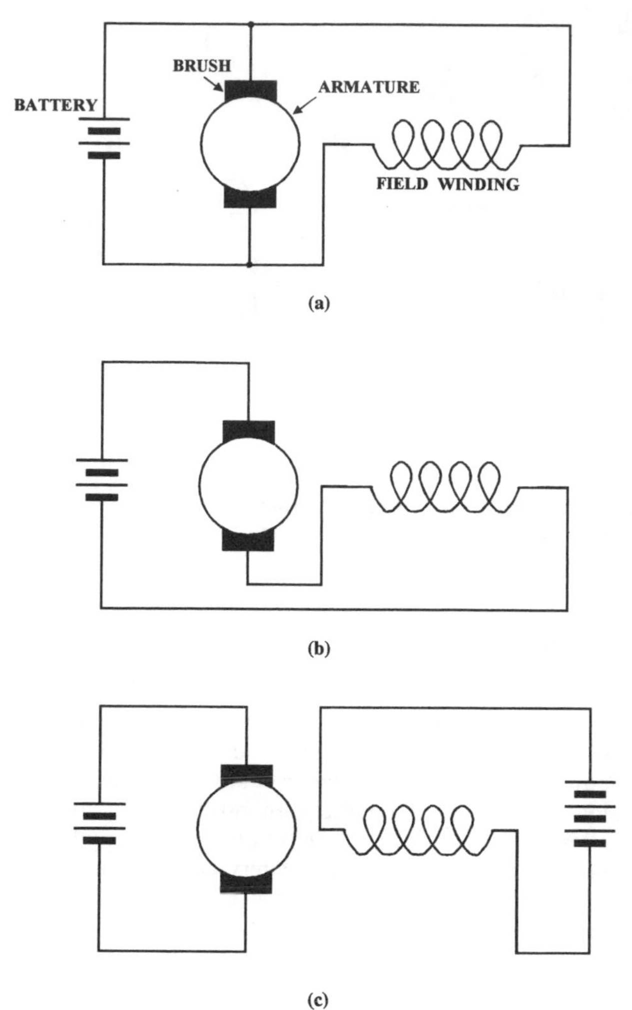

Figure 3.21 The three types of field winding connection used in dc motors: (a) shunt connection, (b) series connection, (c) separate excitation

The stator
The stator field can be produced either by permanent magnets or by an electromagnet fed with an excitation current. The winding for the electromagnet is called the 'field winding' and is usually connected to the battery and armature in one of the three ways shown in Figure 3.21.

Connection (a) has the field and armature wired in parallel and is called the 'shunt connection'. With this arrangement the motor rotates at a more or less constant speed if a steady voltage is applied.

Connection (b) is called the 'series connection'. When power is applied to the motor a large current flows through both the field and armature winding and so a high starting torque, decreasing with speed, is characteristic of this motor. For this reason the series connection is popular for starter motors, but it has the drawback that the motor might accelerate to a dangerous speed and fly apart if it is not driving a load.

Connection (c) shows separate excitation of the armature and the field windings. The currents are separately adjustable to control motor speed.

Permanent magnet stators are used to a great extent in small dc motors, and most of the motors used in vehicles are of this type. Permanent magnets produce a slim, simple, stator construction that requires no current to produce the field flux.

The characteristics of the permanent magnet motor are similar to those of the separately-excited motor, with a constant field current. Motor speed can be controlled by varying the armature voltage, and increases proportionately.

Most permanent magnet motors use a ferrite material ('ceramic magnet') for the stator, but although cheap it only produces a weak field. Where a high torque is required, powerful rare-earth element magnets, such as neodymium-iron-boron or samarium-cobalt, must be used. These are expensive, but allow the manufacture of compact and lightweight permanent magnet starter motors which offer performance superior to that of the traditional series-connected type.

Starter motors

Starting an engine requires that it be turned-over (or 'cranked') at a speed sufficient to cause reasonable turbulence of the incoming air–fuel mixture and so make combustion possible. In addition, the engine's flywheel must be given sufficient momentum that it will keep rotating for the first couple of firing strokes until the engine develops sufficient power to run unassisted. Typically, a petrol engine requires a minimum cranking speed in the region of 50–100 rpm to ensure starting in cold weather.

Early motor cars were invariably hand-cranked by the driver using a 'starting-handle'. Since this was a rather strenuous and at times dangerous exercise, it was not long before electric motors were employed for the task. Some American luxury cars were fitted with electric 'self-starters' as early as 1912, and they were a standard fitment on most prestige cars from the 1920s onward. By the 1960s, even the cheapest of European saloons was fitted with an electric starter.

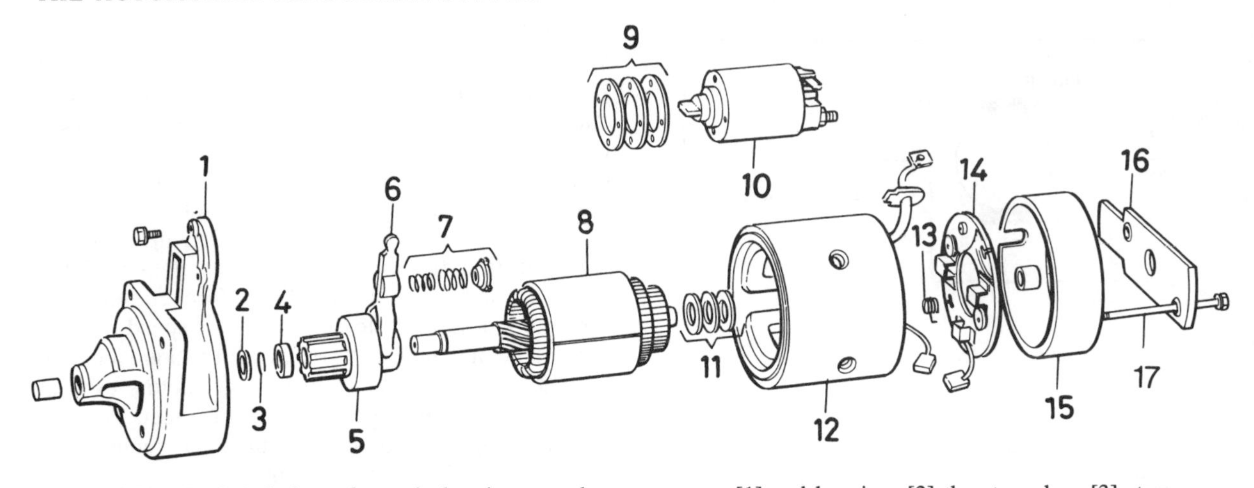

Figure 3.22 Exploded view of a typical series-wound starter motor: [1] end housing; [2] thrust washer; [3] stop ring; [4] stop collar; [5] pinion and one-way clutch; [6] pinion operating lever; [7] return spring; [8] armature; [9] solenoid shims; [10] solenoid; [11] thrust washers; [12] motor body incorporating field windings; [13] brush springs; [14] brushes and holder; [15] rear cover; [16] end bracket; [17] retaining bolt

Starter motor construction

Starters motors are high-torque dc motors, usually of a series-wound construction, which produce the high initial torque required to overcome the inertia of the stationary engine and transmission components. Since they are only operated intermittently, and then only for a very short time, they are designed to operate under a condition of extreme overload at very high efficiency.

In a typical series-wound motor, as illustrated in Figure 3.22, the heavy starting current is supplied via a magnetic switch (the solenoid) to two armature brushes. A further two brushes complete the armature circuit and earth the current through four field windings which are arranged around the armature. As the initial current is very large (hundreds of amps), starting performance is critically dependent on the low resistance of the supply cable and the low internal resistance of the battery. A heavy-duty cable, of as short a length as possible, is always used between the battery and motor.

The engine is turned over by the motor output shaft (armature shaft) which carries a small pinion that meshes with the teeth of a ring-gear fitted around the flywheel. Since the number of teeth on the pinion is only about one-fifteenth that of the number on the ring-gear, a 15:1 reduction drive exists. Thus, for an engine cranking speed of 100 rpm, the starter must turn at 1 500 rpm. Since the torque is multiplied up by a factor of 15, however, a small motor is able to turn a large engine.

Starter solenoid

The starter solenoid, mounted directly on the motor body, provides two functions; (a) it shifts the starter pinion into engagement with the ring-gear before power is applied to the starter, and (b) once the pinion is engaged, it remotely energizes the motor – allowing a heavy-current circuit of minimal length to be constructed between the battery and the motor.

The basic solenoid and motor assembly is shown in Figure 3.23. The solenoid unit consists of a cylindrical plunger surrounded by two windings; a 'pull-in' winding, and a 'hold-in' winding. The plunger is connected to a pivoted shift-lever and also operates a contact bar at its rear.

With no power applied to the starter circuit, the plunger is held out of the solenoid windings by a powerful return spring. When the starter switch is turned, current flows to the solenoid terminal (S) where a proportion flows to ground through the hold-in winding. Some current also flows via the pull-in winding to the motor supply terminal (M), where it flows to ground through the armature and field coils. The magnetic flux created by the currents in the solenoid windings draws the plunger into the core, causing the shift lever to push the pinion forward, into mesh with the ring-gear, and closing the rear contacts (B and M) to complete the circuit between battery and motor. The starter motor is thus energized and turns the engine.

With the contacts, B and M, connected, the pull-in winding is short-circuited and so no current flows

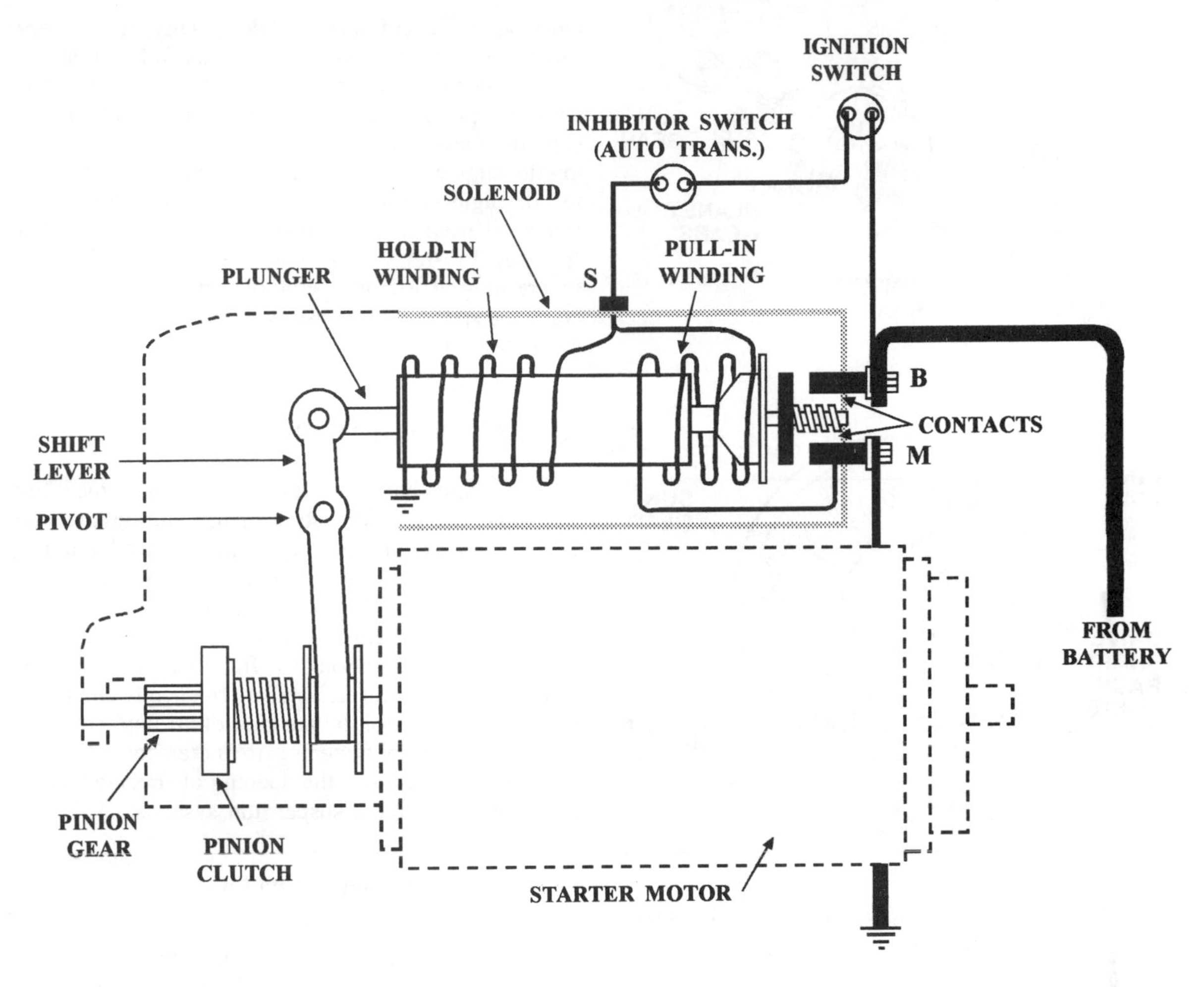

Figure 3.23 Operation of the pre-engaged starter motor

through it. The plunger is thus held in by a small current flowing through the hold-in winding. When the engine fires and the starter switch is released, the current to the hold-in winding is cut and return spring pressure forces the plunger out of the windings and back to its rest position, disengaging the pinion from the ring-gear and de-energizing the motor. The starting cycle is then completed.

Pinion assembly

Since the pinion will remain meshed with the ring-gear for as long as the starter switch is held closed, there is a likelihood that once the engine has started the ring-gear could drive the armature at speeds which might be destructive. In order to prevent this, all starters are fitted with an overrunning (one-way) drive clutch assembly. When the pinion is driven by the ring-gear at a higher speed than the armature shaft, the clutch allows the pinion to overrun the shaft, so avoiding excessive armature speeds.

Permanent magnet starter motors

New developments in magnet technology during the 1980s have allowed the manufacture of starter motors that use rare-earth permanent magnets in place of the current-carrying field coils found in conventional starter motors. The use of permanent magnets offers a number of advantages;

(i) it simplifies the construction and manufacture of the motor.
(ii) it allows a size and weight saving of 30–50% over a conventional series-wound motor.

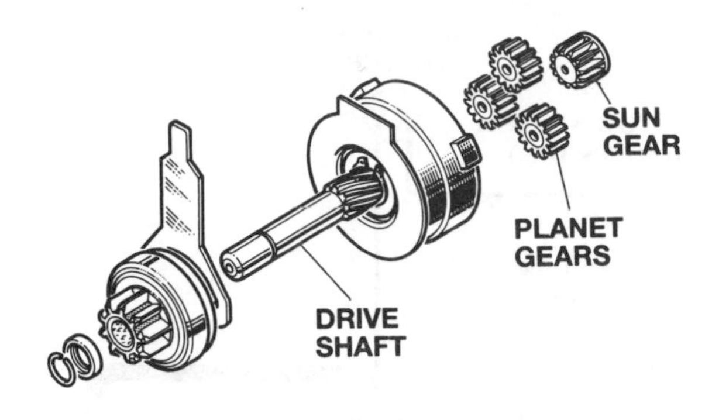

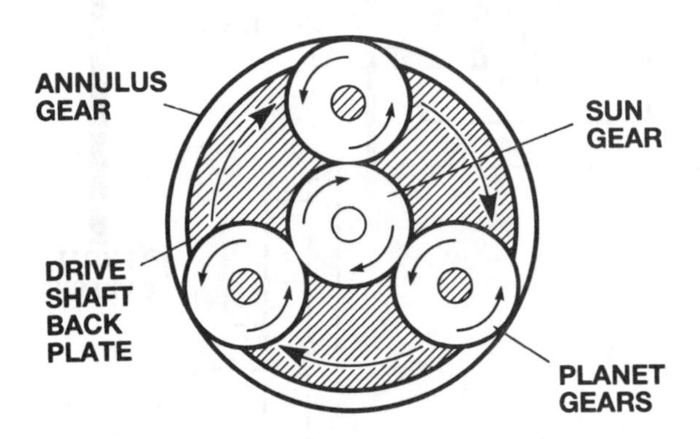

Figure 3.24 Planetary reduction gears used in permanent-magnet starter motors (*Lucas*)

(iii) the electrical resistance and heating arising from the use of field coils is eliminated.

Since the speed-torque characteristics of permanent magnet motors are less favourable than those of series wound motors it is usual to operate them with a reduction drive. This allows the motor to operate at a higher speed and so develop more torque. Typically, an integral planetary gearset is fitted to the motor, as shown in Figure 3.24, giving a reduction ratio in the region 4.5:1. With a ring-gear to starter motor pinion ratio of 16:1 this means that a cranking speed of 100 rpm for the engine corresponds to an armature speed of 7 200 rpm.

Maintenance of the starter circuit

As with most vehicle electrical components, the starter motor is generally highly reliable and will normally last the life of the engine. When starter faults do occur they are generally related to defects in the cables and connections, or to a defective battery. Due to the large currents involved in the starter circuit, all connections must be clean and secure. Additionally, the battery should be at no less than 70% of its full charge.

Mechanical faults are usually associated with failure of the pinion one-way clutch assembly, which can slip and allow the starter motor to spin without turning the engine. The use of an incorrect grade (viscosity) of engine oil may also lead to starting problems. If the oil used in the crankcase is thicker than that specified by the vehicle manufacturer it will cause severe drag at low temperatures, much reducing the cranking speed, and possibly preventing the engine from starting.

3.7 STEPPER MOTORS

Stepper motors have been around for a long time (they were first used in British warships during the late 1920s) but it is only recently, with the introduction of digital control systems, that they have found their way onto automobiles. They are increasingly used in position control mechanisms, where a precisely known number of shaft rotations (or fractions of a shaft rotation) must be performed and are frequently employed to accurately control the position of air flaps (in heating and induction systems), to increment odometer displays and to control the opening of shock-absorber valves in semi-active suspension systems.

Operation of the stepper motor

As the term suggests, a stepper motor rotates in precise steps, the direction, speed and angle of which are determined by the sequence of digital pulses supplied from a electronic drive unit. Simple stepper motors may rotate in 180° steps, whilst more complex ones may rotate in steps of less than 1°. They have a number of useful characteristics, including the ability to start, stop and reverse very quickly. Additionally, the rotor will 'hold' in a chosen stop position, without the need for any kind of brake.

Although there are a number of different types of stepper motor, most of those used in automotive electronics are of the permanent magnet rotor type, illustrated in Figure 3.25. The permanent magnet rotor is free to revolve within four wound stator poles. The step direction and angle is determined by switching the current in the four coils, W_1–W_4.

In general, the control pulses for the motor are generated by a logic circuit (often a microprocessor) and these are then amplified to suitable voltage and current levels by a driver circuit. Figures 3.25(a) and 3.25(b) show how this can be done to achieve 90° increments.

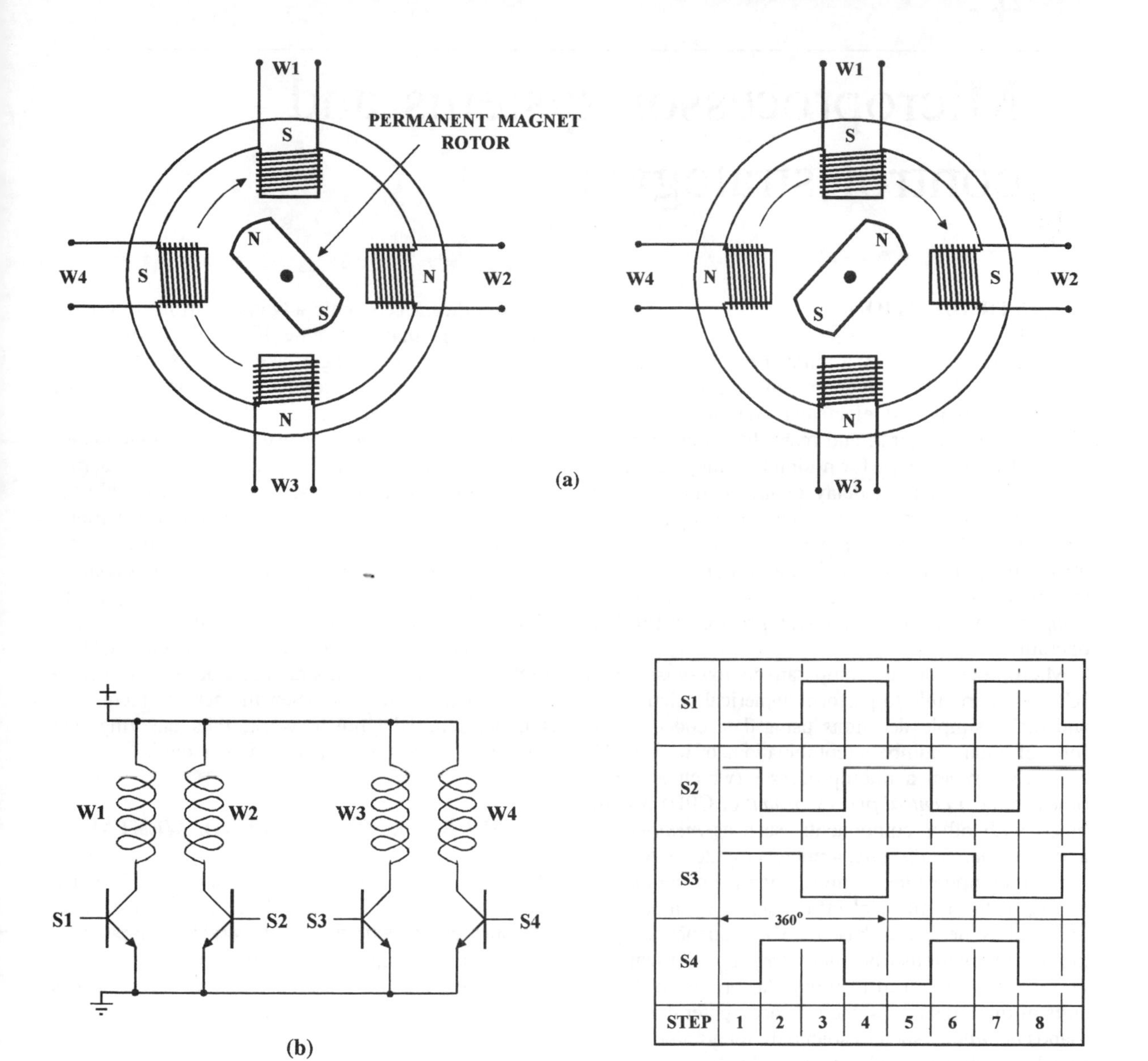

(c)

Figure 3.25 Construction and operation of the permanent magnet stepper motor. (a) Motor construction, (b) drive circuit connection, (c) logic pulse stepping sequence applied to drive transistors

4

Microprocessor systems and control strategies

4.1 INTRODUCTION

Most automobiles carry a number of electronic control units (ECUs) which control the operation of various mechanical and electrical systems. The term 'ECU' is very general, and many ECUs contain just one or two ICs to perform simple timing functions such as courtesy-light delay or heated rear window operation. Other ECUs are much more sophisticated however, and control complex systems such as fuel injection, ignition, automatic transmission and anti-lock brake operation. These ECUs are actually *micro-computers* and so rely upon *microprocessors* for their operation.

Microprocessors are comparatively low-cost digital ICs which are able to perform numerical calculations and make simple decisions using data coded in the form of binary numbers (combinations of logical '1's and '0's). When a microprocessor (which is sometimes called a *central processor unit* or CPU) is combined with other components, such as *memory ICs* and *input/output circuits*, a microcomputer is formed. A microcomputer takes information from the outside world, in the form of electrical signal from *sensors*, and makes decisions based on a sequence of pre-defined instructions (the *computer program*) which is permanently stored in memory. According to the results of these decisions, the microcomputer then commands *actuators*, such as solenoids, relays and motors, to achieve the required outcome. This basic configuration is illustrated in Figure 4.1.

The power of the microcomputer arises from the microprocessor's ability to execute instructions very quickly; for example, a high-performance microprocessor can add two numbers in under one millionth of a second. Microcomputers are thus able to perform many automobile control functions with great speed, accuracy and reliability, and with a precision that cannot be matched by mechanical, hydraulic or vacuum devices. Typical applications include the control of the engine's air/fuel ratio, exhaust gas recirculation (EGR), ignition timing and automatic transmission gear change execution.

A significant economic advantage in using microcomputers is that the same ICs and circuit board (known in computer jargon as the *hardware*) can be used in many different control applications; it is just the computer program (the *software*) that must be changed. For example, it is usual for a vehicle manufacturer to use the same fuel injection system across a wide range of engine types and sizes, it is only the ECU's software that is modified to suit each variant.

Other advantages lie in the field of fault monitoring and diagnostics. Many automobile microcomputers are provided with additional memory space in which the microprocessor can store information relating to any abnormal operation of the electrical or mechanical systems. This data can be retrieved from the computer's memory when the service technician is undertaking diagnostic work, thus ensuring that faults are quickly identified and corrected.

4.2 MICROPROCESSOR FUNDAMENTALS

Microprocessors first became available in 1971 when the Intel Corporation launched a low-cost 4-bit microprocessor, code-named the 4004, fabricated on a single silicon chip. Since that time, microprocessors have undergone very rapid development which has increased their speed and computing power to the extent that they are now indispensable in the control of sophisticated consumer products such as camcorders, washing machines and automobiles. Even the most basic of cars carry at least one microprocessor (in the engine-management ECU) and luxury-specification vehicles can have ten or more located in various ECUs.

Nowadays, there are virtually hundreds of different types of microprocessor available, each optimized for a particular range of tasks, however the basic operational concepts remain the same for all.

Number representation

The manipulation of information within a microcomputer is achieved by coding the data into *binary*

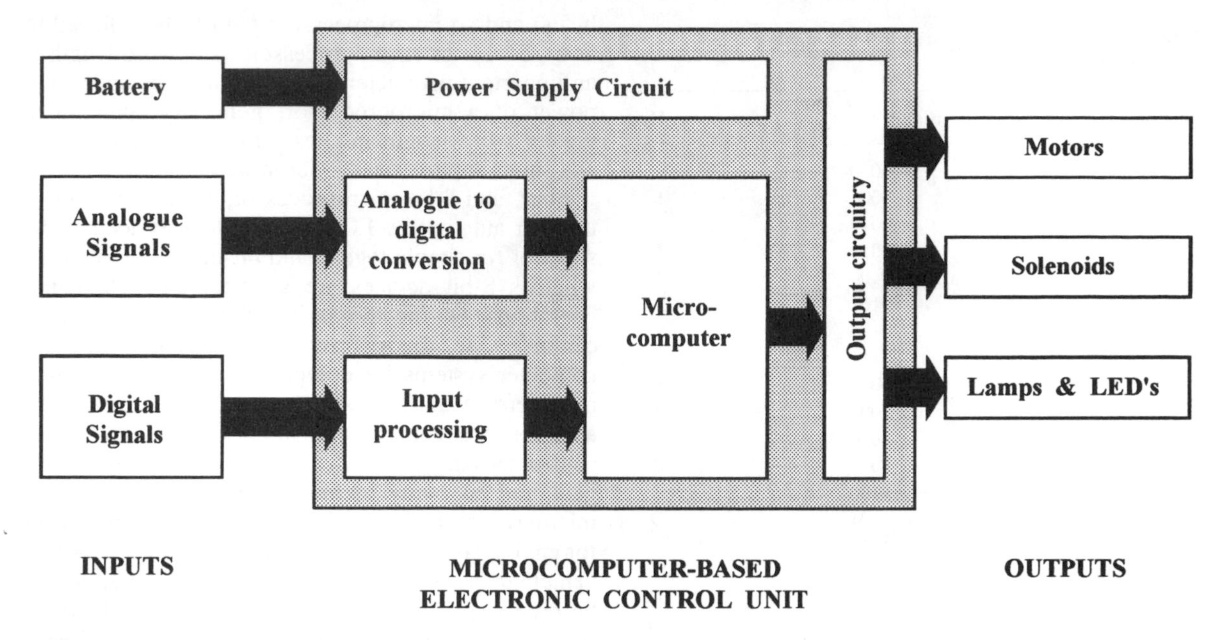

Figure 4.1 Configuration of a microcomputer-based ECU

digits (usually shortened to '*bits*') that can take either a logic '1' (HIGH) or a logic '0' (LOW) value. Groups of bits, in the form of sets of 0s and 1s, are called *words* and can be used to represent numbers. The way in which this is achieved is involved, but an appreciation of the binary number system is necessary when working with computer systems.

All number systems work to a base number. Our everyday counting system, the decimal system, works to base 10 and uses 10 symbols (0, 1, 2, 3, 4, 5, 6, 7, 8 and 9). The importance of each symbol is determined by its position in a number and increases by 10 times (a power of 10) for each added symbol. For example, the number 237 is understood to represent $(2 \times 100) + (3 \times 10) + (7 \times 1)$, whereas the number 2370 represents $(2 \times 1\,000) + (3 \times 100) + (7 \times 10) + (0 \times 1)$.

This is can easily be visualized by inserting the symbols into columns, where each column carries the appropriate decimal weighting:

Column weighting:	1 000	100	10	1
Decimal number:	2	3	7	0

Since the binary number system only uses two symbols (0 and 1) it works to base 2 and numbers must be represented as patterns of 1s and 0s. Again, the importance of each symbol is determined by its position in the number, increasing by 2 times (a power of 2) for each added symbol. For example, the decimal number 163 can be represented in binary by the 8-bit number 10100011. This can be visualized by inserting the digits into columns where each column carries the appropriate binary weighting (i.e. twice the weighting of the column to its right). The right-most bit carries least weighting and is therefore known as the *least significant bit* or LSB. The left-most digit carries most weighting and is known as the *most significant bit* or MSB:

	MSB							LSB
Column weighting:	128	64	32	16	8	4	2	1
Binary number:	1	0	1	0	0	0	1	1

Therefore $10100011 = (1 \times 128) + (0 \times 64) + (1 \times 32) + (0 \times 16) + (0 \times 8) + (0 \times 4) + (1 \times 2) + (1 \times 1) = 163$.

The range of numbers that can be represented by 8 bits is from 00000000 to 11111111, corresponding to the decimal numbers 0 through to 255. To represent numbers larger than 255, more than 8 bits need to be used. 16 bits, for example, can represent decimal numbers up to 65 536 or 64K (note that in computing, the upper-case 'K' represents 1024 rather than 1000, which is represented by a lower-case 'k').

Since binary numbers can become rather long and unwieldy it is usual to simplify their representation by writing them in hexadecimal (base 16) form. Hexadecimal (or 'hex') is a number system in which 16 symbols are used and each symbol position represents successive powers of 16. The decimal symbols 0 to 9 are used in the normal way and the symbols A

Table 4.1 Decimal-binary-hexadecimal number representation

Decimal	Binary	Hex
0	0000	0
1	0001	1
2	0010	2
3	0011	3
4	0100	4
5	0101	5
6	0110	6
7	0111	7
8	1000	8
9	1001	9
10	1010	A
11	1011	B
12	1100	C
13	1101	D
14	1110	E
15	1111	F
16	0001 0000	10
17	0001 0001	11
18	0001 0010	12
19	0001 0011	13

to F are ascribed to the values 10–15 respectively, as illustrated in Table 4.1.

Since a 4-bit binary number covers the range 0–15 it is easy to write a long binary number in hex by separating it into 4-bit groups and then writing the hexadecimal equivalent of each group. For example, the decimal number 3237 may be written in binary as 1100 1010 0101 or in hex as CA5 (1100 = C, 1010 = A, 0101 = 5).

Hexadecimal may appear a slightly bizarre way of counting but it is widely used in computing since microprocessors generally work with binary numbers composed of multiples of 8 bits, called *bytes*, and in hexadecimal each byte can be represented by just two hex digits.

Microprocessor word lengths

The word length used by a particular microprocessor depends upon the number of bits that can be sent in parallel along the *data bus* which runs between the processor and all of the other components that make up the microcomputer. The data bus is a set of parallel circuit tracks which carry the data bits; each track is assigned to carry one particular bit and so the number of tracks sets the word length. Typical word lengths that can be handled by microprocessors are 4 bits, 8 bits (1 byte), 16 bits (2 bytes) or 32 bits (4 bytes) and so microprocessors tend to be referred to as 4-, 8-, 16- or 32-bit processors. The word length is an important parameter in determining the speed and power of a microprocessor; generally speaking a greater word length implies a faster, more powerful and more sophisticated microprocessor.

Four- and 8-bit microprocessors have been widely used in automotive ECUs. 4-bit devices tend to be superior for simple logical and bit-manipulation tasks whereas 8-bit devices are superior for calculating operations. During the early 1980s, 8-bit microprocessors were the mainstay of automotive microcomputer systems for complex tasks such as engine management, and they still provide adequate performance for most control applications. By the late 1980s, control system requirements had become more stringent and so many manufacturers turned to 16-bit microprocessors, which have now become the norm for engine control and some ABS applications. A few vehicle manufacturers have recently moved up to 32-bit processors; these are usually necessary only when there is an additional requirement beyond that of basic control, for example if sophisticated self-monitoring or self-diagnostic functions are needed.

For many applications a cost-effective solution can often be arrived at by combining several microprocessors in one ECU. For example, an ABS controller might use a 16-bit processor to implement the basic anti-lock control strategy, but employ a simpler 8-bit processor to perform peripheral tasks such as wheel-speed calculations.

The microcomputer system

Before looking in detail at how microcomputers operate, it is useful to examine the general layout (or *architecture*) of the microcomputer system.

Although microcomputer technology is continually evolving, the basic architecture has remained unchanged for many years and is as shown in Figure 4.2. At the heart of the microcomputer is the microprocessor, which is the 'brain' of the system. It performs all arithmetical and logical procedures and orchestrates the operation of the other components. In order to function properly, the microprocessor must be connected to memory devices which hold the set of instructions that tell it what to do next. Memory is also used by the microprocessor for the storage of results and other numerical data.

To be able to perform useful control tasks, the microprocessor needs to be able to communicate with the outside world. This is achieved by using input–output (I/O) interface circuits. Input circuits are used to convert electrical signals from sensors into a form

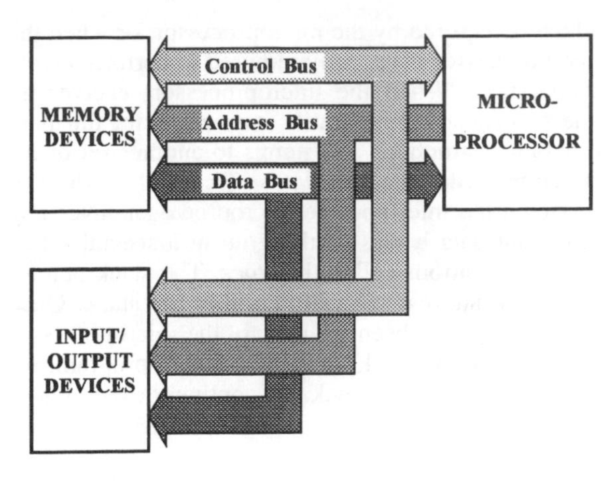

Figure 4.2 Typical microcomputer architecture

that is understandable by the microprocessor. Similarly, output circuits convert the microprocessor's decisions, which emerge in the form of logical 0s and 1s, into signals which can drive actuators such as motors and solenoids.

These three items of hardware, the microprocessor, memory and I/O circuits, usually exist as separate integrated circuits which are connected together by sets of parallel tracks, known as *buses*, on a printed circuit board. Each individual bus track carries a digital signal (in the form of a '0' or '1' voltage level) and three buses are needed, specifically:

(a) *The databus.* The function of the databus is to convey binary numbers between the different components of the microcomputer at great speed. It is a bi-directional bus since data is sent both *to* and *from* the microprocessor. Usually the data bus has an equivalent number of tracks to the microprocessor word length (i.e. a 16-bit microprocessor would use a 16-track data bus).

(b) *The address bus.* Each memory location within the microcomputer is assigned a unique numerical *address*, rather as a house has a unique number in a street. The microprocessor uses the address bus to instruct a specified memory location to connect onto the databus. When a particular memory location is selected, all other locations are automatically deselected, allowing the microprocessor to communicate with just one location at a time. Generally speaking, an 8-bit microprocessor would use a 16-bit wide address bus to give access to 65 536 separate memory locations, each of which could hold a single byte of data, giving '64K bytes' of memory.

(c) *The control bus.* In order for the microcomputer to function efficiently the data bus must only be used by one set of data bits at a time. For example, if two memory locations placed their contents onto the databus at exactly the same time there would be a *bus contention* and data would be corrupted or destroyed. A control bus is therefore needed to carry a group of signals around the microcomputer and precisely synchronize the operations of all the hardware devices.

The most important control signal is the *system clock*, a regular 0,1,0,1, pulse-train to which all actions can be synchronized. The clock circuit is usually built into the microprocessor and uses an external crystal oscillator to generate stable pulses at a high frequency (at least 1 MHz); a higher clock frequency implies a faster-acting microcomputer. Another important control signal is the Read/Write signal. This is a bus line that is used to indicate whether an addressed memory location should transfer its stored data onto the data bus (a *read operation*) or should take new data from the data bus and store it (a *write operation*).

Microprocessors

A microprocessor is an LSI (large scale integration) IC which contains many thousands of transistors, diodes and resistors fabricated on a single chip of silicon just a few millimetres square. The usefulness of a microprocessor comes about from its ability to perform a wide range of operations on binary data, in a sequence determined by instructions held in memory. A microprocessor is therefore not specific to a particular task but can be adapted, using software, to an almost infinite variety of duties; this makes for great economy of production and flexibility in application.

The organization of circuits within the microprocessor chip varies from one device to another, but a typical internal arrangement is shown in Figure 4.3. The two major functional blocks are the *control unit* and the *arithmetic/logic unit* (ALU). In addition, a microprocessor has a set of *registers* that include a program counter, instruction register, general purpose registers and a stack pointer. All of these internal circuits are connected to each other by an internal data bus, and to the other parts of the microcomputer by the three external buses.

The microprocessor's control unit is where all of the synchronizing signals which control the manipulation and transfer of data around the microcomputer are generated. The control unit therefore supervises the execution of the computer program that is held in memory.

The ALU is that part of the microprocessor in which

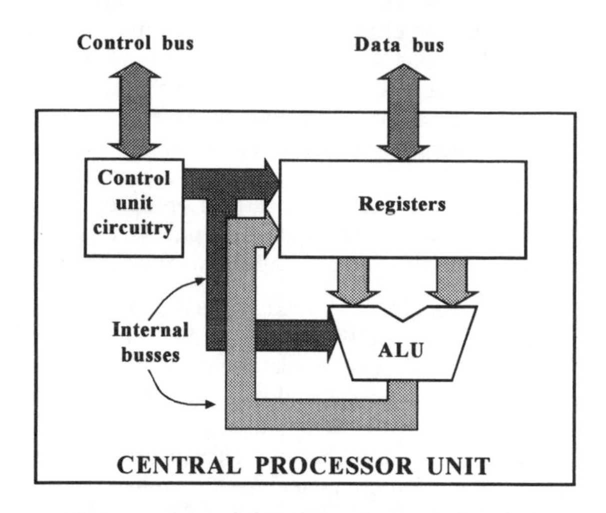

Figure 4.3 Internal organization of a microprocessor

the execution of the computer program actually takes place. It is built up from the types of logic gates described in Chapter Two and can perform a range of operations on binary data, including:

(i) Binary addition and subtraction.
(ii) Complementing of a binary number.
(iii) Left or right rotation of a binary number.
(iv) Boolean logic operations such as AND, OR, NOT and XOR.

In order to execute instructions, the ALU requires access to registers. Registers are memory locations that are built into the microprocessor chip itself and take a direct part in its operation. For example, to keep track of which instruction needs to be executed next, the control unit operates a special register called the *program counter* (PC). The program counter holds the memory address of the data which constitutes the next instruction to be executed. When the control unit decides that it should be transferred to the microprocessor, the instruction data is sent via the data bus to the *instruction register*.

General purpose registers are used by the control unit to temporarily hold data, prior to it being sent elsewhere or used in a computation. To enhance its usefulness, the microprocessor can be instructed to carry out data transfers between registers, or to add or subtract the contents of different registers.

The *stack pointer* (SP) is used during the execution of *subroutines* and *interrupts*. Interrupts are digital signals which originate elsewhere in the microcomputer (usually in the I/O circuits) and are sent along the control bus to arrive at an interrupt pin on the microprocessor IC. An I/O device generates an interrupt signal either when it has a vital piece of data which is required by the microprocessor, or when the microprocessor is urgently required to perform an essential task. When the microprocessor receives an interrupt signal it temporarily stops executing the main program instructions and jumps to another set of instructions called the *interrupt service routine*. Prior to servicing the interrupt, the microprocessor saves any important data it was working on in a special set of memory locations called the *stack*. The stack pointer is used to address a location within the stack. Once the interrupt has been dealt with, the microprocessor returns to its original place in the main program, retrieves data from the stack and continues execution of the main program.

Watchdog timer

An important feature incorporated into many microprocessors used for automotive applications is the *watchdog timer*. The watchdog timer is a simple binary number counter that is loaded with a large number and arranged to decrement (count down) by one digit on each microprocessor clock pulse. The counter is designed to generate a microprocessor RESET pulse when it reaches a count of zero.

Instructions are incorporated into the main program of the microcomputer to periodically re-load the counter before it can reach a count of zero. Therefore, during normal operation of the computer, the watchdog timer is always counting down without ever reaching zero. If an error arises during the operation of the program (perhaps due to an external problem, such as a momentary loss of power) the instructions will not be executed in the correct sequence and so the watchdog timer will not be re-loaded at the appropriate time. This allows the count to reach zero and so a RESET signal is generated, which restarts the microcomputer system and clears the error.

Memory

The purpose of memory is to store instructions and data in the form of binary numbers; each memory location is given a specific address and can store one byte of data. The amount of memory available to a microprocessor is set by the number of address bus bits that it uses. Most 8-bit microprocessors operate with a 16-bit address bus and can therefore address 64K different memory locations. A 16-bit microprocessor can generally address upwards of 1 million different locations. An important distinction is made between memory that the microprocessor can both read and write to (known as *Random Access Memory* or RAM), and memory that can only be read (*Read Only Memory* or ROM).

ROM

ROM is 'long-term' memory, containing data that can be read many times by the microprocessor but cannot be altered or overwritten by it. The data in a ROM is *non-volatile*, meaning that it remains in memory even if the power is removed. ROMs are therefore used to store the program instructions (software) which give the ECU its 'intelligence' and enable it to perform the task to which it is dedicated. Without instructions held in ROM the microcomputer would be useless. Other information held in ROM includes sets of data tables (maps and look-up tables), descriptions of alphanumeric characters for display purposes and fault diagnosis messages. In the production of automobile ECUs many thousands of ROM chips, all with identical data, must be supplied every day. These ROMs are programmed to the car maker's specifications as one of the last stages of their manufacture, prior to shipment to the ECU factory.

During the design of ECUs it is sometimes useful to use ROM which can be easily reprogrammed so that alternative programs can be tried out. For purposes such as these, memory manufacturers can supply *erasable–programmable ROM* (EPROM) chips which are supplied 'blank' and programmed using a desk-top EPROM programmer connected to a personal computer. If a need to change the data should arise, the entire EPROM can be returned to the 'blank' state by unplugging it from the microcomputer and exposing it to strong ultraviolet light for about 25 minutes, thus allowing re-use.

An alternative to the EPROM is the *electrically erasable programmable ROM* (EEPROM). This can be re-programmed on a location-by-location basis by applying a voltage to a special reprogramming pin on the IC whilst it is still connected to the rest of the microcomputer circuitry. EEPROM is sometimes used in production ECUs when the value of data needs to be 'trimmed' during tuning operations to take account of tolerances in mechanical systems.

RAM

Random access memory differs from ROM in that the microprocessor can both read and write data to each memory location. It is therefore very useful for the temporary storage of partial results and for holding data prior to transfer to output circuits.

RAM comes in two varieties; *static RAM* and *dynamic RAM*. In static RAM chips, each memory cell contains a bistable circuit element built-up from two transistors. One of the transistors can be turned on to store a '1' bit or the other turned on to store a '0' bit. The transistors retain their states until overwritten by data from the data bus or the power is removed. Dynamic RAM stores data bits in the form of charge on the gate capacitance of MOSFETs. Depending on whether or not charge is present, each MOS transistor assumes a conducting or non-conducting state (i.e. '1' or '0'). Since the stored charge can slowly leak away, it is important that it is *refreshed* at periodic intervals by having the microprocessor re-write data to each memory location every few milliseconds. Although this is an added burden on the microprocessor, slightly slowing down its computing speed, it is partially offset by the lower cost of dynamic RAM ICs.

Most RAM is *volatile*, which means that all memorized data is lost if the microcomputer power supply is disconnected. Normally this is not a problem, however if the RAM is to hold important 'long-term' data, such as diagnostic fault codes or updated information about engine operating parameters, then *keep-alive memory* (KAM) must be used so that data is retained even when the ECU system power is removed. KAM is provided either by supplying the RAM chips via their own separate power cable which is permanently connected to the vehicle battery, or by using small rechargeable batteries on the microcomputer PCB itself. In the former case, the data stored in KAM is still lost if the vehicle's battery is disconnected.

Input/output circuits

In order to control the operation of a fast mechanical system such as an engine, the microcomputer must be able to communicate with a wide variety of external sensors and transducers in real time (i.e. virtually instantaneously). A common way of *interfacing* to the outside world is to employ sets of I/O circuits, known as *ports*, which the microprocessor addresses in the same way as memory. For convenience, IC manufacturers can supply the I/O port circuits on a single chip which connects directly to the microprocessor's buses. Usually, several I/O ports are provided, together with useful extras such as timer/counter circuits.

When a data read instruction is performed on an I/O port address, data is input from an external sensor and transferred to the data bus for processing. Conversely, if a data write instruction is performed on an I/O port address, then data is transferred from the data bus to the appropriate output circuit.

Input circuits

Although a microcomputer-based ECU must accept input signals from a wide range of sensors, the signals will generally only be of three types, namely,

(i) Switch signals – occasional ON or OFF signals from switches.
(ii) Pulse signals – repeating ON/OFF signals which contain information by virtue of their frequency or timing, for example an engine speed signal.
(iii) Analogue signals – smoothly varying voltage signals from thermistors, potentiometers, airflow meters and so on.

Although signals of type (i) and (ii) are already in a digital form it does not follow that they can be placed directly onto the microcomputer data bus. Whereas most microcomputers operate from a regulated +5 V power supply, many input signals will exceed this voltage and/or contain an excessive amount of electrical noise. The solution is to use an input processing circuit which limits the voltage excursion and filters out high frequency noise; 'clean' 5 V digital signals can then be made available to the microprocessor data bus.

The ability to accurately time digital input signals of type (ii) is vitally important in control applications. For example, the engine management ECU must be able to accurately measure engine rpm and predict the next firing cylinder prior to top dead centre (TDC) using timed digital impulses obtained from the flywheel speed sensor. Typically, engine controller microcomputers need to be able to resolve timed events to within about 3 μs, which translates to less than 0.1 degrees of crankshaft rotation for a V8 engine at 6 000 rpm. While the mechanical aspects of measuring engine speed cannot yet achieve this accuracy, the microcomputer introduces truly negligible error in the fuelling and ignition system timing.

Signals of type (iii) are obviously incompatible with digital systems and must therefore be converted to a binary form before they can be processed by a microcomputer. The device that makes this conversion possible is known as an *analog-to-digital converter* (ADC). A wide range of ADCs are available as single ICs and many are designed for direct connection to a microcomputer data bus. The function of an ADC is to sample an incoming analogue voltage and accurately convert it into a binary number proportional to its amplitude. To do this, the analog signal must be first *quantized* and then *encoded* into binary.

Quantization involves dividing the input signal voltage range into a number of intervals, usually $2^n - 1$ are used, where n is the number of bits used in the binary number. After quantization, a binary number is assigned to each interval and is used to represent any voltage that falls within that interval.

As an example of ADC operation assume that the output voltage of an alternator, which can vary from 0 V to 16 V, is to be monitored by a microcomputer. If a 4-bit ADC is used then each 4-bit binary number output by the ADC will have a relationship to the input voltage as shown in Table 4.2, where each 1 V interval has its own unique code.

Table 4.2 Four-bit binary number representation of 0 V to 16 V analogue input voltage

Input voltage	Bit 3 (MSB)	Bit 2	Bit 1	Bit 0 (LSB)
0 V	0	0	0	0
1 V	0	0	0	1
2 V	0	0	1	0
3 V	0	0	1	1
:	:	:	:	:
13 V	1	1	0	1
14 V	1	1	1	0
15 V	1	1	1	1

In this example the binary representation is crude. Any input voltage in the range 0–1 V is represented by the binary number 0000; for voltages in the range 1–2 V the equivalent number is 0001; for voltages in the range 2–3 V the equivalent number is 0010, and so on until the number 1111, which represents voltages in the range 15–16 V. If the input voltage exceeds 16 V, the ADC will still convert to the code 1111.

If the number of bits is increased, then the *resolution* of the ADC increases and the binary output becomes a more accurate representation of the input voltage. For example an 8-bit ADC would provide 255 intervals from 0–16 V giving 16 V/255 (62.7 mV) per code step. Therefore 00000000 represents input voltages from 0–62.7 mV, 00000001 represents input voltages from 62.8–125.5 mV, and so on. In this way the binary code can give a relatively accurate representation of the analog signal and so the microprocessor is able to detect small changes in input voltage.

In automobile electronics, ADCs of at least 8-bit accuracy are normally used, giving a resolution of 1 part in 255 (about 0.4%). For demanding applications, such as engine management or ABS control, 10-bit or even 12-bit ADCs may be used, giving resolutions of 1 part in 1000 (0.1%) or 1 part in 4000 (0.025%), respectively.

All ADCs require a certain amount of time to perform the conversion of an analog signal to its binary number equivalent and then make the data available to a microcomputer. This time is called the *conversion time*, and it is a key parameter to be considered when an ADC is selected. ADCs which have a short conversion time generally consume more power and cost more than those which have a long conversion

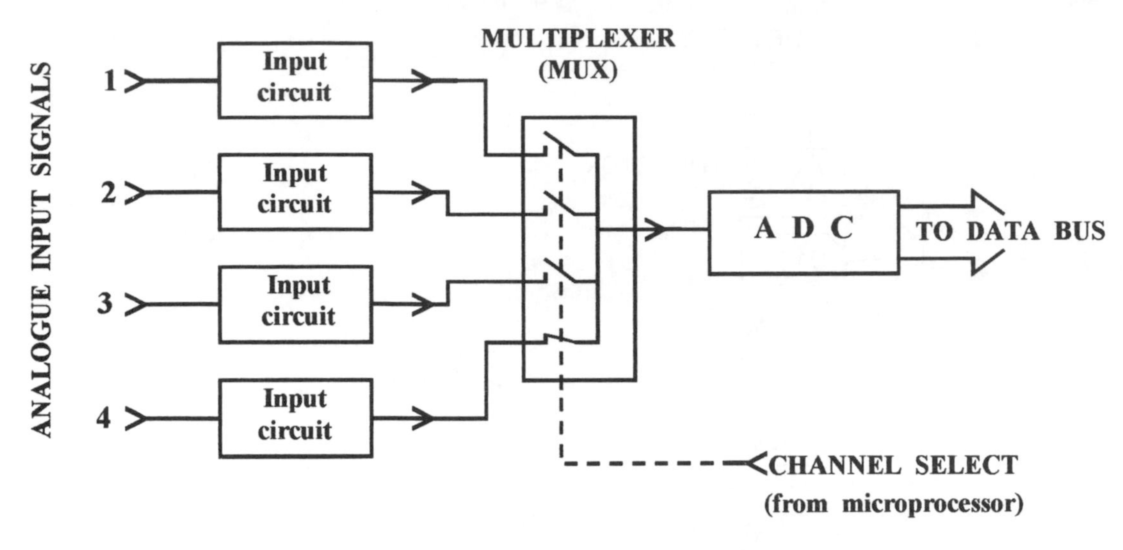

Figure 4.4 Using a multiplexer to connect each input signal to a common ADC

time. The most important factor when considering an ADC's conversion time is the highest frequency input signal that will need to be converted. *Nyquist's criterion* states that the ADC should have a sampling rate which is at least twice the highest-frequency component present in the input signal. For example, if the input signal is changing at a rate of 100 times per second (100 Hz), then the ADC must sample at least 200 times per second (200 Hz) to be able to keep pace with changes in the input. This corresponds to a conversion time of 5 ms, which is typical for automobile applications.

In many ECUs, where a number of analogue signals need to be converted to a digital form, a significant cost-saving can be achieved by using a *multiplexer* (MUX) to connect each analog signal, in turn, to a single ADC (Figure 4.4). The MUX is simply a chip which contains a bank of electronic switches (actually MOSFETs) that are used to select the input channel that is to be sampled and converted by the ADC. Control and synchronization of the MUX and ADC is undertaken by the microprocessor in accordance with the program instructions.

Output circuits

After processing the input signals, the microcomputer needs to command electrical actuators to carry out its decision. For example, it may need to open a solenoid valve, light a lamp or cause a relay to close. Although microcomputer-based ECUs are used in a wide variety of applications, the range of output signal types is small and may be classified as:

(i) occasional ON/OFF pulses, for operating relays or lighting lamps;
(ii) rapid and regular ON/OFF pulse signals of varying duration, for fuel injectors or solenoid valves;
(iii) sets of fixed pulses for operating stepper motors;
(iv) an output voltage which is the equivalent of a digital number; for example, for driving analogue displays.

Most microprocessor ICs operate at a voltage of 5 V and are only able to deliver an output current of a few milliamps, whereas most actuators operate at the vehicle battery voltage (12–14 V) and may require an operating current of up to 1 A or more. Output circuits must therefore be provided so that the actuator loads can be driven from the microprocessor's feeble output signals. The usual way in which this is achieved is to connect one side of the load directly to the unregulated battery supply and use a Darlington pair, controlled by the microprocessor, to switch the other side to ground. Such a 'low-side switching' circuit is illustrated in Figure 4.5. The diode shown in the Figure is a 'quench diode', it must be connected in parallel with an inductive load (such as a solenoid or relay) to protect the transistor from the powerful back-emf spikes that are generated when the load is switched off.

An alternative driver configuration is the 'high-side switch' in which one side of the load is connected permanently to ground and a transistor is used to switch the other side to a regulated supply voltage.

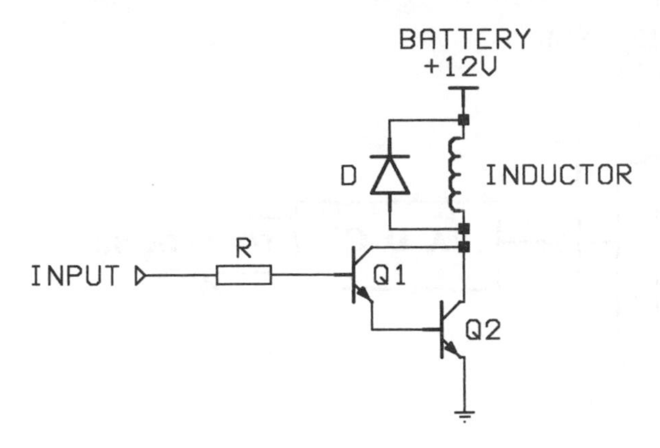

Figure 4.5 'Low-side' switching circuit using a Darlington pair

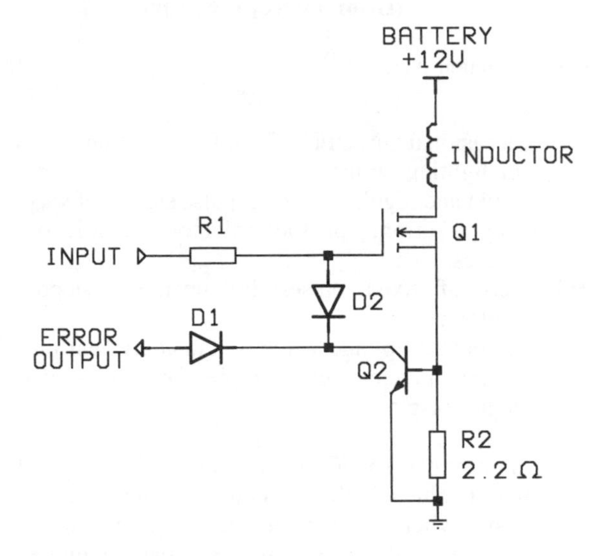

Figure 4.6 'Low-side' switching with current monitoring and error indication

High-side switching is often used when a load requires a voltage other than the battery voltage (e.g. 5 V).

Occasionally, an output circuit is required to have protection against short-circuiting of the solenoid or relay winding. This can be achieved using the circuit configuration shown in Figure 4.6. The load is switched by a field effect transistor (FET) and the load current is monitored by a low-value resistor. If a short occurs in the load windings it causes the voltage drop across the 2.2 Ω resistor to rise, turning on the npn transistor which pulls down the gate of the FET and so switches it off. The'error' terminal is also taken low, informing the ECU of the fault and allowing it to disable the output if required. In some applications the current through a load must be very carefully controlled, a good example is in the operation of fuel injectors. Fuel injectors are small solenoid valves which must open for a precisely defined time (a few milliseconds) to inject an exact quantity of petrol into the inlet manifold. In many injection systems the injector is simply constructed to have a winding resistance of about 14 Ω and is connected in the manner shown in Figure 4.7. When the transistor is switched on, a current of about 1 A flows through the injector windings and the injector opens. There are two drawbacks with this arrangement; (i) the large inductance of the injector windings leads to a slowly rising initial current and so a relatively slow opening response and (ii) the speed of opening varies as the battery voltage changes. These deficiencies lead to an imprecision in the volume of injected fuel.

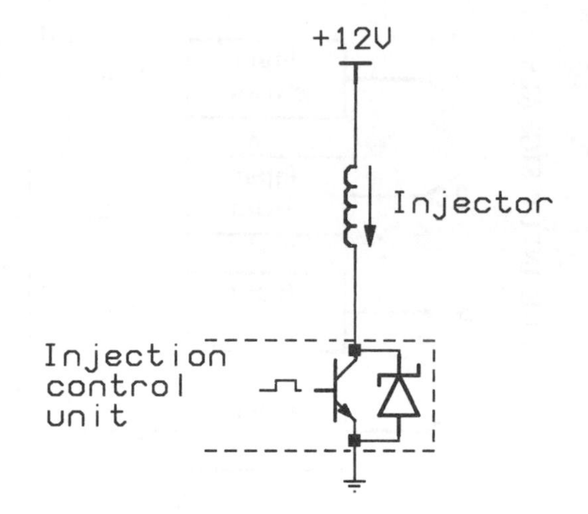

Figure 4.7 Simple injector switching circuit

To overcome the first of these disadvantages many manufacturers use low-inductance injectors which have a fast response time. Since injectors of this type have an inherently low resistance, the current through them must be limited by incorporating a voltage-dropping resistor into the drive circuit (Figure 4.8). Unfortunately this is not a very efficient design because a lot of the drive circuit's output energy is uselessly dissipated in the resistor.

In order to provide an optimal solution, the circuit configuration of Figure 4.9 is now commonly used. A sensing resistor, R_S, is used to develop a control voltage, V_A, at point 'A', which is proportional to the current through the low-inductance injector. The current control circuit constantly monitors V_A and provides feedback control of the injector current by rapidly turning the drive transistor ON and OFF to

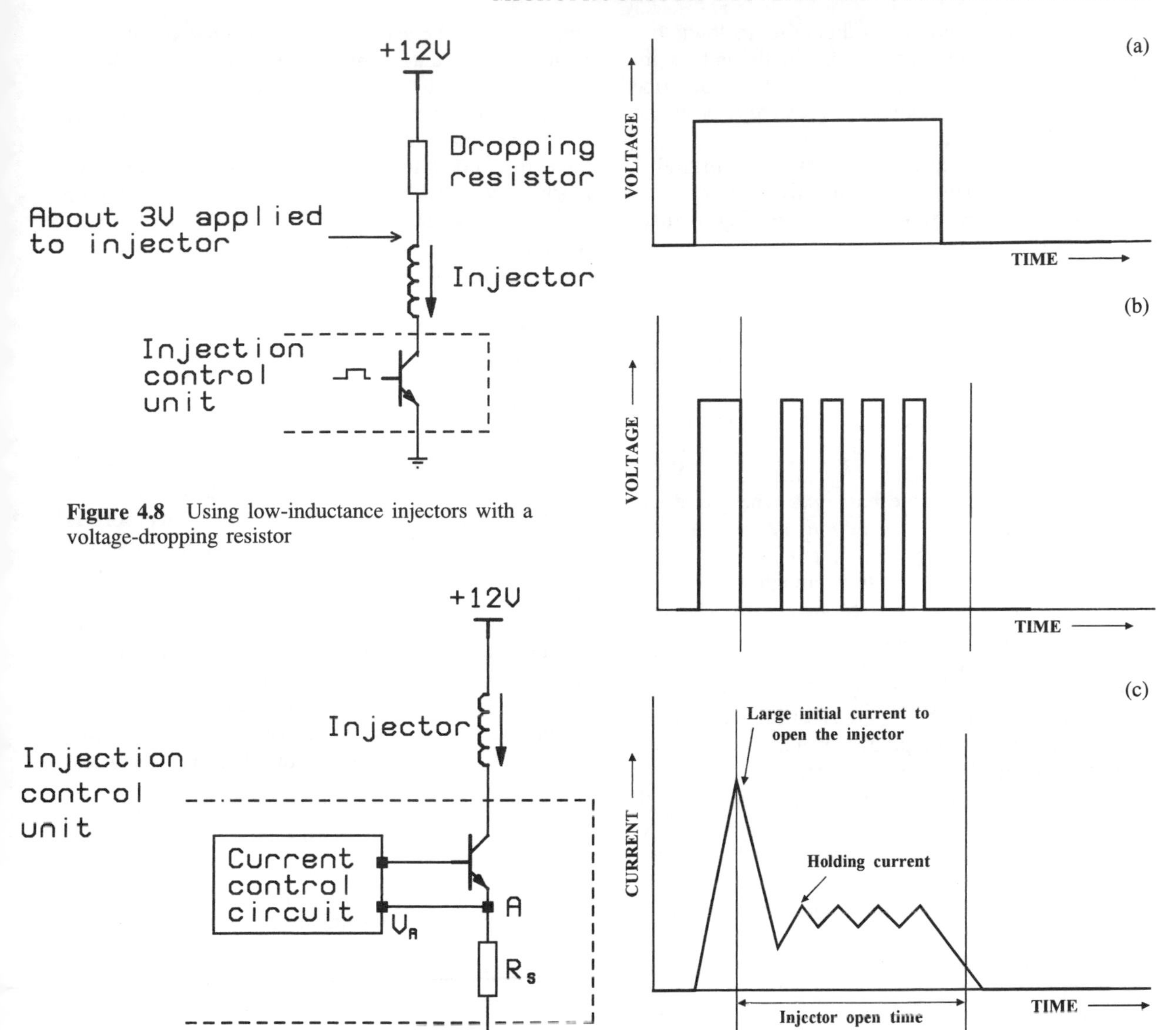

Figure 4.8 Using low-inductance injectors with a voltage-dropping resistor

Figure 4.9 Using a current-control circuit to provide feedback control of current through a low-inductance injector

Figure 4.10 Waveforms for current-controlled injector driver circuit. (a) Injector opening signal from microprocessor, (b) switching pulses applied to base of power transistor, (c) current through injector

maintain the required injector current. Figure 4.10 shows the waveform for the injector current. A large initial current (about 5 A) is used to quickly open the injector. Once open, a smaller 'holding' current (averaging about 1.5 A) is used to keep the injector open for the required length of time. This design provides for a rapid injector opening response and automatically compensates for fluctuations in battery voltage. The current control circuit can be enhanced to include circuitry to monitor for fault conditions (open or short-circuit injectors) and provide automatic shut-down.

Digital-to-analogue converters

There are some situations where digital data needs to be converted to an equivalent analogue voltage or current value. For example, a microprocessor-based ECU may be required to produce a smoothly varying analogue voltage to drive an analogue speedometer mechanism. This calls for a digital-to-analogue converter (DAC) to change the digital 0s and 1s into their analogue equivalent. As with ADCs, described above, DACs are available either as chips or already integrated into a microcontroller IC.

The circuitry required to perform a digital-to-analogue conversion is comparatively simple and therefore fast-converting DACs are readily available at low cost.

Serial input/output

Within the microcomputer system, data is transferred from one place to another along a parallel data bus in which an individual wire is dedicated to each data bit. Whilst parallel data communication is acceptable over very short distances (usually less than a metre) it requires cumbersome cable harnesses when used over longer distances. For this reason data transmission over distances of more than a metre or so is accomplished using a *serial data link* through which data is sent one bit at a time along a single wire. The serial data output circuit accepts parallel data bytes from the data bus and arranges them into a corresponding serial strings of '1's and '0's. The data then has 'START' and 'STOP' bits added at the ends of each byte to give 10-bit groupings which are sent one after the other along the wire to the receiver.

The sending and receiving circuits must use the same fixed bit rate which will typically be 2 400, 4 800, 9 600 or 19 200 baud (one baud is one data clock period per second). The data is sent *asynchronously* which means that the two circuits do not have to operate with highly accurate timing, instead they can resynchronize on the START and STOP bits of each byte. Serial data communication circuits are frequently encountered on automotive electronic systems where they are used to connect an off-board diagnostic scan tool to an ECU for the purposes of checking the system function and reading stored diagnostic trouble codes.

Microcontrollers

A microcontroller is a single-chip microcomputer which integrates the microprocessor, memory (RAM and ROM), input/output circuits and timers into a single LSI device. A typical automotive microcontroller, the Motorola 68HC05, is illustrated in Chapter Two.

Although microcontrollers are often more restricted with regard to memory capacity and computational power than a microcomputer system, they have the advantages of lower cost, lower power consumption, smaller size and greater reliability. Microcontrollers are widely used in automobile ECUs since they are ideally suited to dedicated real-time control tasks which require many input and output circuits but relatively little memory. Furthermore, many IC manufacturers have responded to the demands of the automotive market by developing customized microcontrollers with built-in input/output functions specially designed for automotive use. For example, for many years the Intel 8031/51 8-bit microcontroller family was used in a range of engine, transmission and ABS controllers from a variety of manufacturers. Its big brother, the Intel 8096 16-bit microcontroller, is currently popular and is used in the Rover MEMS (modular engine management system) ECU.

Motorola have long held a large slice of the automotive microcontroller market, their MC 68705R3 is the popular 8-bit microcontroller used in the Bosch Motronic engine management system. It incorporates 112 bytes of RAM, 4 Kbytes of EPROM, four I/O ports, a timer and a four-channel ADC integrated into one IC.

A very powerful and sophisticated Motorola microcontroller, the 32-bit MC 68332, was designed specifically for the automotive market, it is currently used by Saab in their Trionic engine management system and is due to be incorporated into forthcoming ECUs from a number of other manufacturers.

4.3 COMPUTER PROGRAMS

To enable a microcomputer to control the operation of an electrical or mechanical system it must be provided with a sequence of instructions, the program, that tells it what to do and how to do it. Since the operation of the microcomputer is based on the binary number system the instructions must also use binary numbers, which are permanently stored in ROM and therefore always available to the CPU. Each instruction causes the microprocessor to perform a single, simple, arithmetical or logical process; for example the binary addition of two numbers or the transfer of data from one memory location to another, and so on. Taken alone, a single instruction achieves very little, but when hundreds of thousands of instructions can be executed each second the microprocessor becomes a very powerful tool.

A typical 8-bit microprocessor can execute about 50 different *types* of instruction, and together with the various combinations of addressing that are available for use by each instruction type this gives an *instruction set* of 100–200 executable instructions from which to construct a program.

Program execution

A microprocessor executes a program by fetching an instruction from memory, executing it, and then fetching the next instruction. The program counter is automatically incremented after each memory access so that it always addresses the next consecutive memory location. Since the program counter is 16 bits wide (on an 8-bit microprocessor), it can address any location in the 64 Kbytes of address space.

Instructions are generally comprised of one, two or three bytes. The first byte always holds the *operation code* (op code) that tells the microprocessors what to do and so it is directed to the microprocessor's internal instruction decode logic. This circuit then issues appropriate internal and external control signals to all of the other elements of the microcomputer. The second and third bytes, if the instruction has them, are routed to the arithmetic/logic unit (ALU) if they represent data, or into the program counter if they represent an address from which to fetch data.

The speed with which a microprocessor can process data depends on the instructions that are executed. Each instruction requires a certain number of clock cycles in order to be executed, so the speed of program execution depends on the speed of the processor that is being used. A typical low-cost 8-bit microprocessor with a 1 MHz clock will execute its simplest instruction in about 2 μs and the most complicated instruction will require about 7 μs.

Programming languages

The set of binary numbers which comprise the microprocessor's instruction set is called the *machine code*. It is very difficult for the average user to write a program in machine code, errors are likely to arise and the resulting listing of 0s and 1s is very difficult for other programmers to understand. For these reasons *programming languages* have been developed which are much easier for the programmer to under stand and can subsequently be converted into machine code for use by the microprocessor.

Programming languages can be classified into two main types, *assembly languages* and *high-level languages*.

Assembly languages
Assembly languages use English-like *mnemonic codes* (i.e. memory aids) to represent machine code instructions on a one-to-one basis. For example, the binary number 10101001 might be machine code for 'load data from memory into accumulator'; in assembly language this instruction would be represented by the mnemonic 'LDA'.

Once a program has been written in assembly language it must be converted into machine code using a special program called an *assembler*. Each microprocessor family has its own set of instructions and mnemonics and therefore requires its own specific assembler.

High-level languages
Commonly used high-level languages include Pascal, C and BASIC, they are sophisticated programming languages which use instructions that closely resemble English words.

Since each high-level language statement may correspond to several machine language instructions a complex *compiler* program must be used to convert them into machine code. As a result of this conversion process some inefficiency in the number of machine code instructions and memory usage appears in the compiled program.

Despite these drawbacks high-level languages are frequently used for the development of automotive microcomputer programs since they are easy to understand and the same program can easily be recompiled for use with different types of microprocessor.

Preparation of the high-level or assembly language program is usually undertaken on an IBM-compatible PC, which is also used to generate the final machine code for the microprocessor.

After completion of this task, the code must be tested on prototype hardware to ensure that it functions correctly and any errors rectified, a process known as *debugging*.

To perform debugging, the machine code is transferred from the PC's memory into an EPROM chip using an EPROM programmer. The EPROM is then installed into the prototype ECU.

Once the ECU designers are confident that the program is operating to its specification, the final version of the machine code is supplied to the memory chip manufacturer for incorporation into mass-produced ROMs.

4.4 CONTROL STRATEGIES

As its name implies, the function of an electronic control unit is to control the operation of a system and provide an optimum response to counter the influence of external disturbances. The *controlled system* can be entirely mechanical, for example an engine, or it can be electromechanical, for example an automatic heater/air-conditioner system. Whatever the case, the ECU outputs a *control signal* which commands a *controlling element* (usually an actuator) so as to keep the system operating within a given specification. The signal applied to the controlling element is determined by the control strategy, which takes the form of a *control algorithm* that is incorporated into the ECU's software. Control algorithms are designed to quickly respond to disturbances in the controlled system's operating conditions and so re-establish stable operation.

Control systems divide into two main categories; *open-loop* and *closed-loop*.

Open-loop control

An open-loop control system is illustrated diagrammatically in Figure 4.11(a). The *input variables* to the system are the *reference variable*, or *set-point*, w, and the *disturbance variables*, z_1 and z_2 The control unit monitors w and z_1 and uses its *control algorithm* to generate an *output variable*, y, which modifies the operation of the controlled system.

The major characteristic of this type of control system is that it is only able to respond to the disturbances, z_1, that are *directly measured* by the control unit. There is no compensation for disturbances of type z_2 that are not measured. For example, in a control unit designed to control the air–fuel (A/F) ratio of a non-catalyst engine, w would be the ideal A/F ratio, say 15:1, and the z_1 disturbance inputs would be air and coolant temperatures, and engine speed and load. The control unit would process this information to provide an output signal, y, to open a fuel injector and deliver the appropriate amount of fuel. Any variation in engine condition, perhaps due to a build-up of combustion chamber deposits, would not be detected by the control unit and therefore could lead to a deviation from ideal operation and a consequent drop in performance.

Although open-loop systems can be used when relatively coarse control is required they are not suited to systems that require accurate control over a long service life. For this reason they have been largely superseded by closed-loop control systems.

Closed-loop control

Figure 4.11(b) illustrates the configuration of a closed-loop control system. It differs from the open-loop system in that the control unit is placed within a closed loop and so can act to compensate for *all* types of disturbance (z_1 *and* z_2) that affect the controlled system.

The control unit accepts a *feedback signal*, x, which represents the output from the controlled system, and then performs the function of comparing this value with the set-point value, w, to establish an *error value*, $(x - w)$, between the two. A control algorithm is then applied to the error value to determine the appropriate corrective action which would eliminate the error. The controller then outputs the required corrective signal value, y.

The sophistication of closed-loop systems means that they can respond quickly and accurately to any changes in operating conditions, leading to smooth, stable and precise control over the life of the vehicle. It is for these reasons that closed-loop strategies have become almost universally adopted for automobile control tasks. Table 4.3 gives simplified examples of some common closed-loop systems.

Feedforward control

The principle of closed-loop feedback control (described above) is that the controller implements corrective action *after* the external disturbances, z_2, have influenced the output, x, of the controlled system. In contrast, a feedforward control system measures the external disturbances directly and takes corrective action *before* they can influence the output.

Feedforward control is useful when there are just a few external disturbances that can easily be measured. A good example is in an engine management system, where the use of a feedforward signal from an airflow meter is combined with a feedback control system based around an exhaust-gas oxygen sensor (λ sensor) to give a sharp engine response and optimum control of fuelling.

Closed-loop control algorithms

The function of a closed-loop control algorithm is to provide the control system with a means to:

(i) minimize the value of the error between the feedback signal, x, and the set point value, w;
(ii) minimize the response time of the system for sudden changes in load.

For simple control tasks ON/OFF control can be used; it is best suited to processes which have a slow response time, such as temperature or fluid level control. Usually the output signal, y, is inactive (i.e. 'OFF') when the input signal, x, lies within an acceptable range. However, when the input exceeds a predetermined threshold the output switches fully 'ON'. A good example is the control of a radiator cooling fan using a temperature-sensitive thermo-switch. If the

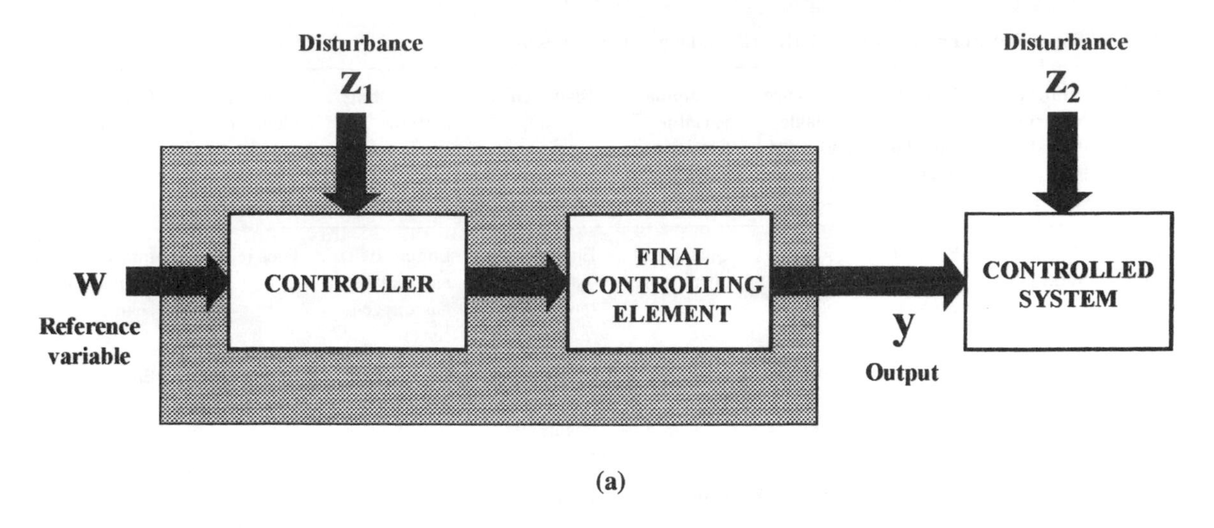

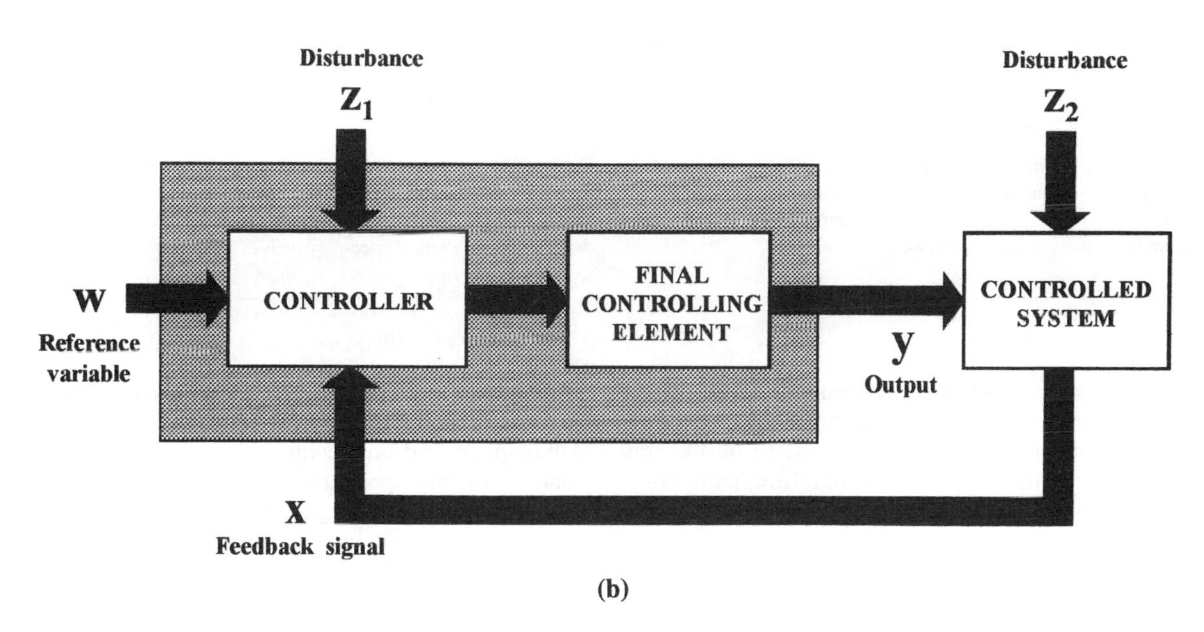

Figure 4.11 The two types of control system: (a) open-loop control; (b) closed-loop control

engine coolant temperature is above the radiator thermo-switch's ON temperature, the radiator cooling fan is energized and the coolant temperature falls. When it drops below the ON temperature, the thermo-switch opens and the cooling fan stops. This type of control therefore provides for either full cooling or no cooling and results in a coolant temperature that fluctuates about a set point (which is fixed by the characteristics of the thermo-switch). The fan's ON/OFF switching cycle is referred to as the *limit cycle* of the system, and the response time of the limit cycle depends upon the characteristics of the sensor and controller.

Limit cycle control is used in a number of other automobile systems, including engine air–fuel ratio control, hydraulic pressure control and adaptive suspension height control. Although the strategy is apparently very simple, a full analysis of its behaviour in a real application is in fact very involved.

For more complicated control tasks other control strategies must be used. The most common is the so-called *PID control* strategy, where PID is an acronym

Table 4.3 Simplified examples of automotive closed-loop control systems

Control system	Indirectly controlled variable	Directly controlled variable (feedback signal, *x*)	Reference variable (set-point, *w*)	Manipulated variable (output variable, *y*)	Disturbances (z_1, z_2)	Controlling system	Control element	Controlled system
Fuelling control	Air/fuel ratio (λ)	Exhaust oxygen content	λ = 1.0	Quantity of injected fuel	Engine wear, leaks, temperature	Engine ECU and exhaust gas oxygen sensor	Fuel injector	Intake tract and combustion chamber
Idle speed control	Idle speed	Idle speed	Preset idle speed	Intake air volume	Idle load, air temperature, engine temperature	Engine ECU and speed sensor	Idle air control valve	Engine
Knock control	Knock	Knock sensor signal	Zero knock	Ignition timing	Engine wear, temperature, fuel quality	Engine ECU and knock sensor	Ignition coil switching transistor	Combustion chamber
Anti-lock braking system (ABS)	Wheelslip limit	Wheel speed	Wheelslip limit	Brake line pressure	Road and tyre condition	ABS ECU and wheel speed sensor	ABS solenoid valve	Road/tyre
Cabin air-conditioning control	Cabin temperature	Heater and cooler output	Cabin preset temperature	Hot/cold air mixing	Outside air temperature, sunlight level, driving speed	Temperature control ECU and temperature sensors	Air-con. compressor and heater valves and flaps	Cabin

for *Proportional, Integral, Derivative.* These control algorithms are highly effective in minimizing the influences of the external disturbances, z_2, on the controlled system and most automobile controllers therefore use P, P + I or P + I + D control.

Proportional control

The fundamental principle of PID control is *proportional correction* in which the controller output signal, y, is made proportional to the error between the set point, w, and the feedback signal, x. Mathematically,

$$y = K_P \times (x - w),$$

where K_P is the *proportional gain* of the controller. An example of proportional action is the closed-loop control of engine idle speed. Normally a car engine is designed to idle at a set point speed of around 750 rpm. If the idle speed is greater than this, then the proportional control algorithm will close the engine's idle air control valve by an amount proportional to the speed error. Conversely, if the idle speed is below the set point then the valve is opened by an amount proportional to the speed error. After a short time the idle speed should settle close to the set point value. Although proportional control is superior to ON/OFF control in many applications, it still has a number of limitations. In particular, if the controller is required to operate over a wide range of error values then the system can be rather slow to respond. Furthermore, poorly designed proportional controllers can oscillate if the response time of the controller is not well matched to that of the controlled system.

Integral control

To overcome the deficiencies of proportional control, and increase control accuracy, *integral action* may be added. The nature of integral action is to steadily increase the value of the output signal, y, for as long as the error $(x - w)$ exists. y is therefore the *integral* of $(x - w)$.

Returning to the example of engine idle speed control, the response of an integral control algorithm is to progressively close off the idle air control valve for as long as the idle speed remains above the set point value.

Derivative control
A derivative control algorithm is added to a control system to increase stability and minimize any tendency to overshoot. It produces an output signal, y, proportional to the *rate of change* of the error signal, $(x - w)$. Mathematically,

$$y = K_D \times [d(x - w)/dt],$$

where K_D is the *derivative gain* of the controller.

Derivative action therefore provides a *lead function* and acts to predict the effects of changing disturbances on the controlled system, speeding up the response of the controller. The amount of derivative action that is applied to the system depends on how fast the error signal is changing and so its greatest effect is therefore in the early part of a system's transient response when there is a large, rapidly varying error due to a sudden change in load.

Returning, once more, to the example of the engine idle speed controller, a derivative response provides a large and rapid change in the idle air control valve setting when the idling engine is subjected to a sudden load change. For example the engagement of 'DRIVE' on an automatic gearbox, or the operation of the air-conditioner.

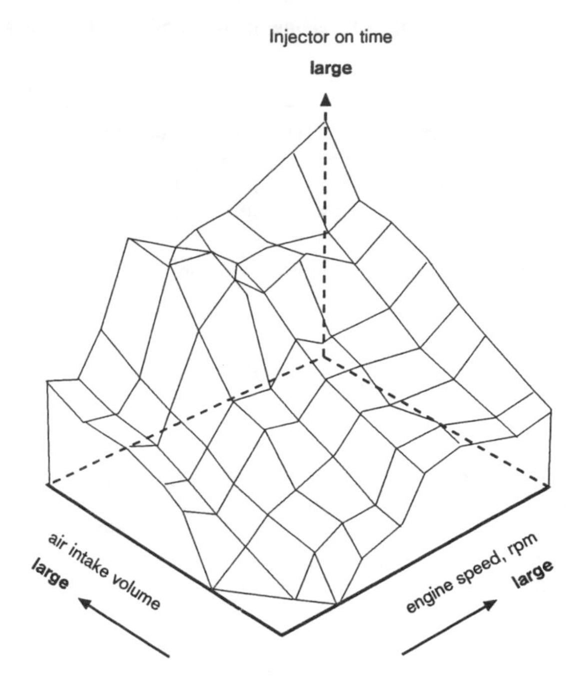

Figure 4.12 Simplified example of three-dimensional 'map' relating two input quantities to single output quantity

Look-up tables and maps

Microcomputers are used to implement control strategies in various ways. One way is to provide the microprocessor with a mathematical *control equation*, permanently held in ROM, that is repeatedly evaluated using the most recent input data. The result of the calculation is then used to generate an appropriate actuator drive signal. Many ABS controllers use this approach and work by continually measuring each wheel speed and then calculating wheel acceleration or deceleration. When the calculated deceleration exceeds a predetermined threshold the controller actuates an electromechanical brake pressure reduction valve to prevent the wheel from locking. Whilst this strategy is acceptable for simple and relatively slow systems it is not suitable for very complex, high-speed systems. For example, the control of an engine would require the evaluation of many complicated equations, each with a large number of variables. Even a sophisticated microprocessor could not calculate quickly enough to keep up with an engine.

The solution to this problem is for the ECU designer to provide the microprocessor with sets of predefined results that cover all of the engine's possible operating conditions. These results, in the form of *look-up tables* permanently held in ROM, are then available to the microprocessor in the time taken to execute a memory 'fetch' instruction.

During engine operation the ECU monitors sensor signals which indicate the engine's speed and load conditions. These signals are then digitized, and the resulting binary numbers are used as *pointers* to indicate the memory address locations that hold output data appropriate to the instantaneous speed/load conditions. The microprocessor fetches this data and passes it to the output circuits.

Since the look-up tables generally relate two input variables to a single output variable, they can be visualized as three-dimensional surfaces or *maps* (Figure 4.12). The low cost of digital memory means that modern engine management ECUs are able to hold numerous detailed maps relating spark timing and fuel injection volume to variables such as engine speed, load, temperature, turbo boost and ambient air temperature.

Table 4.4 shows typical map data, in this case relating ignition timing to engine speed and inlet manifold pressure (i.e. engine load). Ignition timing is controlled by selecting the advance value determined by the engine load and engine rpm from the data stored in memory. This particular map is referred to as a 16×16 map since it contains data for 16 values

Table 4.4 Look-up table giving ignition timing (BTDC) as a function of engine speed and load

Engine speed (rpm)

Load (bar)	800	1 000	1 200	1 400	1 600	1 800	2 000	2 400	2 800	3 200	3 600	4 000	4 400	4 800	5 200	5 600
1.0	5	7	9	10	11	13	15	17	19	20	22	22	22	24	25	25
0.96	5	7	9	10	11	13	15	17	19	20	22	22	22	24	25	25
0.92	8	8	9	11	12	14	17	21	24	27	29	29	28	29	30	30
0.88	9	10	11	12	14	18	22	25	30	32	32	33	35	33	33	33
0.84	12	13	15	18	18	18	25	30	32	34	34	34	34	34	34	34
0.80	14	15	17	18	18	18	25	33	34	34	35	35	35	35	35	35
0.76	17	18	18	18	18	18	27	35	35	37	38	38	37	37	37	37
0.72	18	18	19	19	20	23	31	36	38	39	40	40	39	39	39	39
0.68	20	20	20	20	24	28	37	38	40	41	42	42	43	42	42	42
0.64	22	23	24	25	28	32	40	41	41	43	45	45	45	45	45	45
0.60	23	25	26	27	31	35	40	42	42	44	45	45	45	45	45	45
0.56	25	25	29	29	32	37	40	41	42	43	45	45	45	45	45	45
0.52	25	28	30	32	35	37	40	40	41	42	45	45	45	45	45	45
0.48	25	28	35	35	35	38	38	40	40	40	40	45	45	45	45	45
0.44	23	24	35	35	35	35	35	38	38	38	38	40	40	40	40	40
0.40	20	23	27	28	28	28	35	35	35	35	35	35	35	35	35	35

of engine speed and 16 values of engine load, giving a possible 256 advance settings that can be directly read out. If the engine is operating at a load/speed condition not directly given in the table, then the microprocessor must calculate an appropriate timing value using a mathematical process called *interpolation*.

For example, if the engine is running at 2400 rpm with an inlet manifold vacuum of 0.52 bar, then from Table 4.4 it can be seen that the ignition timing takes the value 40° BTDC. However, if the engine speed is 1 300 rpm and the inlet vacuum is 0.90 bar then a result cannot be read directly from the look-up table and the microprocessor must interpolate.

First, the timing angles at 0.88 bar and 0.92 bar are calculated for an engine speed of 1 300 rpm. Referring to Table 4.4, at 0.88 bar the advance at 1 200 rpm is 11° and the advance at 1 400 rpm is 12°, therefore at 1 300 rpm the advance is calculated as (11° + 12°)/2 = 11.5° BTDC. Similarly, at 0.92 bar the advance is (9° + 11°)/2 = 10° BTDC. Therefore the advance at 0.90 bar is the mean of the values at 0.88 bar and 0.92 bar, that is 10.75° BTDC.

Engine mapping

The process of obtaining engine map data is called *mapping*. Mapping involves running a fully-instrumented pre-production test engine on a dynamometer and appying varying levels of load whilst adjusting fuelling and timing for maximum power and lowest emissions, taking into account likely production tolerances. Once the preliminary map data has been obtained, the engine is run in a test vehicle to assess overall drivability and responsiveness. At this stage, final tweaks can be made to the maps using portable computer equipment carried in the test car.

When the development engineers are satisfied with the prototype ECU's performance, the maps are programmed into mass-produced ROMs that are installed in production ECUs. Often the map data ROM IC is designed to plug into a socket on the ECU circuit board, a feature that allows a standardized ECU design to be configured for a specific engine variant simply by inserting the appropriate ROM at the final stage of production.

Modern control theories

The availability of very powerful and low-cost microcontrollers has led to the development of new control theories that offer improvement over conventional PID control. Two of these new control strategies, *fuzzy-logic control* and *adaptive control*, have recently appeared on production vehicles.

Fuzzy-logic control

Fuzzy-logic is a control concept that was developed in Japan during the early 1970s. Since that time fuzzy control systems have been used to supervise the operation of trains, lifts, video cameras, cars and even helicopters. Many Japanese engineers remain committed to fuzzy control and it has been extensively promoted by the Mitsubishi car company for the control of vehicle systems, including automatic transmissions, traction control systems and semi-active suspension systems.

The basis of fuzzy control theory is the use of *multivalue logic*; i.e. rather than using logic with just two values, 'YES' or 'NO' or '1' or '0', fuzzy-logic can have many values. For example, using conventional logic a room thermostat might operate on the basis of '0' = 'cold' and '1' = 'hot', whereas using fuzzy logic the thermostat would return a range of values between 0 and 1 corresponding to temperatures ranging through 'very cold', 'cold', 'a bit cold' 'a bit warm', 'warm', 'hot' and 'very hot'. The fuzzy controller then applies a set of pre-programmed *rules* to this input data to control the operation of a heater, for example:

(i) If the room temperature is hot, turn off the heater.
(ii) If the room temperature is warm and the outside temperature is mild, decrease the heat a little.
(iii) If the room temperature is about right, don't change the heater output.
(iv) If the temperature is cool and the outside temperature is cool, increase the heat output a little.
(v) If the temperature is cold, greatly increase the heater output temperature.

By inputting fuzzy data from a large number of sensors, and then applying a comprehensive set of rules, it is possible for a fuzzy controller to recognize particular patterns in a system's operating conditions and thereby provide an ideal output for any set of circumstances.

Fuzzy controllers use standard microprocessors programmed with special fuzzy control software. Since fuzzy control demands the execution of many multiply and accumulate instructions it is the calculating speed of the microprocessor that limits the performance and accuracy of the control system.

Adaptive control

Adaptive control, as the term suggests, has the ability to alter the pre-programmed controller output value to suit the instantaneous operating conditions. Such systems have been successfully commercialized for both fuelling and ignition timing. For example, an adaptive ignition timing controller contains a pre-programmed ignition map, giving the basic advance, but it has the added capability to experiment with small temporary timing alterations, either advance or retard, to see what effect they have on engine operation.

By continuously monitoring engine rpm the controller is able to detect whether an alteration from the mapped timing value leads to increased speed, indicating improved combustion. If speed does increase, then the controller permanently modifies the map value to the altered value.

The advantage of adaptive control is that the control map is continuously updated to take account of engine wear, fuel variability and changing driver behaviour.

5

Engine management systems

5.1 INTRODUCTION

Today's car engines must offer low exhaust emissions, good fuel economy and excellent driving performance under all driving conditions. Many factors are important in achieving this aim. Improvements in the mechanical design of the engine, such as the shape of the combustion chamber, location of the spark plug and number of intake valves are very significant. However, precise control of the air–fuel mixture ratio and spark timing have become of central importance in maximizing an engine's power and efficiency, and minimizing its emissions. For a modern engine this task is now considered to be beyond the capabilities of simple mechanical control systems and electronic engine management must be used. Such a system consists of a microprocessor-based electronic control unit (ECU) and a large number of electronic and electromechanical sensors and actuators. It is the job of this system to:

(i) Provide accurate control of the air–fuel mixture ratio via a fuel injection system.
(ii) Assure accurate and precise ignition timing for all engine operating conditions.
(iii) Monitor and control numerous additional parameters such as idle speed, exhaust-gas recirculation, air conditioner operation and fuel evaporative emissions to ensure consistently good performance under all circumstances.

5.2 COMBUSTION PROCESSES IN THE SPARK-IGNITION ENGINE

Introduction

To understand the operation of the engine management system it is first of all vital to develop an appreciation of the combustion process itself and the factors that influence it. A vast amount of research has been done in this area, perhaps most notably by the British engineer Sir Harry Ricardo who was responsible for many of this century's remarkable advances in combustion engineering. Since 1913, when Ricardo started his pioneering research work at Cambridge, the mechanical design and construction of engines has been enormously improved. But perhaps the greatest improvements have come about quite recently, as a result of the precise control afforded by electronic systems.

In a four-stroke spark-ignition (SI) engine the fuel and air are normally mixed in the engine intake passages and drawn into the engine during the intake stroke. Turbulence within the cylinder causes further air–fuel mixing to occur as the piston rises on its compression stroke. As the piston nears the top of its compression stroke, a precisely timed electrical spark starts the combustion process. A 'flame-front' of burning gas then propagates away from the spark with a velocity of 20–40 ms^{-1}, progressively consuming the unburnt mixture until it is eventually extinguished on contact with the cold cylinder wall. The hot gases produced by this combustion process raise the cylinder pressure to about 30 bar (about 450 psi) and so force the piston downwards, turning the crankshaft and hence the roadwheels. The burnt gases are subsequently expelled to atmosphere on the exhaust stroke and the cylinder is ready to be refilled.

It is the composition of the inducted air–fuel mixture and the timing of the ignition spark which so radically influences this combustion process, and hence the performance and economy of the engine and the quantity of pollutants in its exhaust.

Ignition timing

Since a typical air–fuel mixture takes a few thousandths of a second to completely burn, engine designers must arrange for the ignition spark to occur just before the piston reaches the top of its compression stroke. This is known as 'ignition advance' since the spark occurs in advance of the piston reaching top dead centre (i.e. it occurs before top dead centre – BTDC). Ignition advance gives time for combustion to occur and so allows the cylider pressure to reach a maximum just as the piston starts down on its expansion (power) stroke. To illustrate this, consider an engine in which the mixture burning time is 3 ms. To obtain maximum power from the engine, the mixture must be fully burned by the time the crankshaft has turned between 10° and 20° past top dead centre. At 1 000 rpm the crankshaft turns through 18° in 3 ms, whereas at 2 000 rpm it turns through 36°. Since the point of complete combustion is fixed, it is clear that the spark must occur earlier in the compression cycle as the engine speed increases.

The amount of advance required varies considerably between different engine models and is established by running the engine on a dynamometer at various speeds and loads. For maximum power and torque it is usually found that the cylinder pressure should reach a maximum at about 16° of crankshaft rotation after top dead centre (16° ATDC), this maximizes the transfer of energy from the fuel to the piston. The amount of spark advance required to achieve this condition depends upon the speed of flame propagation, which is in turn influenced by a large number of factors, including:

(i) The shape of the combustion chamber. Modern combustion chambers are designed to give a short flame propagation time.
(ii) The air–fuel ratio of the intake mixture. Combustion is most rapid when there is just sufficient air available to burn all of the fuel. Fuel-rich or fuel-lean mixtures burn more slowly.
(iii) The pressure of the vapour in the intake manifold. A low intake pressure gives a less dense air–fuel mixture which burns slowly. A higher pressure gives a more dense mixture which burns more quickly.
(iv) The amount of turbulence (swirl) present in the mixture. A high level of turbulence accelerates the combustion process.
(v) The chemical composition of the fuel used. Poor quality fuel may either burn slowly or 'knock'.
(vi) The amount of recirculated exhaust gas present in the mixture. Flame speed falls as the proportion of exhaust gas in the intake mixture is increased.

The graphs shown in Figure 5.1 illustrate how engine output torque alters as ignition timing is varied for one particular engine operating at constant speed with a fixed air–fuel ratio. With the timing set at 30° of spark advance (30° BTDC) the cylinder pressure rises smoothly to a maximum of about 25 bar at about 15° ATDC. This represents the best timing for the particular operating conditions and is known as the maximum brake torque (MBT) timing. A primary function of the engine management system is to continually assess the engine operating circumstances and compute the amount of ignition advance required to maintain MBT timing.

If the spark timing is advanced beyond MBT timing, say to 50° BTDC, then the cylinder pressure rise starts earlier in the compression stroke. The rising piston must therefore work against the rapidly expanding cylinder gases, resulting in a loss of efficiency and an excessively steep rise in cylinder pressure. The maximum pressure of about 33 bar is achieved close to TDC and is accompanied by intense pressure oscillations due to spontaneous combustion in portions of the unburned mixture (the 'end-gas'). These pressure oscillations are audible as 'knock' (a metallic rattling sound) and can cause engine damage. Ignition timing of 50° BTDC therefore represents overadvance for this engine.

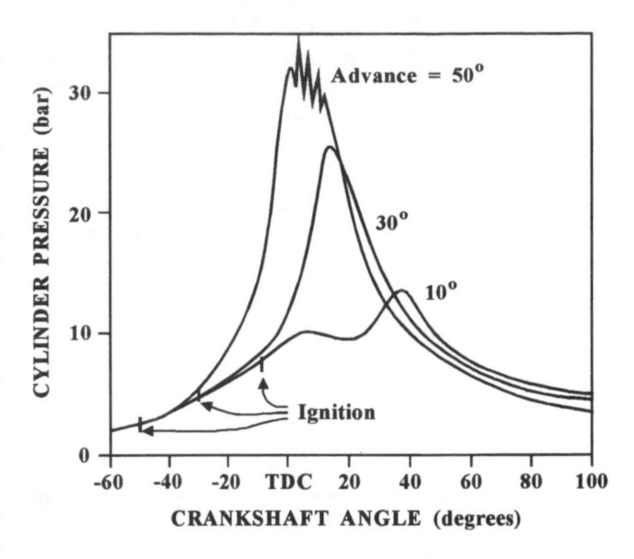

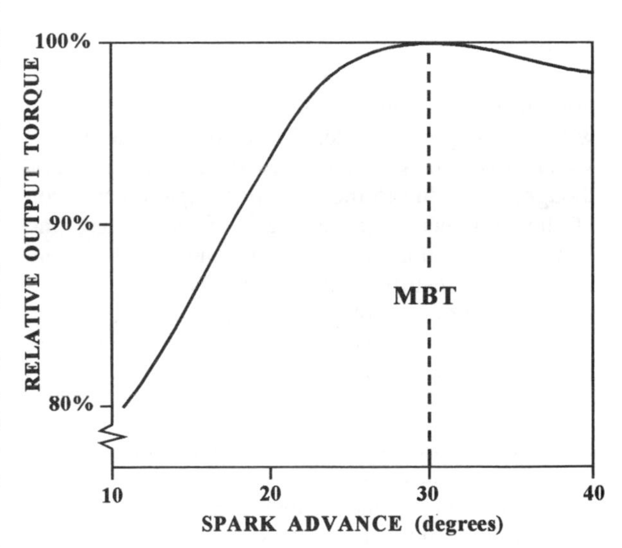

Figure 5.1 Cylinder pressure and relative output torque as a function of spark advance in a typical engine

With the spark timing retarded to 10° BTDC, the start of the combustion process is considerably delayed and so peak cylinder pressure occurs comparatively late in the expansion stroke. The amount of energy transferred from the fuel to the piston is therefore diminished. Since combustion may not be completed by the time the exhaust valve is opened it is possible that retarded timing may lead to an increase in exhaust emissions.

Abnormal combustion

Under normal combustion conditions the flame starts at the spark plug and propagates smoothly across the combustion chamber until the air–fuel mixture is completely burned. However, under some circumstances abnormal combustion can occur, resulting in engine inefficiency and possible damage. The two most commonly encountered types of abnormal combustion are (i) knock an (ii) surface ignition.

Knock

The term 'knock' derives from the metallic rattling noise that accompanies abnormal pressure oscillations within the engine cylinder. The theory of exactly how knock arises is still not fully developed, but it is thought to be due to the extremely rapid combustion of the unburned mixture ahead of the propagating flame. When this mixture (the end-gas) suddenly ignites, the resulting uneven pressure rise causes the combustion chamber to resonate and the engine structure to vibrate. There are two theories to explain this sudden release of energy, (i) detonation and (ii) autoignition.

(i) *Detonation.* Certain factors such as unstable fuel compounds or an excessive cylinder pressure can cause the advancing flame front to increase its velocity to a supersonic speed. Under these circumstances the end-gas burns at a rate much faster than normal, leading to knock.
(ii) *Autoignition.* This theory suggests that as the air–fuel mixture in the end-gas is heated to very high temperatures and pressures by the advancing flame front it starts to undergo a process of spontaneous oxidation. This results in a sudden and explosive combustion of the end-gas, leading to knock.

Short duration knocking on acceleration is quite common in many engine designs and is not likely to cause damage. Long term knock, especially at high speeds, can cause severe engine damage. Pressure pulses of up to 150 bar at frequencies in the 6–12 kHz range can cause head gasket failure, piston erosion, holing and piston ring breakage. Very heavy knock transfers extra heat to the combustion chamber surfaces, leading to overheating and possible piston seizure.

To safeguard against knock damage most engine management systems incorporate some means of detecting knock so that the ignition timing can be retarded to protect the engine.

Surface ignition

As the term implies, surface ignition refers to ignition of the mixture by a 'hot-spot' on the combustion chamber rather than by an electrical spark. Usually an overheated spark plug or a glowing fragment of combustion-chamber deposit is responsible. When surface ignition occurs after the spark it is referred to as postignition and may go unnoticed. Surface ignition prior to the spark is called preignition and can usually be heard as 'pinking', since it represents over-advanced ignition and may result in knock. Surface ignition is beyond the control of the engine management system.

Mechanism of pollutant formation

Petrol is a complex blend of hydrocarbons; chemical compounds such as paraffins, cycloparaffins, olefins and aromatics, containing mainly carbon and hydrogen atoms. In a car engine it is these hydrocarbon compounds which are burnt with air to release energy and hence propel the vehicle.

Burning is the chemical process of oxidization; the combining of the hydrocarbon with the oxygen in air. If sufficient oxygen is available then the hydrocarbon will be completely oxidized – all of the carbon atoms will combine with oxygen to produce CO_2 (carbon dioxide) and all of the hydrogen atoms will combine to produce H_2O (water), both of which are harmless substances,

$$HC + O_2 + N_2 \rightarrow CO_2 + H_2O + N_2.$$

Stoichiometric (chemically perfect) combustion occurs with an air–fuel ratio of about 14.7:1. Thus to completely burn 1 gm of petrol, consisting mainly of heptane (C_7H_{16}) and hexane (C_6H_{14}), about 14.7 gm of air are required (a volume of about 12.2 litres). In practice, air–fuel mixtures may be either fuel-rich or fuel-lean and so the ratio of the *actual* air–fuel ratio to the *stoichiometric* air–fuel ratio is a useful parameter for describing mixture composition. This parameter, termed the relative air–fuel ratio or excess air factor, is given the symbol λ (the Greek letter 'lambda'), and is defined as,

λ = (actual air–fuel ratio)/(stoichiometric air–fuel ratio)

For fuel-lean mixtures, λ is greater than 1
For stoichiometric mixtures, λ equals 1
For fuel rich mixtures, λ is less than 1

Unfortunately complete oxidization rarely occurs in a spark-ignition engine (even when operated with a stoichiometric mixture) since the hydrocarbons are constrained to burn for a short period of time wth a fixed volume of air. Atmospheric nitrogen, which is normally chemically inert, starts to react with oxygen under the conditions of high temperature and pressure which prevail in the combustion chamber. The nitrogen thus consumes oxygen which would ordinarily react with the hydrocarbons, leading to their incomplete combustion and resulting in the formation of undesirable combustion products, specifically the pollutants CO (carbon monoxide), HC (unburned hydrocarbon) and various oxides of nitrogen (NO, NO_2 and N_2O – collectively termed NO_X), i.e.

$$HC + O_2 + N_2 \rightarrow CO_2 + CO + H_2O + HC + N_2 + NO_X.$$

Even with a stoichiometric mixture ($\lambda = 1$) pollutant gases are produced and up to 2% of the exhaust gas will be composed of HC, CO and NO_X. Unfortunately these compounds have very harmful effects:

(i) CO (carbon monoxide) combines with haemoglobin in blood and interferes with its oxygen-carrying capability. This is a very dangerous condition which can result in toxicosis. Mild poisoning effects are felt when breathing air with a CO concentration as low as 0.03%, depending on exposure time.

(ii) HC (unburned hydrocarbon) is not only emitted from the exhaust, but also from the crankcase and fuel system. It does not present a direct threat to human health, although there is evidence that some compounds are carcinogenic (cause cancer). The major problem associated with HC is that it reacts with NO_X and atmospheric ozone to produce an unpleasant photochemical smog. Such smogs have become a particular problem in large cities, especially where there is a sunny climate.

(iii) NO_X (oxides of nitrogen) combine with haemoglobin in blood in a similar way to that described for CO, potentially leading to anoxia. Their acidic properties can lead to the formation of acid rain, and they are known to trigger respiratory irritation and asthma.

The relative quantities of the pollutants emitted by the engine depend principally upon the relative air–fuel mixture ratio, λ, and vary in the manner illustrated in Figure 5.2. Maximum combustion temperature occurs slightly lean of stoichiometric at about $\lambda = 1.05$ (an air–fuel ratio of about 15.5:1) and decreases rapidly to either side. Since NO_X emissions are temperature dependent they tend to follow this characteristic and so decline rapidly as λ increases.

Conversely, HC emissions increase away from stoichiometric. On the rich side ($\lambda < 1$) this arises as a consequence of incomplete combustion due to an inadequate oxygen supply. On the lean side ($\lambda > 1$) slow or incomplete burning means that the exhaust valve is opened before combustion is entirely completed, allowing some unburned HC into the atmosphere. With very lean mixtures, misfire may set in and HC emissions will rise sharply.

5.3 EMISSION CONTROL STRATEGIES

Introduction

Pressure to limit the amount of toxic gases emitted by automobiles first came about in the USA during the 1960s when the US government became concerned at the deteriorating quality of air in many large cities. The problem was particularly acute in Southern California where a combination of high traffic density and a sunny climate led to the formation of photochemical smog.

The first US emission limits were set by the 1968 Clean Air Act which required that vehicles sold in 1970 should have emission levels some two-thirds less than the uncontrolled 1967 levels (Figure 5.3). Subsequently, in 1970, Senator Muskie gained the approval of Congress for further steep reductions in automobile emissions. The 'Muskie Laws' required that by 1975–76 all vehicles should have emission levels of about 10% of the 1970 levels and in order to enact this legislation the US Environmental Protection Agency (EPA) was established to test and Type Approve vehicles.

Faced with such a daunting task and a short timescale the US automobile industry immediately looked at all possible technologies and concluded that only exhaust aftertreatment using catalytic converters would be available in time. Initially, oxidation catalysts were used to limit HC and CO emissions. Then exhaust gas recirculation (EGR) was added to give a reduction in NO_X emissions. Subsequently, as increasingly stringent NO_X limits were introdced in the late 1970s, the three-way catalytic converter was developed to reduce all three pollutants. By the mid-1980s virtually the entire US petrol-engined car fleet was fitted with this emission-control system and it has since become a standard throughout most of the developed world.

European emission control legislation has lagged about a decade behind the US, enabling European car

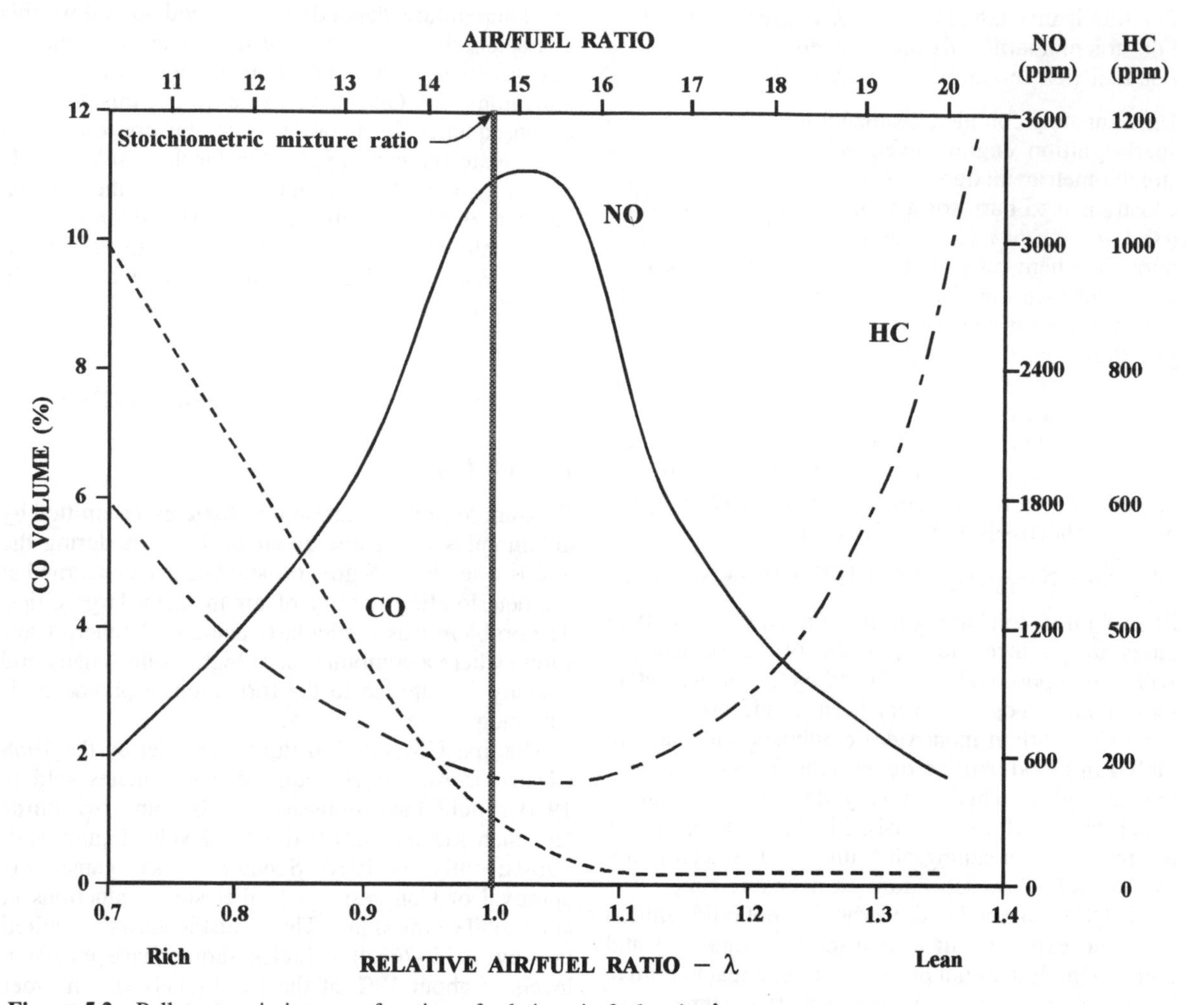

Figure 5.2 Pollutant emission as a function of relative air–fuel ratio, λ

makers to learn from mistakes made in the US and exploit more advanced electronic control techniques. Many European manufacturers, including Rover, Volkswagen and Peugeot-Citroën, developed emission-control strategies based on lean-burn engine technology which enables low emission levels to be met using simpler catalytic converters and with greater fuel economy. Unfortunately political pressure from some European governments has led to the imposition of EU-wide legislation which requires the use of US-style catalytic converters, and since January 1993 all new cars have been fitted with such systems (Table 5.1).

Although these vehicles emit less pollution than those without a catalyst they nevertheless still produce enormous quantities of toxic gasses. Researchers have estimated that over a ten year period a medium-size catalyst-equipped car covering 8 000 miles per year at an average fuel consumption of 35 miles per gallon will produce approximately;

- 44 tonnes of carbon dioxide
- 5 kg of sulphur dioxide
- 47 kg of nitrogen dioxide
- 325 kg of carbon monoxide
- 36 kg of unburned hydrocarbons.

This amount of exhaust gas emission will create about 1 billion cubic metres of polluted air. It is one of the reasons why many car makers (especially in Japan) are still working on lean-burn engines, and it is likely that forthcoming legislation to limit CO_2 emission will result in a revival of interest in lean-burn technology.

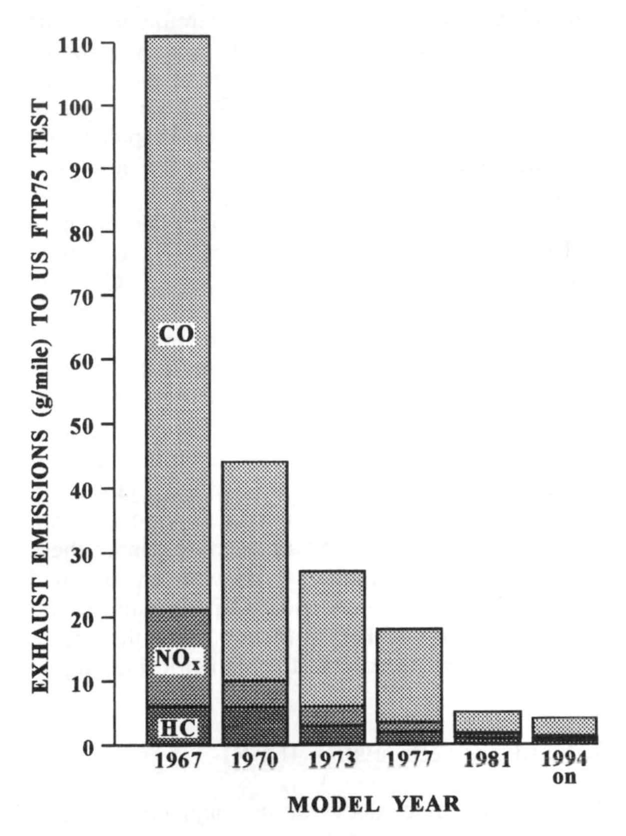

Figure 5.3 Reductions in exhaust emissions since 1967 (USA)

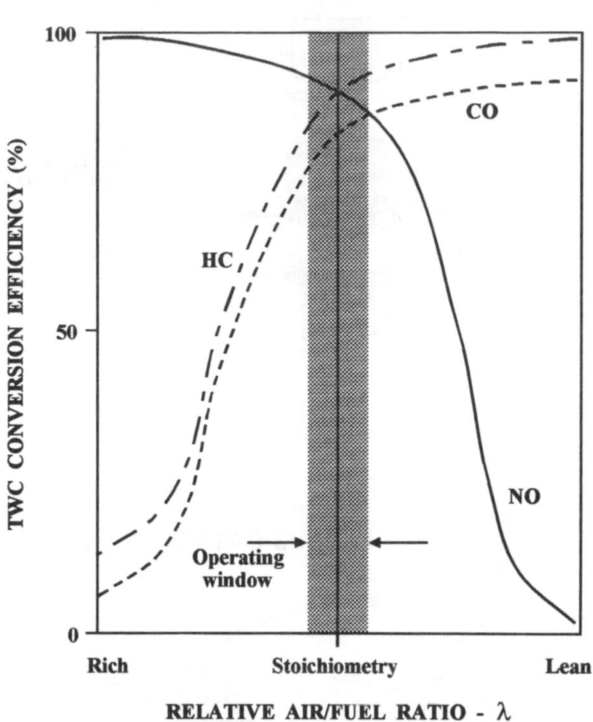

Figure 5.4 Conversion efficiency of the three-way catalytic converter as a function of relative air–fuel ratio

Table 5.1 European emissions regulations for petrol engines

Model year	$HC + NO_X$ (gm per km)	CO (gm per km)
1992/93	0.97	2.72
1996/97	0.5	2.2
1999/2000 (Proposed)	0.1 + 0.1	1.5

Three-way catalytic converters

A catalyst is a substance which promotes a chemical reaction without chemically changing itself. For example, the CO and HC in exhaust gas will not normally oxidize unless exposed to oxygen at a temperature in excess of 700°C (well above the typical exhaust-gas temperature range of 400–600°C). In the presence of a platinum catalyst, however, this oxidation process occurs at only 300°C without having any effect on the platinum itself.

At stoichiometry there is just the right amount of oxygen available to burn all of the hydrocarbon. If an engine can be operated at all times with an air–fuel ratio close to stoichiometry then it is possible to use a catalytic converter to clean the exhaust gases by first removing the oxygen from the NO_X (a process of chemical reduction) and then using this liberated oxygen to oxidize the CO and HC. The vehicle's exhaust gases will then contain only carbon dioxide (CO_2), water (H_2O) and nitrogen (N_2), all of which are harmless constituents of air. Such a catalytic converter is called a three-way catalyst (TWC) since it removes all three pollutants.

The purification characteristics of the TWC vary widely depending on the operating air–fuel ratio of the engine (Figure 5.4). Optimum pollutant elimination occurs for mixtures very close to stoichiometry; within a window of about ± 0.1 A/F ratios ($\lambda = 1.00 \pm 0.01$ approximately) the conversion efficiency may reach 90%. The narrowness of this widow is such that it is beyond the control range of mechanical systems such as ordinary carburettors. A TWC emission control system therefore requires a 'closed-loop' electronic air–fuel ratio control system which uses an exhaust-gas oxygen sensor (EGO sensor) to provide an electrical feedback signal indicating the composition of

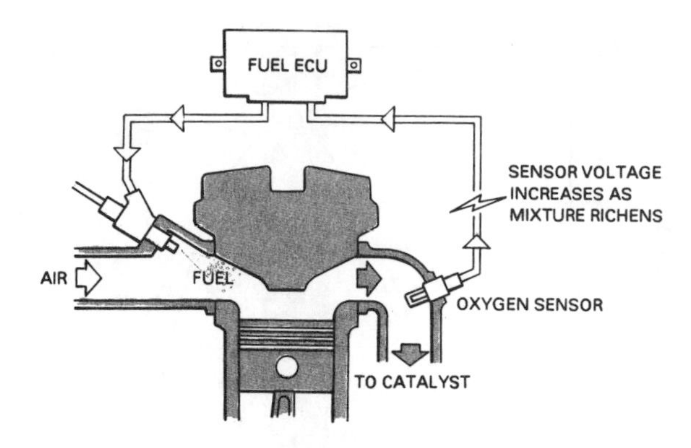

Figure 5.5 Use of closed-loop control system to maintain air–fuel ratio at 14.7:1 ($\lambda = 1$) (*Rover*)

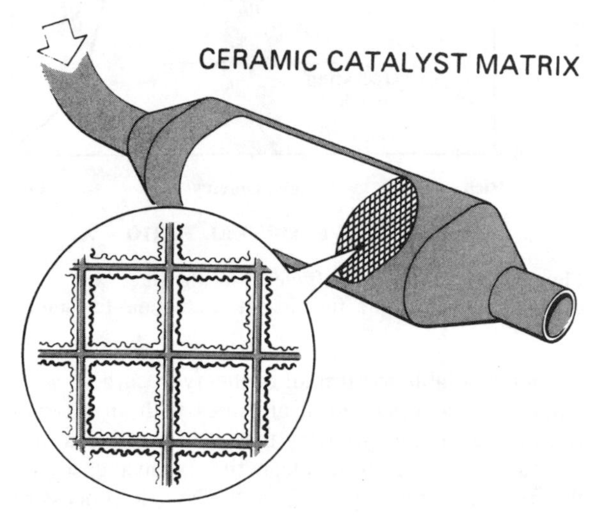

Figure 5.6 Structure of typical catalytic converter (*Rover*)

the exhaust gas. An electronic control unit then adjusts the amount of injected fuel to achieve the required air–fuel ratio (Figure 5.5). In this way the mixture can be maintained at stoichiometry for all driving conditions and so emissions are held to a minimum.

Structure of the three-way converter

A typical TWC structure is illustrated in Figure 5.6. It consists of a catalytic element mounted in a metal casing which forms part of the exhaust system. The position, size and amount of catalyst used depend on the engine type and operating conditions.

Since all of the exhaust gas must come into contact with the catalyst surface in order to react, it is very important that the active material is thinly applied over a large area. Most converters therefore use a fine ceramic honeycomb structure, with passageways about 1 mm^2, onto which a thin washcoat of the catalytic metal is deposited. Only about 10% of the weight of the element is catalytic material, but it is spread over a surface area of several hundred square metres with a gas capacity of 1–2 litres.

Most converters use platinum (Pt) for the oxidation of HC and CO, and rhodium (Rh) for NO_X reduction. Additives such as nickel (Ni) and cerium (Ce) may also be incorporated to improve performance. Alumina (Al_2O_3) is used as the supporting ceramic.

Catalytic action does not occur until the exhaust gases heat the element to about 350°C ('light-off' temperature) and so a cold vehicle must be driven for several minutes before emission control can commence. Even then, it is possible that the catalyst will have warmed-up before the cold engine can run smoothly with a stoichiometric mixture. Under these circumstances the engine is fed with an enriched mixture and extra air ('secondary air') is pumped into the converter to enable it to function until the engine is warm enough to accept stoichiometric fuelling.

Exhaust gas recirculation (EGR)

EGR can be used alone, or in conjunction with a three-way catalytic converter, to reduce NO_X emissions. A small fraction of the previously burned exhaust gas is recirculated to the intake manifold to act as an inert dilutant in the intake mixture. The air–fuel ratio is therefore unchanged by EGR, but the quantity of burnable vapour in the cylinder is reduced. The effect of this is to reduce the maximum combustion temperature and hence limit the formation of NO_X. EGR is generally accomplished by connecting the exhaust manifold to the intake manifold via a small passage. The flow of recirculated exhaust gas through the passage is controlled by an EGR valve whose opening is determined by the engine management ECU in accordance with preprogrammed parameters. EGR must be used with caution because adding exhaust gas to the intake mixture slows down its burn-rate; a high level of EGR can therefore result in increased HC emissions and possible misfiring. The amount of EGR an engine can tolerate depends on many factors, including the combustion chamber design and engine load. Very large reductions in NO_X emissions are possible with EGR levels of 20–30% of the intake mixture, but for satisfactory combustion 15% EGR is about the maximum that can be used provided the ignition timing is suitably retarded to take account of the longer combustion time.

EGR control

An engine which is cold or at idle requires a full cylinder charge for smooth running, and so the EGR valve is closed under these circumstances. During light acceleration or low-speed cruising a small amount of EGR may be used to reduce NO_X concentration whilst maintaining good drivability. At moderate engine loads NO_X levels are high and so the maximum possible amount of EGR is used; a large reduction in NO_X emission is then possible. When maximum power is required from the engine a full fuel charge must be delivered into the cylinders and so EGR is shut off.

Lean-burn engines

Lean-burn engines achieve low emissions by operating at air–fuel ratios lean of stoichiometric. Referring to Figure 5.2 again, it can be seen that at air–fuel ratios greater than about 19:1 ($\lambda > 1.3$) CO and NO_X emissions fall to very low levels. An additional benefit is that fuel consumption improves by at least 10%, reducing the emission of CO_2 (the 'greenhouse gas') by the same amount. HC emissions, however, start to rise steeply as a result of incomplete or irregular combustion and so a lean-burn engine may require an oxidizing catalyst to purify its exhaust.

Evaporative emission control systems

Many vehicles are fitted with an evaporative emission control system to prevent fuel vapour escaping to the atmosphere from the fuel tank. Evaporative emissions consist of volatile hydrocarbons (HC) which vaporize at relatively low temperatures, the most hazardous of these is benzene. Benzene is classified as a 'genotoxic carcinogen', which means that at the most minute concentrations it is able to cause leukaemia by directly attacking DNA in the blood-forming cells to which it is attracted after being inhaled into the lungs.

In order to reduce evaporative emissions a charcoal-filled canister is connected to the fuel tank vent line to store fuel vapour. When conditions allow, the engine controller opens a solenoid valve to allow intake manifold vacuum to 'purge' the charcoal canister by drawing the stored vapour into the engine, where it is burned.

5.4 IGNITION SYSTEMS

The ignition system must provide an adequate voltage to initiate a discharge across the spark plug electrodes and supply sufficient energy to ignite the air–fuel mixture. This must occur for all engine operating conditions and at the appropriate time on the compression stroke.

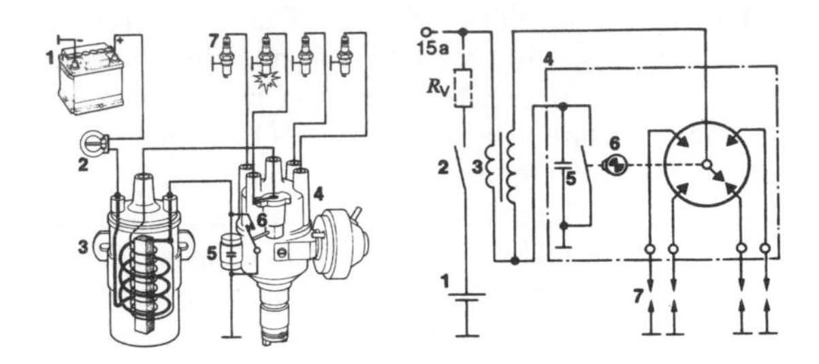

Figure 5.7 Conventional ignition system (components at left, circuit diagram at right): [1] battery; [2] ignition switch; [3] coil; [4] distributor; [5] capacitor (condenser); [6] contact breaker points; [7] spark plugs; [R_V] ballast resistor (not always fitted) (*Bosch*)

On modern engines it is normal for the ignition system to form a subsection of an integrated engine management system, sharing sensors and circuits with the fuelling and (occasionally) transmission control systems. For clarity, however, it is useful to consider the ignition system separately.

Ignition fundamentals

With a stoichiometric mixture, and under ideal circumstances, it is possible to achieve ignition with spark energies as low as 1 mJ and spark durations as short as 50 μs. In practice, however, ignition must be successful under even the most unfavourable driving conditions and so a spark energy of the order of 30 mJ is required for about 1 ms.

The voltage required to enable a spark to occur also varies. Carbon deposits, electrode erosion and insulator leakage all increase the voltage demand as does the use of a lean mixture. Typically at least 15 kV (15 000 volts) is required, with some lean-burn engines demanding 40 kV or more.

To meet these requirements a variety of ignition systems have been developed.

Contact-breaker ignition system

This ignition system (also known as the Kettering system) was first patented in 1908 and remained the standard fitment for most vehicles up until the introduction of electronic ignition systems in the early 1980s. Although now obsolete, it provides a relevant introduction to the topic since many fundamental principles are common to more modern systems.

Figure 5.7 shows the typical contact-breaker ignition

system found on older vehicles. The circuit consists of a battery, ignition switch, coil and distributor assembly (cam, contact-breakers and rotor arm). With the engine running, a steady current flows from the battery, through the switch to the coil primary's positive terminal. The coil negative terminal is connected to the contact-breaker points, which act as a switch to earth. A cam mounted on the rotating distributor shaft causes the points to open and close in sequence with the compression stroke of each of the cylinders. Since each cylinder fires on every other revolution of a four-stroke engine, the distributor cam rotates at half engine speed. When the contacts are opened, the current in the coil primary winding decreases very rapidly, and a high voltage pulse is induced in the secondary. This high voltage is routed via the high tension (HT) leads to the distributor rotor arm which directs it to the appropriate spark plug. The contact points then close again and the cycle repeats. The next plug to fire will be the one that is connected to the distributor cap insert that is aligned with the rotor arm when the points next separate.

When the contacts first separate, self-induction causes a brief continuation of the primary current, leading to the appearance of a high voltage at the coil negative terminal. This voltage is in the region of 300 V and causes arcing across the opened points. To bring the primary current to a controlled stop, and greatly reduce the size of the arc, a capacitor is installed inside the distributor and connected across the contacts. When the points first open, induced primary voltage causes the capacitor to charge and current is diverted away from the arc.

Ignition coil design

The ignition coil is a pulse transformer which steps up the low battery voltage to a high ignition voltage. Typically the coil is constructed with a primary of a few hundred turns of 1 mm diameter insulated copper wire, and a secondary winding of 10 000–30 000 turns of 0.1 mm wire. These windings are assembled over a laminated soft iron core and inserted into a one-piece steel or aluminium case which is filled with oil to prevent internal arc-over. Finally, the case is sealed with a cap of insulating material which houses the primary and secondary terminals.

In the design of the ignition coil it is imperative that sufficient primary resistance is present to protect the contact-breaker assembly from burning and pitting due to excessive current. Acceptable points life is obtained with primary currents up to about 4 A (a primary resistance of about 3 Ω).

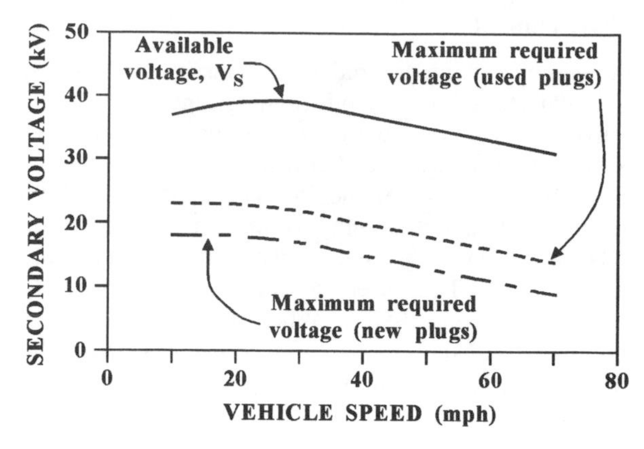

Figure 5.8 Available and required secondary voltage as a function of speed

Secondary voltage requirements

When the contacts open, a large proportion of the input energy to the coil primary is transferred to the secondary circuit and appears as a high voltage (see Section 2.9). This voltage must charge the capacitances associated with the secondary system and ionize the vapour between the spark-plug electrodes. The maximum available voltage in the secondary circuit, V_S, is,

$$V_S = (2E_S/C_S)^{1/2} \tag{5.1}$$

where E_S is the energy transferred from the primary to the secondary and C_S is the secondary capacitance. Clearly, a higher secondary voltage can be obtained if the secondary capacitance is kept small.

The secondary capacitance is mainly determined by the length and routing of the HT leads and their proximity to ground, as well as spark plug and rotor arm gaps. Overall capacitance is typically in the range 20–100 pF.

Dirt on insulating components also adds a shunt capacitance to the circuit and it is for this reason that the secondary components should always be kept clean and neatly routed.

The voltage required to fire a spark plug depends on many factors including the engine compression ratio, load and speed, and the spark plug gap and operating temperature. The actual voltage developed in the secondary circuit is fixed by the requirements of the spark plug. This may vary from under 10 kV to over 40 kV but misfiring will occur when it exceeds V_S, the maximum voltage available from the ignition system.

Figure 5.8 shows typical curves for the available voltage and maximum required voltage for new and used plugs; the used plugs require an extra 5 kV to fire because of the increase in plug gap and erosion of

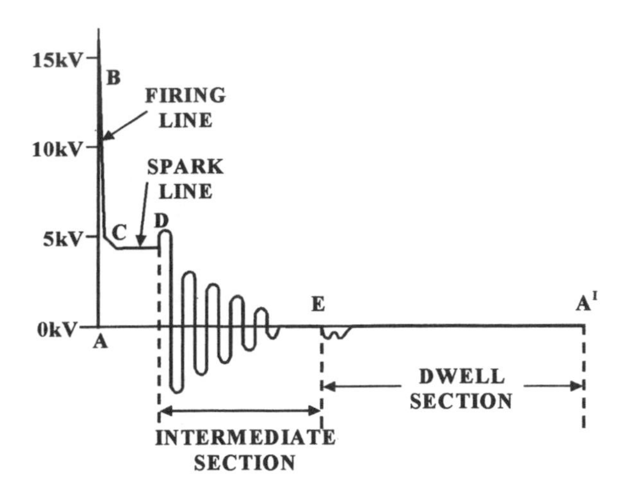

Figure 5.9 Typical secondary voltage waveform

the electrodes. If a plug is severely worn, or defective, the required firing voltage will exceed the available voltage and misfiring sets in, usually first noticed during part-throttle acceleration in the medium speed ranges.

Secondary voltage waveform

Figure 5.9 illustrates a typical secondary voltage pattern. For simplicity the waveform is divided into four parts. The firing line, A–B, occurs at the instant that the points open and represents the rise of the secondary voltage from zero to the value required to break down the gap between the plug electrodes and cause ionization of the air–fuel mixture. The spark voltage is very high (15 kV in this example) as is the spark current (about 200 A), but it flows for only an instant (about 10 ns).

Once the gas between the spark plug electrodes has been ionized, much less voltage is required to maintain the spark current than was required to cause ionization. The voltage therefore falls from about 15 kV to about 4 kV, and stabilizes at this level for about 2 ms. This is called the spark line (C–D) and during this time energy from the coil is dissipated in the arc between the plug electrodes. The arc current is typically in the region of 100 mA and the arc temperature is about 3 000°C.

At the point D the spark ceases because most of the energy in the coil has been exhausted and there is insufficient current available to maintain ionization. The small amount of remaining energy is dissipated in the circuit as a damped oscillation, seen between points D and E.

At point E the contact-breakers reclose and so the period E–A′ represents the 'dwell' time for the distributor, typically about 60% of the complete cycle time A–A′. The dwell time is the time for which the coil's primary circuit is closed and so the primary current shows an inductive rise from zero (at E) to a few amps (at A′). For maximum spark energy it is desirable that this current should saturate during the dwell period, high engine speeds may not permit this.

Spark polarity

Since electrons leave a hot surface more readily than a cold one, the secondary system is usually wired so that the hottest spark plug electrode is the negative one. For most spark plugs this means that the centre electrode is given the negative polarity and the outer electrode is earthed. If a spark plug is used with a positive voltage at the centre electrode it usually takes about 5 000 volts more to fire and there is greater erosion of the earth (outer) electrode.

Mechanical advance control

In the contact-breaker ignition system the required speed/load variation in ignition advance is obtained by moving the relative positions of the contact-breaker assembly and the cam. This movement is performed by two separate mechanisms: the centrifugal advance mechanism and the vacuum advance mechanism.

(i) The centrifugal advance mechanism uses a pair of pivoted advance weights, fitted to the distributor shaft, that move against calibrated springs connecting them to the cam assembly. As the distributor shaft rotates, the outward centrifugal movement of the weights advances the cam assembly in relation to the shaft as the engine speed increases. Timing therefore varies from no advance at idle to full advance at a high engine speed, where the weights reach the full extent of their travel.

(ii) During part load operation the engine intake pressure is low (partial vacuum) and so a smaller mass of air–fuel mixture enters the cylinder. Under these conditions a greater ignition advance is required to allow for the slower burning time of the mixture. Conversely, at high loads (open throttle) the intake pressure is high (little or no vacuum) and so the cylinder receives a full charge of mixture, requiring less advance. To provide an advance based on load the distributor is equipped with a vacuum advance mechanism. This usually consists of a spring-loaded diaphragm assembly linked to the contact-breaker base-plate by a short rod. The intake vacuum is directed onto the diaphragm and therefore causes it to

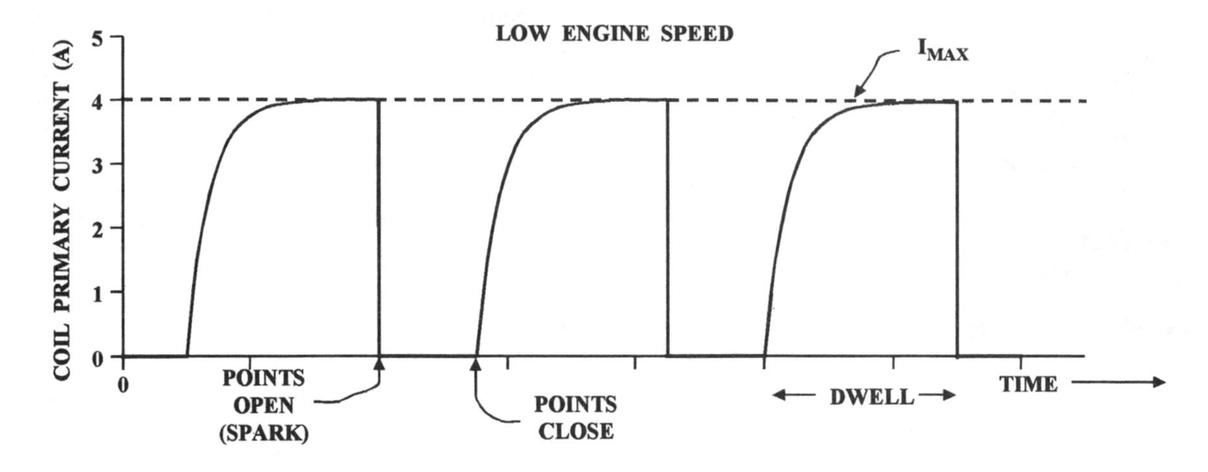

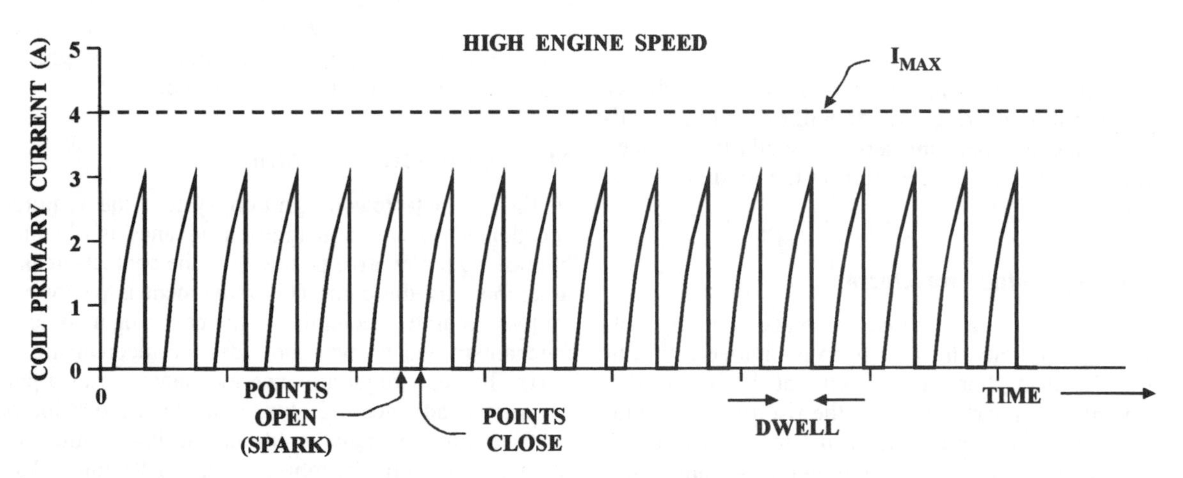

Figure 5.10 Fall in maximum coil primary current, at high engine speed

move against the spring, rotating the points relative to the cam and advancing the spark timing.

At any given engine speed there will be a set advance resulting from the operation of the centrifugal mechanism, plus a possible additional advance resulting from the operation of the vacuum mechanism.

Shortcomings of the contact-breaker ignition system

Although the contact-breaker ignition system has been used for over 80 years it suffers from a number of well-known shortcomings and is therefore no longer fitted as original equipment. Drawbacks include:

(i) Rapid wear of the contact-breaker rubbing block, necessitating frequent readjustment of the contact gap.

(ii) Rapid erosion of the contact faces leading to a comparatively short points life (10 000–20 000 km).

(iii) Points that are able to switch primary currents only up to about 4 A, placing a lower limit on the primary winding resistance (and hence inductance) and an upper limit on the spark energy.

The last point is particularly significant with modern high-revving engines since it restricts the rate of growth of the primary current as engine speed increases, limiting the build-up of coil energy. This problem is illustrated in Figure 5.10 which shows how a system designed to operate with a maximum primary current, I_{MAX}, at low rpm is unable to achieve that current at high rpm due to the combined effect of a high coil inductance and reduced dwell time.

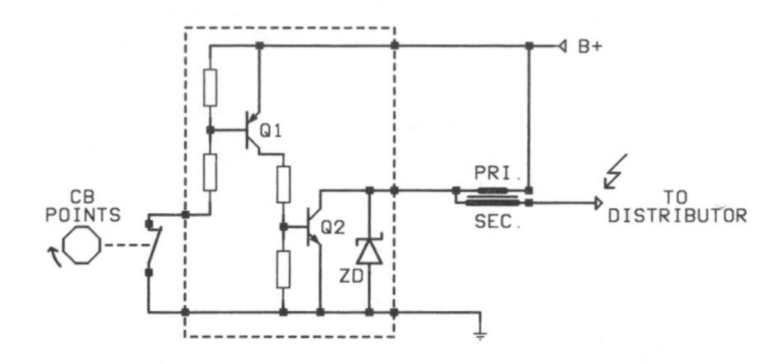

Figure 5.11 Elementary transistor-assisted points ignition system

With the introduction of wider spark-plug gaps (exceeding 1 mm) and higher compression ratios, modern ignition systems must provide voltages of around 35 kV and a spark duration of over 2 ms to ensure good ignition over extended service intervals. To meet these requirements automobile manufacturers have introduced a variety of electronic ignition systems.

Transistor-assisted contact-breaker system

The simplest eletronic ignition systems retain the points as a control device, but use a power transistor to switch the coil primary circuit, permitting a higher current to be used (about 8 A rather than 4 A). This leads to a higher spark voltage and a longer spark duration. Maximum primary current can be achieved even with the short dwell times that apply at high engine speeds. A bonus is that the points only have to pass a current of a few hundred milliamps and so contact arcing is greatly reduced and erosion is minimal.

Figure 5.11 shows an elementary transistor-assisted ignition circuit. A conventional distributor assembly is retained; the contact-breaker points are used to switch the driver transistor, Q1, which then switches the power transistor, Q2. With the points closed, Q1 is biased on and so Q2 conducts to complete the primary circuit. When the points open, Q1 turns off which in turn switches Q2 off to stop the primary current and induce a high secondary voltage. A 100 V zener diode is installed between Q2's collector and emitter to protect it from the induced voltage spike that appears on the primary terminal when the current ceases.

Although the transistor-assisted ignition system represents a great improvement over the Kettering system, some deficiencies of the latter remain, in particular,

(i) rapid wear of the contact-breaker rubbing block leads to frequent maintenance intervals; and
(ii) the points may 'bounce' at high speed as the return-spring becomes weak through extended use.

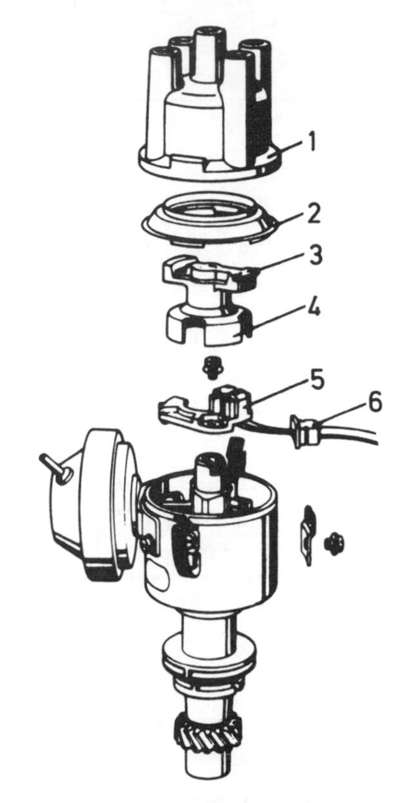

Figure 5.12 Distributor incorporating Hall-effect sensor [1] distributor cap [2] dust cover [3] rotor arm [4] flux screening 'trigger wheel' [5] Hall sensor assembly [6] grommet

For these reasons vehicle manufacturers have retained the electronic circuitry of the transistor-assisted system, but have replaced the points with non-contact switching devices.

Breakerless transistorised ignition

In breakerless electronic ignition systems the coil primary current is switched by a power transistor that is controlled by a non-contact trigger device mounted in the distributor in place of the contact-breaker points. A variety of trigger devices can be used, although Hall-effect sensors and optical switches are most common. The primary switching circuit is similar in both cases.

Hall-effect switching

The operation of the Hall-effect sensor was described in Section 2.10 (Figure 2.23) and is further illustrated in Figure 5.12. Circuit details are shown in Figure

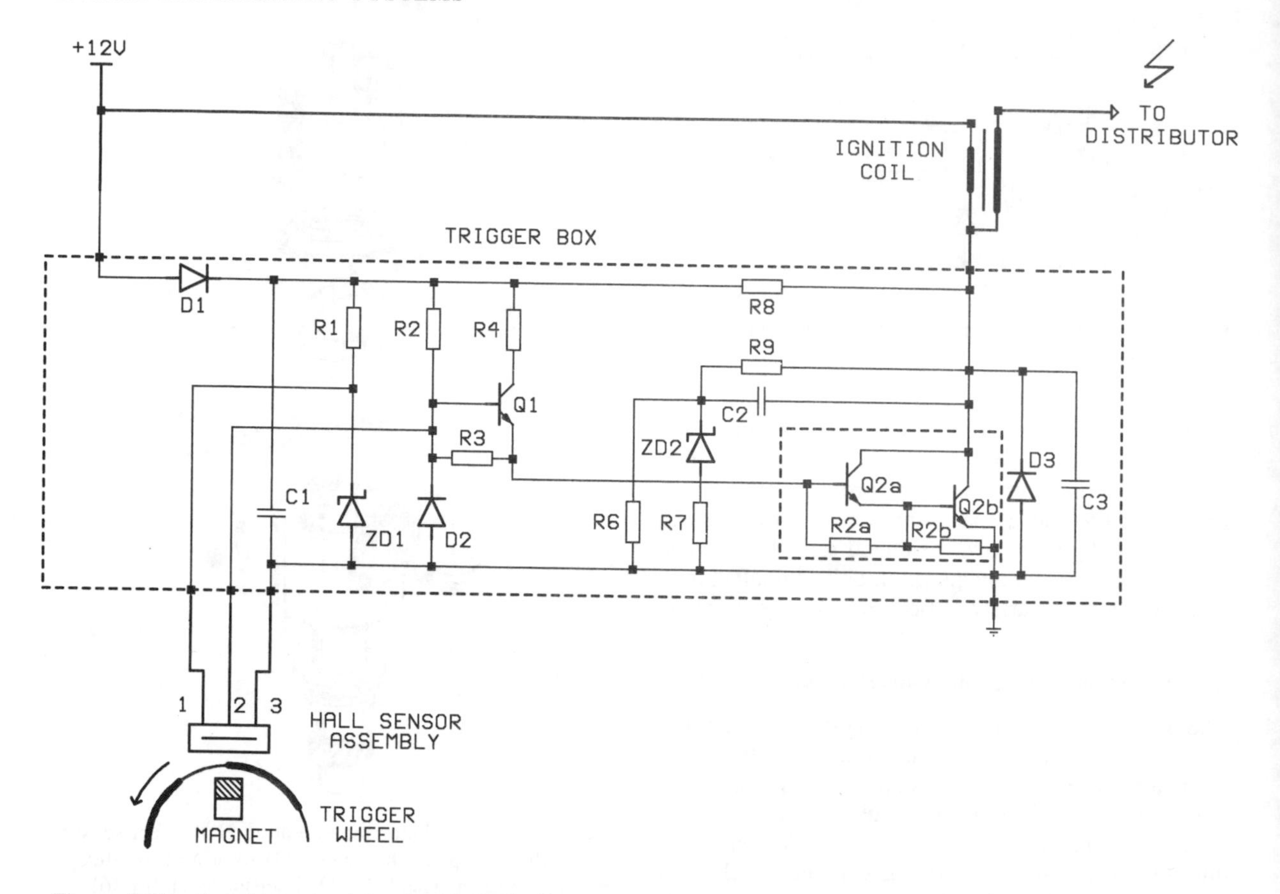

Figure 5.13 Switching circuit used with Hall effect sensor

5.13. The Hall IC is provided with a stabilized supply voltage on terminal 1 and a ground connection on terminal 3. When a trigger-wheel vane passes in front of the Hall IC assembly, the IC output on terminal 2 is pulled-up to a small positive voltage by the resistor R2. This holds transistor Q1 ON which in turn allows the Darlington-pair output stage, Q2, to switch ON and enables the primary current to flow. When the trigger-wheel vane moves away from the Hall IC, the IC output on terminal 2 falls to 0 V and so Q1 switches OFF. This causes Q2a and Q2b to stop conducting and so the ignition coil's primary current is interrupted, generating the ignition spark.

Optical switching

An alternative to the Hall-effect sensor is the optical trigger wheel. Such systems were very popular at one time and frequently fitted as an aftermarket modification to a conventional distributor.

The optical trigger unit consists of an infra-red LED and matching phototransistor. The phototransistor is located opposite the LED so that a trigger wheel can pass between the two and interrupt the infra-red light beam. The trigger wheel is similar to that used in Hall effect systems and has as many vanes as there are cylinders. Each time the light beam is interrupted by a vane, a trigger pulse is sent to the electronic switching circuit to switch off the primary current and generate a spark.

A variation of this principle is used in the Lucas distributor fitted to some Nissan vehicles (Figure 5.14). These distributors use a precision rotor with 360 slots at 1° intervals. Four larger slots are located at a smaller diameter and represent TDC for each of the four cylinders; number one cylinder being identified by having a greater slot width than the other three.

The breakerless systems described so far have the advantage that the contact-breaker points are eliminated, offering the prospect of reliable and virtually maintenance-free operation over high milage. They do however, still suffer from the drawback of having a dwell time that varies with engine speed. At low speeds the coil is switched on for far too long between each spark and wastes energy, whilst at high

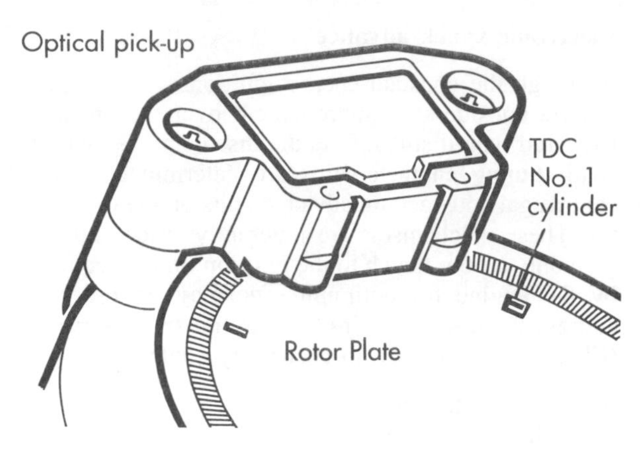

Figure 5.14 Optically triggered distributor (*Lucas*)

speed it may not be switched on for long enough and so the primary current does not have time to reach its maximum possible value. Spark energy is therefore reduced at high rpm.

Constant-energy electronic ignition

The objective of a constant-energy ignition system is to ensure that a constant secondary voltage is available right up to maximum engine speed by maintaining the dwell time at a fixed value. Figure 5.15 shows a simple constant-energy system incorporating a special distributor, an electronic switching module (which is sometimes built into the distributor) and a low-resistance ignition coil.

The distributor shaft is fitted with a gear-shaped iron rotor which has as many teeth as there are cylinders. The rotor teeth move past a stationary pickup which comprises a small coil wound around a permanent magnet core. As each tooth passes the pickup core it causes an increase and then a decrease in the magnetic flux through the pickup coil, inducing a voltage signal which has an amplitude in proportion to the tooth speed (Figure 5.16a,b). As a tooth approaches the pickup, the voltage signal rises above a reference voltage, V_{REF}, and the control module turns on the coil primary current to start the dwell period (Figure 5.16b).

The signal voltage rises to a maximum as the tooth and pole piece come into alignment, falling abruptly as they move apart. When it drops below the 0 V threshold, the control module switches off the coil primary current to generate a spark. Since the coil switch-off point is referenced to 0 V the spark timing remains fixed with increasing engine speed, however the commencement of the dwell period occurs earlier due to the higher signal voltage (Figure 5.16c,d). The consequence of this is that the dwell angle increases with engine speed and so the dwell time remains more-or-less fixed, fulfilling the 'constant-energy' requirement.

A refinement of the constant-energy principle is the closed-loop ignition controller which ensures that the desired primary current is attained, even under conditions of varying battery voltage and coil resistance. The system uses a resistor connected in the coil primary circuit to monitor current. When the coil primary is switched on, the voltage developed across the monitoring resistance rises in proportion to the current through the primary circuit and can therefore be used to trigger a current-limiting circuit when a preset maximum value is reached. This means that the coil primary resistance has little influence on the maximum current reached, enabling the use of compact designs which have a very low resistance and a high efficiency.

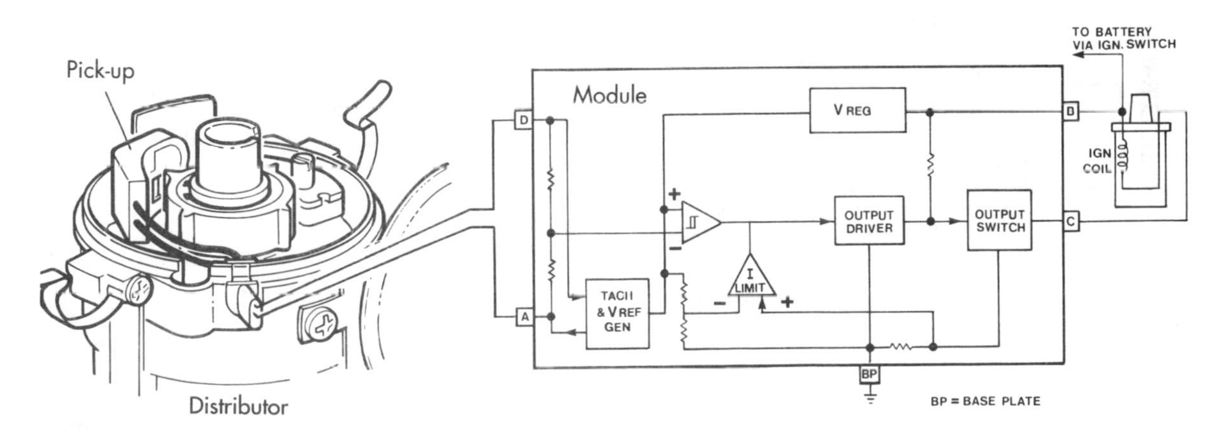

Figure 5.15 Constant energy ignition system (*Lucas*)

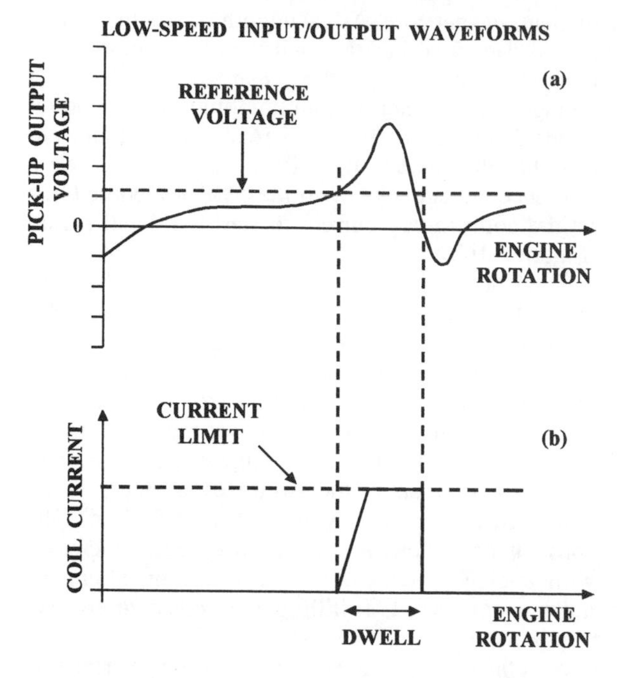

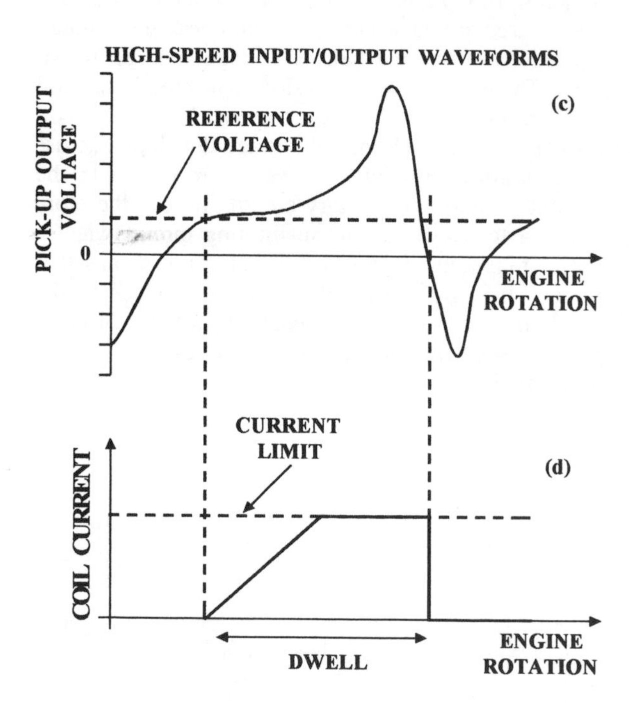

Figure 5.16 Output of the induction-type distributor pulse generator at low and high engine speeds

Electronic spark advance

Although the constant-energy ignition system represents a considerable improvement over the breaker-triggered type it still suffers the disadvantage that the spark timing characteristics are determined by the centrifugal and vacuum advance units on the distributor. These mechanisms are inherently crude and can give only a poor match to the ignition timing required by the engine for optimum efficiency. In order to address these shortcomings electronic spark advance (ESA) was developed to accurately control

(i) the ignition timing;
(ii) the dwell time;
(iii) the coil primary current.

ESA is now standard on virtually all new vehicles and usually forms an integrated part of a single engine management system. However, for clarity of explanation it is treated separately here.

In ESA (which is also called *programmed ignition* and *digital ignition*) the basic spark timing for various engine operating conditions is stored in the microcomputer memory of the ignition ECU as a map, and selected on the basis of information obtained from a number of sensors. Corrections can then be made to take account of the particular driving circumstances. In this manner very complex timing characteristics are possible and the control accuracy of the ignition system is vastly improved (Figure 5.17).

The basic input parameters to the ESA controller are engine speed and a signal related to engine load (intake manifold pressure, intake air-flow, or injected fuel quantity); this information is then used to extract a basic advance angle from the map. The map data is obtained from an ideal test engine and will typically contain information relating to at least 16 values of engine speed and 16 values of load and therefore 16×16 (256) timing values. A process of mathematical interpolation can be used to improve the resolution of the system. Several maps may be available within memory to suit different fuel grades and these can usually be selected using external links ('octane adjust' links).

ESA timing calculation

The ignition timing is calculated as

$$\text{Ignition timing} = \text{Initial advance} + \text{Basic advance} + \text{Correction advance}$$

The *initial advance* represents the baseline timing for the engine and is the setting used during cranking,

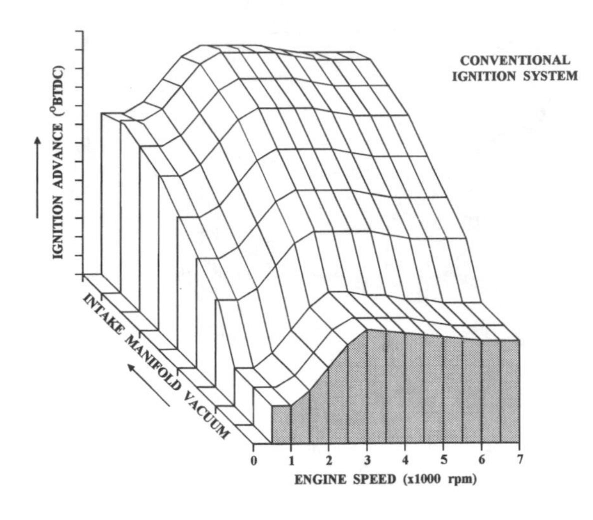

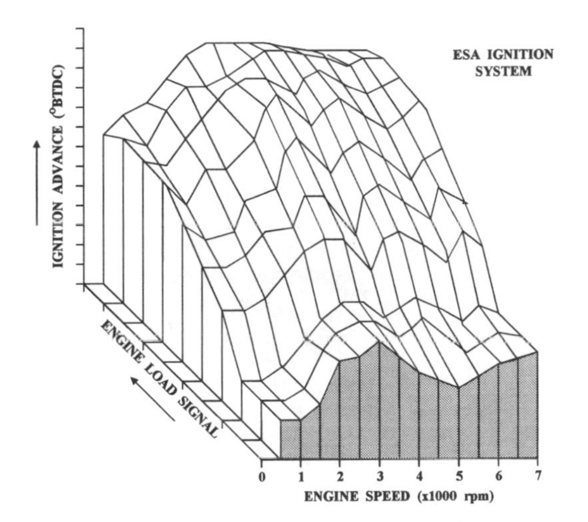

Figure 5.17 Comparison of ignition advance map provided by conventional and electronic spark advance (ESA) systems

when the engine speed fluctuates markedly. The ignition timing is never less than the initial advance.

The *basic advance* is the advance value obtained from the timing map in accordance with the instantaneous engine load and speed conditions.

The *correction advance* is an amount of advance which is added to the initial advance and basic advance to take account of special engine operating circumstances, for example,

(a) *Warm-up*. A low coolant temperature signal indicates to the ECU that the engine is in the warm-up phase and so requires additional ignition advance to ensure good drivability. The amount of warm-up correction advance is progressively reduced as the coolant temperature rises.
(b) *Overheating*. When the coolant temperature is excessive a negative correction advance (i.e. a retard) is applied to reduce the combustion temperature and safeguard the engine.
(c) *Idling stabilization*. A correction advance is applied if, at idle, an additional load is placed on the engine. This helps stabilize the idle speed against large variations due to operation of the air-conditioner, power-steering, transmission engagement or heavy electrical loads.
(d) *Knock control*. When combustion knock is detected the correction advance must take a negative value (ignition retard) to stop the knocking and safeguard the engine.

The result of the ignition calculation is a precisely timed switching signal which is applied to the base of the power transistor that controls the coil primary current. The resulting HT pulse is routed to the distributor, whose only task is HT distribution to the appropriate cylinder. Some ESA ignition systems dispense with the distributor entirely by using either double-ended coils to serve pairs of cylinders or having one coil for each spark-plug.

Distributorless ESA ignition system

Figure 5.18 illustrates the arrangement of components for a simple distributorless ESA ignition system. It includes:

(i) An ignition ECU incorporating a built-in intake manifold absolute pressure sensor (MAP sensor) and an 8-bit microprocessor with timing and other data stored in 4 KBytes of ROM.
(ii) Two sensors for the determination of engine speed and crankshaft position.
(iii) A knock sensor.
(iv) An overboost pressure switch (for turbo applications).
(v) Two ignition primary current switching modules.
(vi) Two double-ended ignition coils.

Figure 5.19 shows a connection diagram for these components. The ECU (A) houses the microprocessor (P) and all of its associated interfacing circuitry. Some circuits, including memory and the two analogue-to-digital converters (A/D1 and A/D2) are integrated into the microprocessor chip itself. The system functions are as follows:

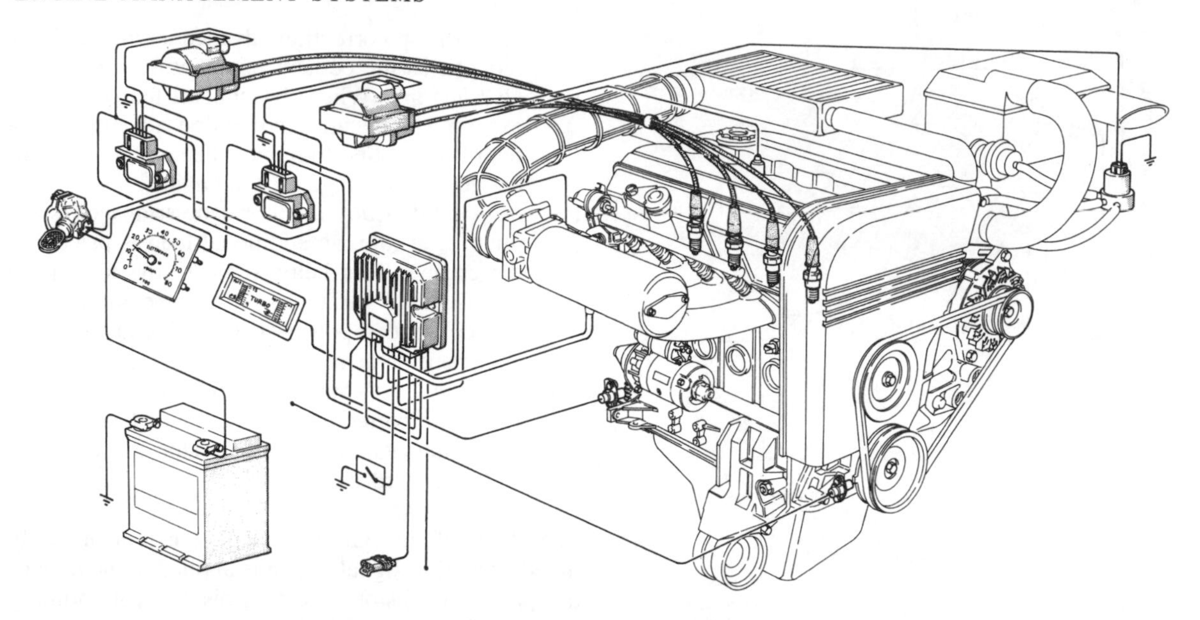

Figure 5.18 Distributorless ESA ignition system using two 'double-ended' coils (*Magneti Marelli*)

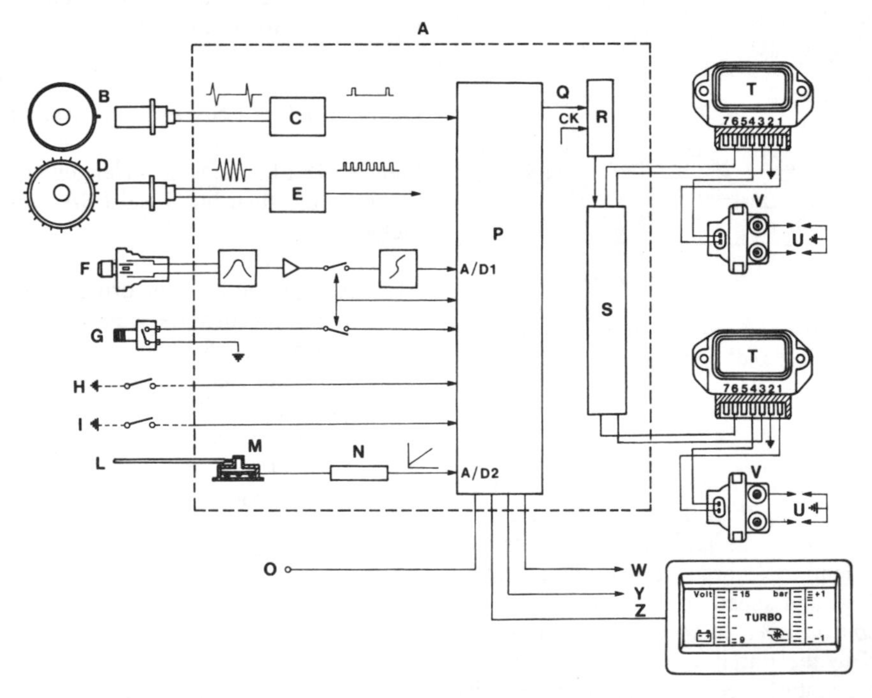

Figure 5.19 Connection diagram for distributorless ESA ignition system: [A] ECU; [B] TDC sensor; [C] comparator; [D] flywheel teeth sensor; [E] comparator; [F] knock sensor; [G] boost pressure switch (turbo models); [H] octane adjust switch; [I] full load switch; [L] intake manifold vacuum pipe; [M] MAP sensor; [N] amplifier; [O] diagnostic connector; [P] microprocessor; [Q] data output; [R] clocked register; [S] cylinder pair selector (1 + 4 or 2 + 3); [T] primary current switches; [U] spark plugs; [V] ignition coils; [W] output to injection ECU; [Y] diagnostic connector; [Z] turbo gauge output (turbo models)

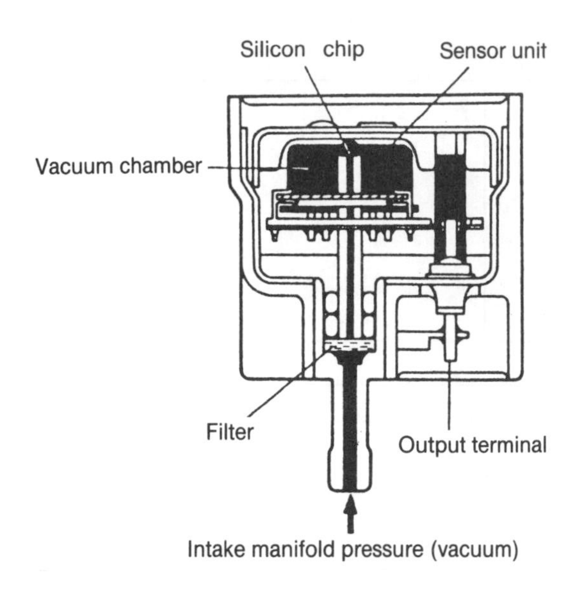

Figure 5.20 MAP sensor construction

(i) Engine speed and TDC information is obtained from the magnetic reluctance sensors (B and D) in the form of sinusoidal pulses which increase in frequency and amplitude as engine speed increases. The signal conditioning circuits (C and E) incorporate comparators which trigger on each input pulse and produce digital output pulses of fixed amplitude.

(ii) The MAP sensor measures the inlet manifold vacuum as an indication of engine load. Together with the engine speed signal, the MAP signal is used to read out a basic advance value from the microprocessor's ROM. MAP sensors vary in design, but most are based on the piezoresistive properties of thick-film resistors. The sensing element is a Wheatstone bridge fabricated on a thin alumina (Al_2O_3) diaphragm which is bonded to a rigid base-plate of the same material. More compact sensor structures are also possible using the same technique, but with silicon as the sensing material. In both cases the element is encapsulated in an air-tight enclosure and connected to the inlet manifold via a pressure pipe (Figure 5.20). As the inlet vacuum changes, the diaphragm is deflected and so the piezoresistors are subjected to either tension or compression. This changes their resistance and therefore the bridge output voltage, which is amplified and supplied to the ECU where it is converted to a digital value proportionate to pressure (A/D2). Since the bridge resistors are temperature sensitive it is important that the sensor incorporates some form of temperature compensation circuitry, as shown in Figure 5.21.

(iii) Monitoring of combustion knock is provided by the knock sensor (F). The control of knock is particularly important in modern high-compression engines because the advance required for maximum efficiency is often at, or beyond, the knock limit for the engine.

Knock sensing may be achieved in a variety of ways, including direct measurement of cylinder pressure or measurement of the spark-plug ionization current just after ignition. The most popular method is the measurement of structure-borne cylinder vibrations using a piezoelectric sensor (Figure 5.22). The sensor is designed to be mechanically resonant at the engine's knock frequency (generally in the range 6–12 kHz), and is bolted to a relatively flexible location on the cylinder block or head. When knocking takes place, combustion pressure oscillations cause vibrations in the engine structure which, in turn, vibrate the knock sensor at its resonant frequency. The piezoelectric element is thus subjected to a cyclic stress from the vibration plate, resulting in the generation of a small voltage across the faces of the element (Figure 5.23).

The ECU filters the knock signal and samples it for a short time, just after each spark, to assess the knock condition. If the knock signal exceeds a comparison level (Figure 5.24) then the ignition timing is immediately retarded until knocking ceases. Thereafter, the full advance is progressively reinstated until knocking occurs once again and the control cycle is repeated. In this manner the engine is consistently operated close to its knock limit, but with no danger of damage. Note that when knocking is detected in a cylinder, the ECU may either retard the spark timing for just that cylinder, or for all cylinders.

In the event that the sensor fails, the ECU detects a change in the circuit resistance and sets the timing to a heavily retarded value to protect the engine.

(iv) Additional sensors are used to enable correction for changes in other parameters. The octane adjust switch (H) is used to select between alternative timing maps for high and low octane fuels. The full load switch (I) causes the microprocessor to switch to a different advance map when it detects wide-open throttle, ensuring maximum engine power.

(v) The corrected ignition advance is output from the microprocessor as a logic pulse of duration equal to the coil dwell period. The pulse is fed to the cylinder pairs selector (S) which switches either the 1 + 4 or 3 + 2 primary current control module (T). The two coils (V) are designed to store enough energy to cause a spark over two gaps simultaneously so that both ends of each secondary winding can be connected to a plug. Sparks thus appear at plugs 1 + 4 or 3 + 2, one

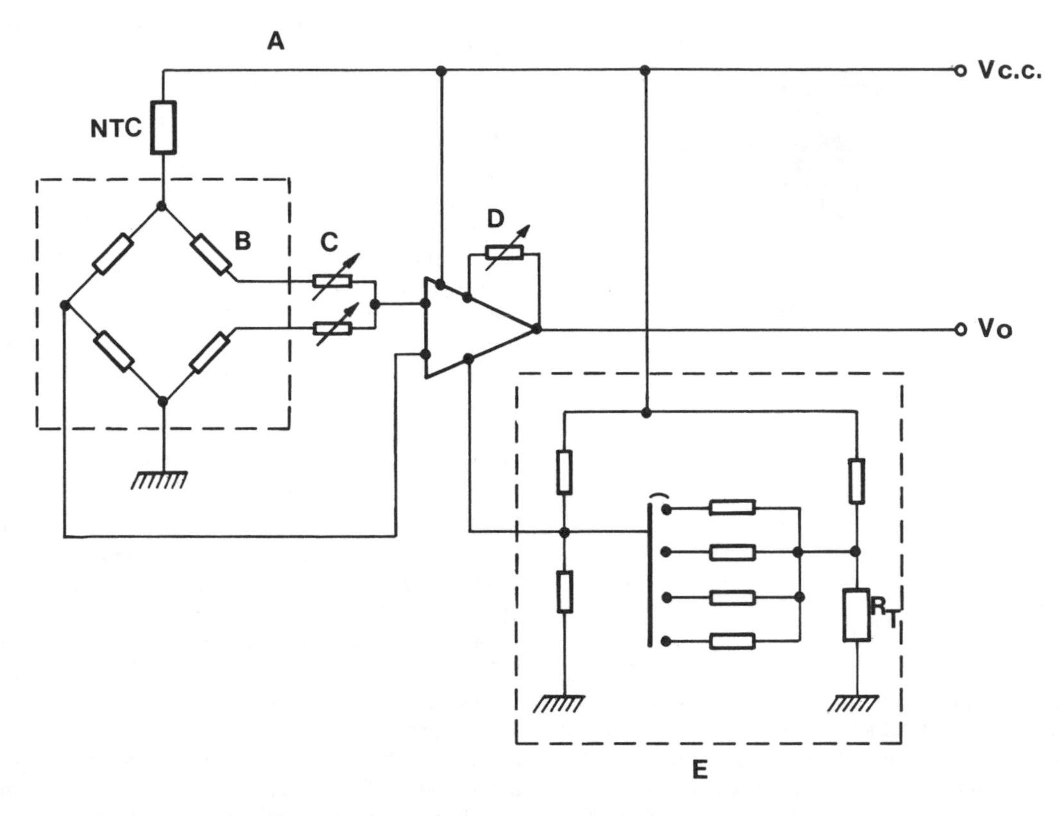

Figure 5.21 MAP sensor with temperature compensation circuit: [A] temperature compensated circuit; [B] bridge resistors; [C] zero trim; [D] gain trim; [E] thermal compensation trim (*Magneti Marelli*)

spark having a negative polarity and the other a positive polarity. Only one of the sparks is actually used to cause combustion, the other occurs on another cylinder's exhaust stroke and is therefore termed a 'wasted spark'.

Although double-ended ignition coils eliminate the distributor they tend to lead to accelerated spark-plug erosion because of the occurrence of the wasted spark. For this reason some manufacturers have adopted designs in which each plug is fitted with its own 'plug-top' coil. This is obviously an expensive solution, but it can be justified on high performance vehicles.

Other ignition systems

Capacitive discharge ignition

Whereas the ignition systems described so far use a coil to store electrical energy, the capacitive discharge ignition (CDI) system uses a capacitor (Figure 5.25). The capacitor is charged to high voltage (300–500 V) by a charging circuit; either a single charging pulse or a series of pulses may be used. When a spark is required the thyristor is fired and the capacitor quickly discharges through the coil primary. In a CDI system the coil serves only as a step-up transformer and so has a very low inductance, giving a spark of very short duration (0.1–0.5 ms) with a very fast rise-time and a high voltage.

The main application of CDI systems is on high-performance cars which, when driven gently, are susceptible to plug fouling. The short, strong, spark burns off any deposits on the electrodes.

Saab CDI system

An interesting development of CDI is used on some Saab vehicles. It consists of four 'plug-top' ignition coils and associated CDI circuitry built into a module which is mounted on the cylinder head and triggered from the engine management ECU (Figure 5.26). When the engine is cranked the ignition discharge module uses information from a crankshaft position sensor to fire the plugs on those cylinders which are approaching TDC. When the engine starts it detects the rise in plug ionization current due to combustion and so is able to accurately synchronize ignition to

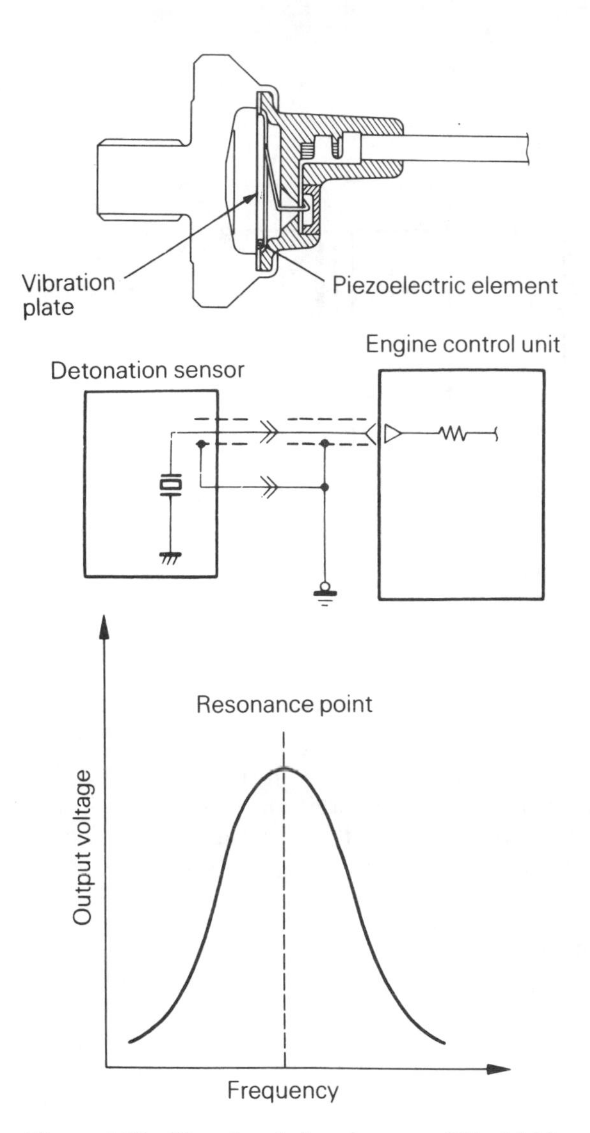

Figure 5.22 Piezoelectric knock sensor (*Mitsubishi*)

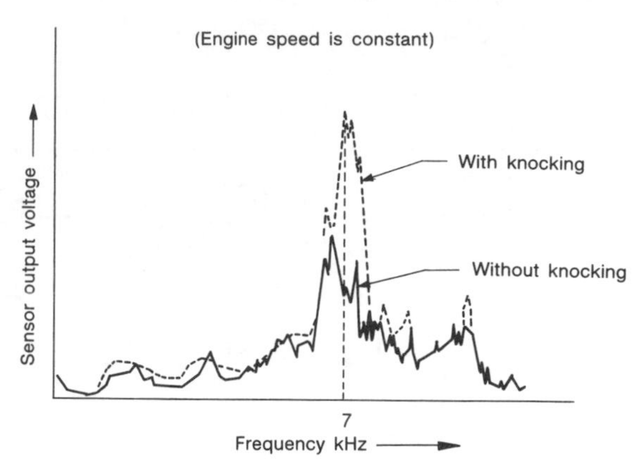

Figure 5.23 Output of knock sensor (*Nissan*)

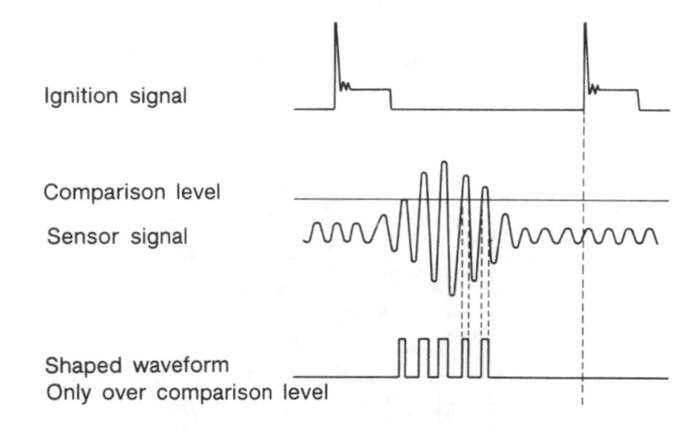

Figure 5.24 Determination of knock intensity (*Nissan*)

the appropriate cylinders. Good cold starting is assured by having the system fire a quick succession of sparks ('multisparking') from 10° BTDC to 60° ATDC when the starter motor is operated at a coolant temperature below 0°C.

Knock detection is performed by applying a bias voltage to the spark-plug electrodes immediately after ignition and monitoring the resulting ionization current. Knocking produces pressure spikes in the cylinder, which cause corresponding spikes in the ionization current. These spikes are detected and the ignition timing or fuel injection duration is then adjusted to prevent reoccurrence.

Twin-spark ignition systems

Ignition systems using two spark plugs per cylinder have been used by several Japanese manufacturers (largely to help meet strict emission regulations in their home-market) and by Alfa Romeo in Europe. The two plugs are fitted either side of the combustion chamber and are fired simultaneously to give two benefits:

(i) The initial flame area is almost doubled, giving a faster burn rate which is useful when the engine is operated with lean mixtures or high levels of EGR.

(ii) Variations in the mixture composition in the region around the plugs have less impact on the burning rate, giving a more repeatable burn.

5.5 SPARK PLUGS

Although a spark plug may appear a simple component, its design is no less sophisticated than that of

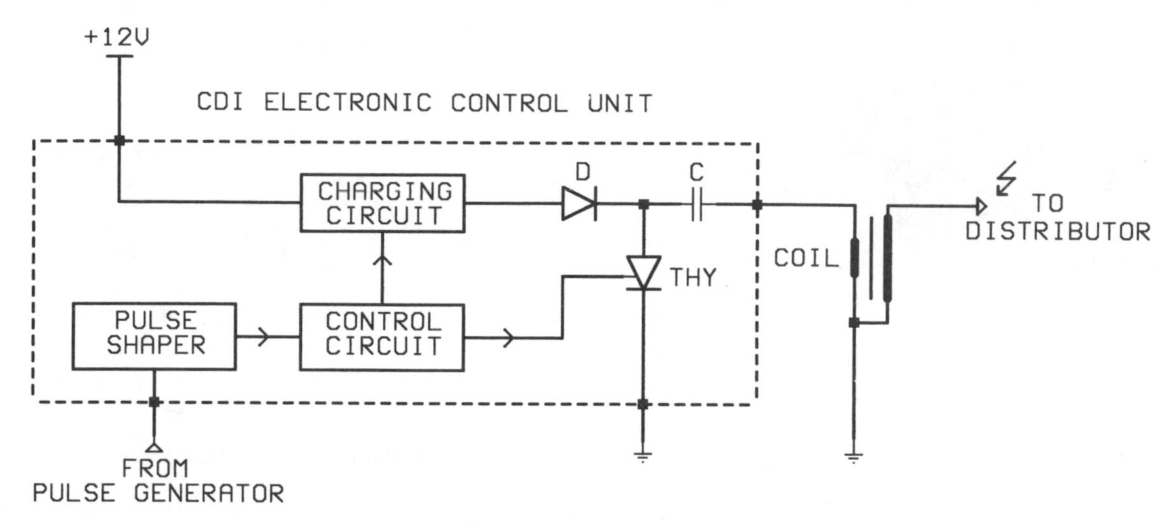

Figure 5.25 Block diagram of capacitive discharge ignition (CDI) system

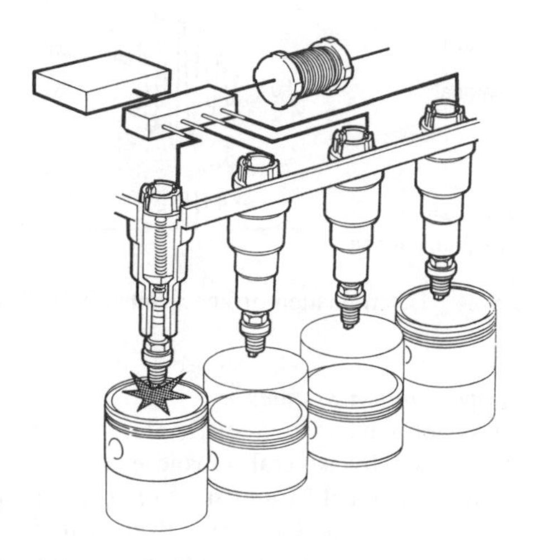

Figure 5.26 Arrangement of Saab CDI system (*Saab*)

the other electrical elements of the vehicle. Its functions are to provide a gas-tight current path from the HT lead into the cylinder and an electrode gap across which the high-voltage spark may jump. There are many different types of spark-plug, each designed for specific engine applications, however most passenger cars use the design typified in Figure 5.27. Major components are:

(i) The insulator.
(ii) The inner conductor (terminal nut, terminal stud, conductive seal and centre electrode).
(iii) The outer shell, incorporating the ground electrode and fixing thread.

The insulator is made from specially-formulated alumina, which has a low porosity, a high electrical resistance and good heat conduction properties. The outer surface is glazed to prevent the adhesion of dirt, and ribbed to give a longer HT leakage-current path.

The inner conductor comprises a terminal stud which locates in a special conductive glass seal. Normally the seal is designed to have a significant electrical resistance which serves to damp oscillations that would otherwise occur due to the interaction between the spark current and the capacitance of the HT leads. The seal connects the stud to the centre electrode, which is housed in the insulator nose (Figure 5.28) and has a slightly smaller diameter than the bore in the nose to take account of the greater expansion of the electrode when the plug heats up.

The construction of the centre electrode has a fundamental influence on the electrical and thermal properties of the spark plug. When the insulator nose is short the thermal path from the centre electrode to the outer shell is also short and so the plug has a high heat dissipation, giving a 'cold' spark plug. A 'hot' spark plug has a long nose, and therefore a long thermal path from the centre electrode. The heat rating of the plug must be carefully chosen to suit the engine. At idle speed the centre electrode should operate above 400°C to ensure adequate self-cleaning and so prevent fouling. At high engine speed the temperature of

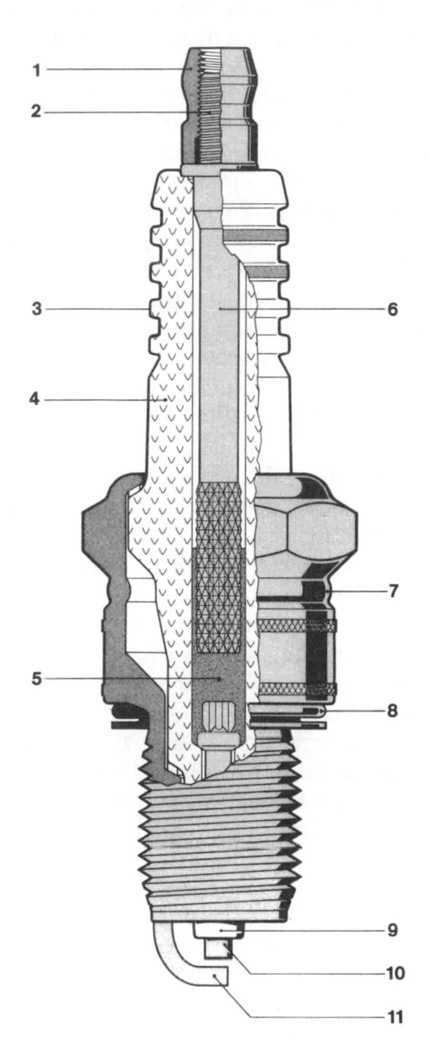

Figure 5.27 Typical spark-plug design. [1] Terminal; [2] threaded connection; [3] leakage-current barrier; [4] Al_2O_3 insulator; [5] conductive seal; [6] terminal stud; [7] swaged and heat-shrunk fitting; [8] captive sealing washer; [9] insulator tip; [10] centre electrode; [11] ground electrode (*Bosch*)

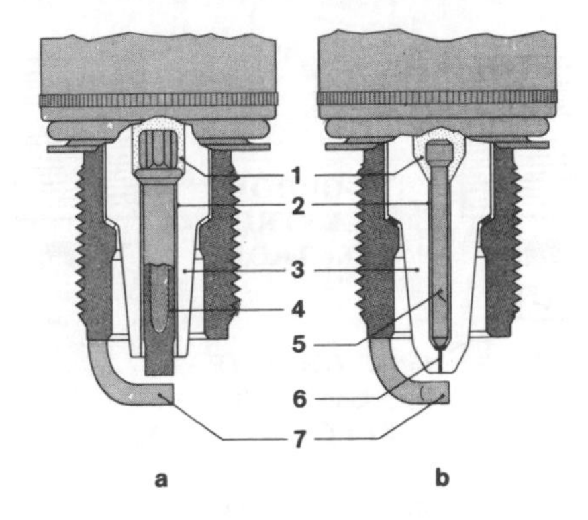

Figure 5.28 Insulator and electrode detail for (a) standard centre electrode, (b) platinum centre electrode. [1] Conductive seal; [2] air gap for thermal expansion; [3] insulator nose; [4] compound centre electrode; [5] metal contact pin; [6] platinum centre electrode – sintered in without gap; [7] ground electrode (*Bosch*)

the plug rises, but should remain below 850°C otherwise surface ignition (plug pre-ignition) may occur.

Most spark plugs use a compound centre electrode consisting of a copper core, plated with a nickel-chromium alloy. The copper gives good thermal conductivity whilst the nickel-chromium alloy resists chemical attack from the combustion products.

The gap between the electrodes is determined by the ignition voltage, engine compression ratio and air–fuel mixture ratio for the engine. Generally, an electrode gap of 0.8–1 mm is used, but a difficult-to-ignite lean mixture may require a gap up to 1.5 mm whereas a high-compression racing engine needs a gap of less than 0.6 mm.

In recent years small-diameter centre electrodes, made of silver or platinum, have become popular. The narrow tip makes it easier for a discharge to occur and so the rise in the required ignition voltage during the service life is smaller than with a conventional plug. Plug change intervals can often be extended to 50 000 km or more.

Other specially shaped electrodes are also used, including 'V' grooved and 'U' grooved designs (Figure 5.29). These are claimed to improve sparking performance by causing an electric field concentration at the edges of the electrode contour and so may reliably ignite a weaker mixture.

5.6 FUEL CONTROL SYSTEMS

Introduction

The function of the fuel control system is to introduce fuel into the incoming air stream in precise accordance with the engine operating conditions, and then uniformly distribute it to the individual cylinders. For most of the history of the motor car, the carburettor has been the most common device used to achieve these requirements. It uses a restriction ('venturi') in the intake passage upstream of the throttle valve to

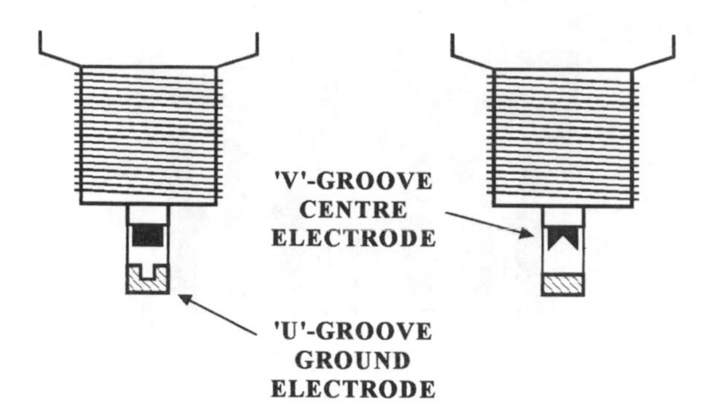

Figure 5.29 'U' and 'V' grooved spark-plug designs

create a pressure drop which sucks fuel from a small nozzle called a 'jet'. The pressure difference between the air inlet and the throat of the venturi depends upon the intake air flow rate and so is used to meter the appropriate amount of fuel to intake air. By careful selection of jet size and throat diameter it is possible to achieve a suitable air–fuel ratio. Traditionally this is set at about 13:1 ($\lambda = 0.9$) when maximum power is required, or about 16:1 ($\lambda = 1.1$) for maximum economy.

Although carburettors can be designed to meter fuel with great accuracy, the introduction of emission legislation requiring the use of three-way catalysts soon highlighted certain shortcomings, specifically:

(i) Difficulty of A/F ratio feedback control from an exhaust gas oxygen sensor.
(ii) Many moving parts, leading to rapid wear (80 000 km lifetime).
(iii) Poor mixture distribution through the intake manifold.
(iv) Poor mixture control during cold running and transient conditions such as acceleration and deceleration.

In an attempt to overcome these deficiencies some manufacturers designed carburettors that were fitted with electronic control systems, for example the Bosch Ecotronic, however the problem of rapid wear-out and poor transient response remained. The more precise and controllable method of fuel injection has therefore completely superseded the carburettor.

Fuel injection

The idea of injecting fuel into an engine first arose in the aircraft industry, when the need to deliver a constant air–fuel mixture at varying altitudes and engine angles led to the development of a variety of novel fuelling systems. In the period prior to World War II, the Bosch company of Germany designed a mechanical 'direct injection' system in which petrol was sprayed directly into each combustion cylinder. The system drew on Bosch's expertise in diesel injection and worked on a similar principle, with a plunger-type high-pressure pump and spring-load injector nozzles. Its good performance led to it becoming a standard fitment on German warplanes and, after the war, on Mercedes sports cars. In the mid-1950s the modern practice of indirect injection was developed, whereby fuel is injected into the intake manifold just behind the inlet valves. This had the great advantage that the injector was no longer exposed to the combustion process and a much lower fuel pressure could be used.

The desirability of continuously controlling the quantity of injected fuel soon led manufacturers to consider electronically controlled systems. The first truly electronic petrol injection system used valve circuits and was developed in 1957 by the Chrysler Corporation. Subsequently many other companies worked on electronic systems using transistors, most notable amongst these was the Bendix Corporation which pioneered *intermittent indirect injection* using solenoid valve injectors. This is the system which was brought to market by Bosch in the early-1970s and is now fitted as standard on virtually all cars. It is simply referred to as electronic fuel injection (EFI).

Electronic fuel injection

The great advantage of EFI is that it allows the amount of fuel injected into each cylinder to be precisely controlled on a cycle by cycle basis in response to information obtained from sensors which continuously report the engine operating circumstances. The result is greater engine efficiency and controllability than would ever be possible with a carburettor.

The amount of fuel delivered into each cylinder is controlled by a timed current pulse sent to solenoid valve 'injectors'. Each injector is connected to a delivery pipe that is supplied with fuel from the tank and kept at a constant pressure differential (usually 2–3 bar) with respect to the intake manifold by means of a vacuum-diaphragm pressure regulator valve (Figure 5.30).

The fuel pump may be submerged within the tank, or mounted externally part-way along the fuel line. In either case it consists of a centrifugal roller cell pump driven by a permanent-magnet electric motor (Figure 5.31) and is specified such that it can always supply more fuel than the engine requires. In order to cool and lubricate the electric motor, the entire pump assembly

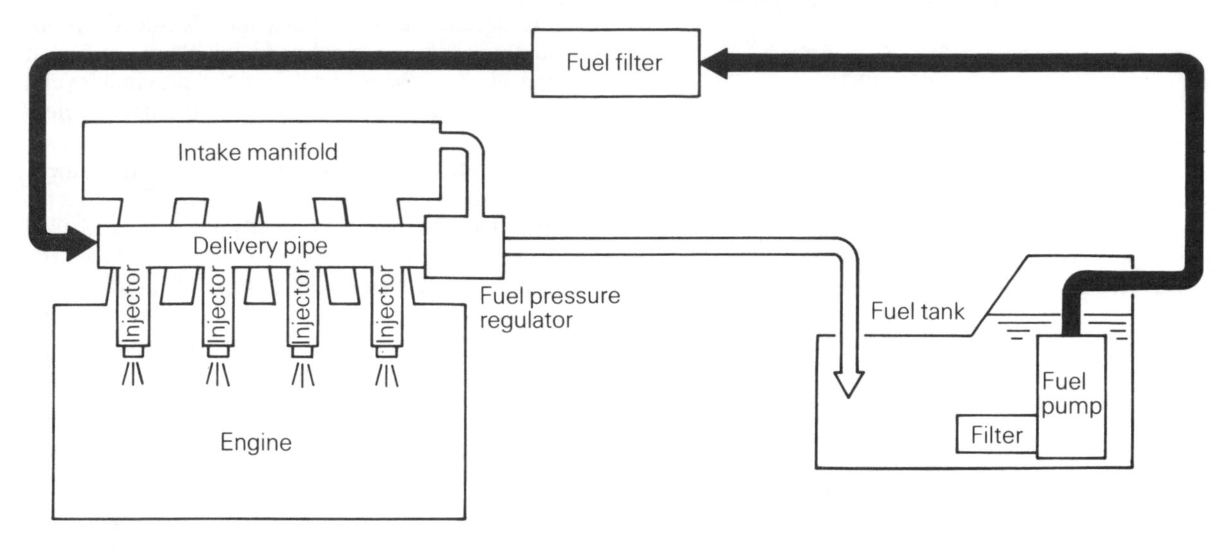

Figure 5.30 Injection system fuel delivery (*Mitsubishi*)

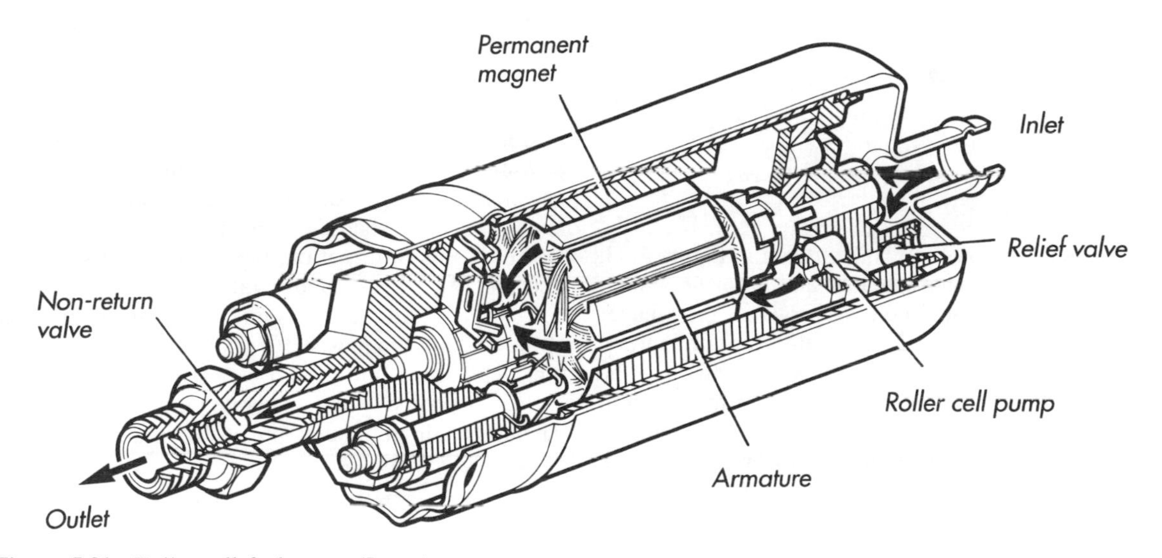

Figure 5.31 Roller cell fuel pump (*Lucas*)

is filled with fuel and therefore it should not be operated 'dry'.

Fuel from the pump is passed through a replaceable filter element which traps dirt particles larger than about 10 μm, protecting the injectors from possible blockage or abrasion. The injector operating pressure is determined by the fuel pressure regulator (Figure 5.32). The lower regulator chamber is connected to the fuel delivery pipe and isolated from the fuel return pipe by a ball valve held closed under spring pressure. If the fuel pressure in the delivery pipe rises above a pre-set level, it lifts the diaphragm and unseats the valve, so allowing some fuel to return to the tank and reducing the delivery pressure until the valve re-closes. With the pump operating, the valve is continuously opened and closed to keep the fuel at the prescribed pressure. To ensure a constant fuel flow rate through the injectors, the fuel pressure in the delivery pipe must be kept constant *with respect to intake manifold pressure*. This is achieved by connecting the regulator's upper chamber to the inlet manifold so that the spring force is moderated in proportion to inlet vacuum.

The design of the fuel injectors varies somewhat,

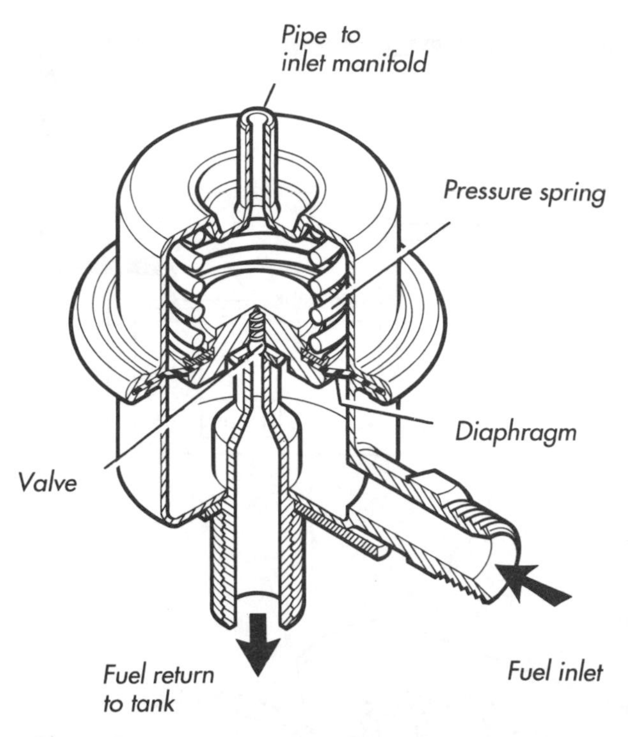

Figure 5.32 Fuel pressure regulator (*Lucas*)

but a typical construction is illustrated in Figure 5.33; it comprises a plunger-solenoid and a nozzle assembly. When the solenoid coil is energized, the plunger is pulled back by 0.1–0.2 mm against a return spring, allowing pressurized fuel to escape as a mist via the plunger tip and discharge orifices. When the solenoid current ceases, the return spring closes the nozzle. Opening and closing response times lie in the range 0.5–1 ms and the injector opening duration is controlled within the range 1–10 ms, depending upon engine speed and load. The injector actuation process may be performed in one of two ways: *group injection* or *sequential injection* (Figure 5.34).

Group injection
With this strategy all of the injectors are opened simultaneously and fuel is injected into the intake manifold once at each crankshaft revolution (360°). Half of the required fuel is therefore delivered on each occasion and so there are two injection pulses for each full engine cycle (720° of crankshaft rotation).

Sequential Injection
This is the preferred strategy for modern designs, it requires more complex control than group injection but delivers greater engine efficiency through improved mixture homogeneity. Each cylinder's injector is individually controlled to deliver a full charge of fuel at the end of the exhaust stroke of the previous cycle, just prior to the induction stroke of the next cycle.

The operation of all modern electronic fuel injection systems is fundamentally similar. The injector opening duration is electronically timed and phased by a microprocessor-based ECU in response to a large number of input signals. The primary input signals are engine speed and intake air mass, and it is the derivation of the latter signal that distinguishes the two fundamental types of EFI system; *speed-density* EFI and *mass air-flow* EFI.

Speed density EFI

Speed-density EFI was first introduced by Bosch, in 1967, with the product name 'D-Jetronic', and it has since been adopted by many other manufacturers.

A modern speed-density engine management system featuring sequential fuel injection and distributorless ignition control is illustrated in Figure 5.35. The primary input signals to the ECU [1] are engine position and speed [14,15] and manifold absolute pressure (MAP) [10]. Additional sensors are used to detect throttle position [8], water temperature [12], knock [11] and exhaust-gas oxygen concentration [13]. Based on this information, the ECU controls the fuel injectors [4], idle air control valve [7] and ignition coil current [16]. Additional connections are included to allow for fault diagnosis [19] and communication with other electronic units, such as a trip computer [20].

Speed-density systems operate on the principle that the mass of air, m_a, drawn into each cylinder per engine cycle is given by,

$$m_a = (V_d n_v p_i)/(RT_i)$$

where V_d is the displacement of the cylinder, n_v is the volumetric efficiency (i.e. the fraction of V_d actually filled on each intake stroke), p_i is manifold absolute pressure, R is a constant and T_i is the intake air temperature. Since m_a is proportional to p_i (MAP) it can be indirectly measured by detecting this pressure. V_d and R are constants, but n_v varies with engine speed; thus engine speed, N, and MAP are used as the two input variables to a fuelling map (look-up table) which contains the basic injector opening pulse width data.

Mass air-flow EFI

Air-flow EFI was introduced by Bosch in 1973 with the product name 'L-Jetronic'; it works on the principle of precisely measuring the quantity of air drawn

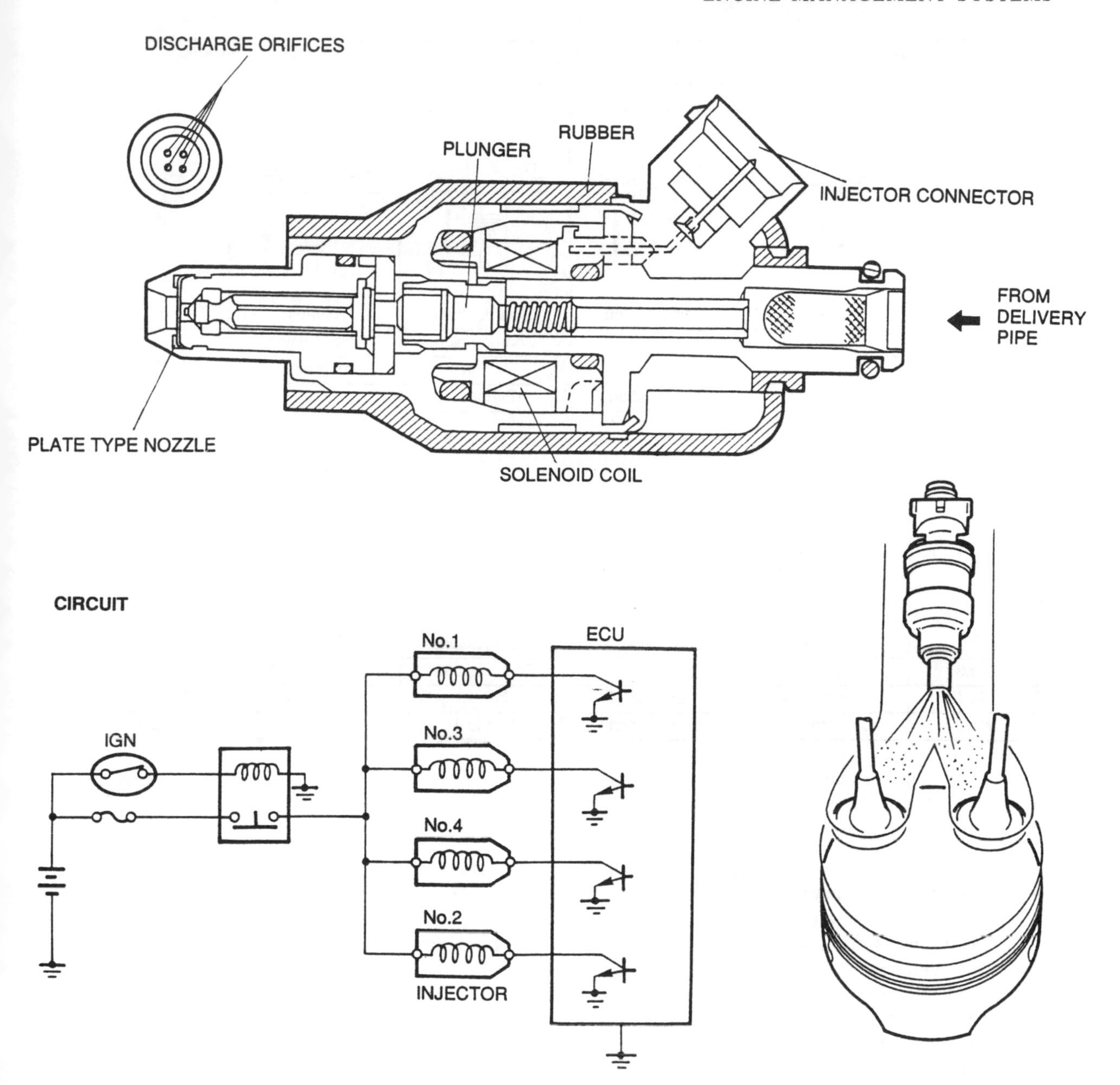

Figure 5.33 Fuel injector (*Mazda*)

into the engine on each intake stroke using an *air-flow sensor* (AFS).

The general arrangement of the L-Jetronic system is very similar to that of the speed-density system of Figure 5.35, but with the MAP sensor replaced by an air-flow sensor located in the inlet path, just after the air filter, and consisting of a hinged flap connected to a potentiometer (Figure 5.36). The incoming air pushes the flap open against a return spring and so rotates the potentiometer. The potentiometer's output voltage to the ECU therefore varies in proportion to the flap deflection which is, in turn, dependent on the air-flow volume (Figure 5.37).

Air-flow EFI is more costly and complicated than speed-density EFI but offers several advantages, including:

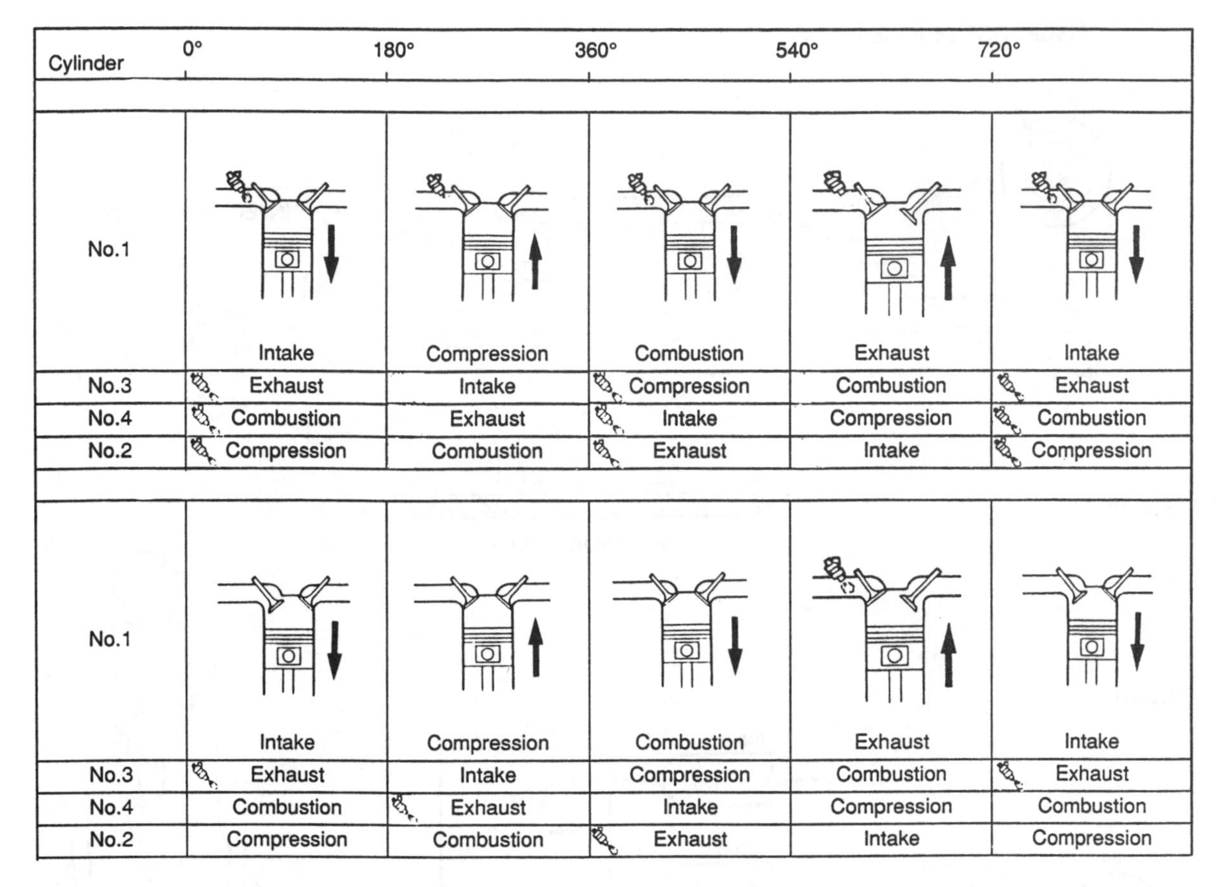

Figure 5.34 Injector timing options; *group injection* (upper diagrams) and *sequential injection* (lower diagrams) (*Mazda*)

(i) Variation in volumetric efficiency, n_v, with engine speed, N, is automatically compensated for.
(ii) Since the intake air-flow is measured directly, the system is insensitive to variations in engine displacement, V_d, due to internal deposits and build tolerances.
(iii) The system inherently provides compensation for the addition of EGR gases to the inlet manifold.

The basic injector pulse width in a mass air-flow EFI system is determined on the basis,

$$\text{Basic pulse width} = KQ/N$$

where Q is the measured intake air quantity, N is the engine speed and K is a constant for the particular engine. Thus Q and N are the inputs to the fuelling map, and compensations for other factors are made in the exactly same manner as described for the speed-density system.

Since the introduction of the L-Jetronic EFI system, a variety of alternative techniques for air-flow sensing have been developed and these are summarized in Table 5.2. The most widely used are hot-wire AFS and the Karman vortex AFS.

Air-flow sensor construction and operation

Flap-type air-flow sensor

Since its introduction in 1973 the flap-type AFS has been steadily improved. A major drawback of this type of sensor is its poor transient response due to the inertia of the flap; when the throttle is snapped open there is a slight delay as the flap strives to keep up with the increasing air flow, similarly there can be an overshoot as the flap reaches the appropriate opening angle. In order to overcome these deficiencies, recently developed flap-type sensors use a lightweight magnesium flap together with a carefully designed damping volume to provide a fast response and freedom from jitter due to intake air pulsations.

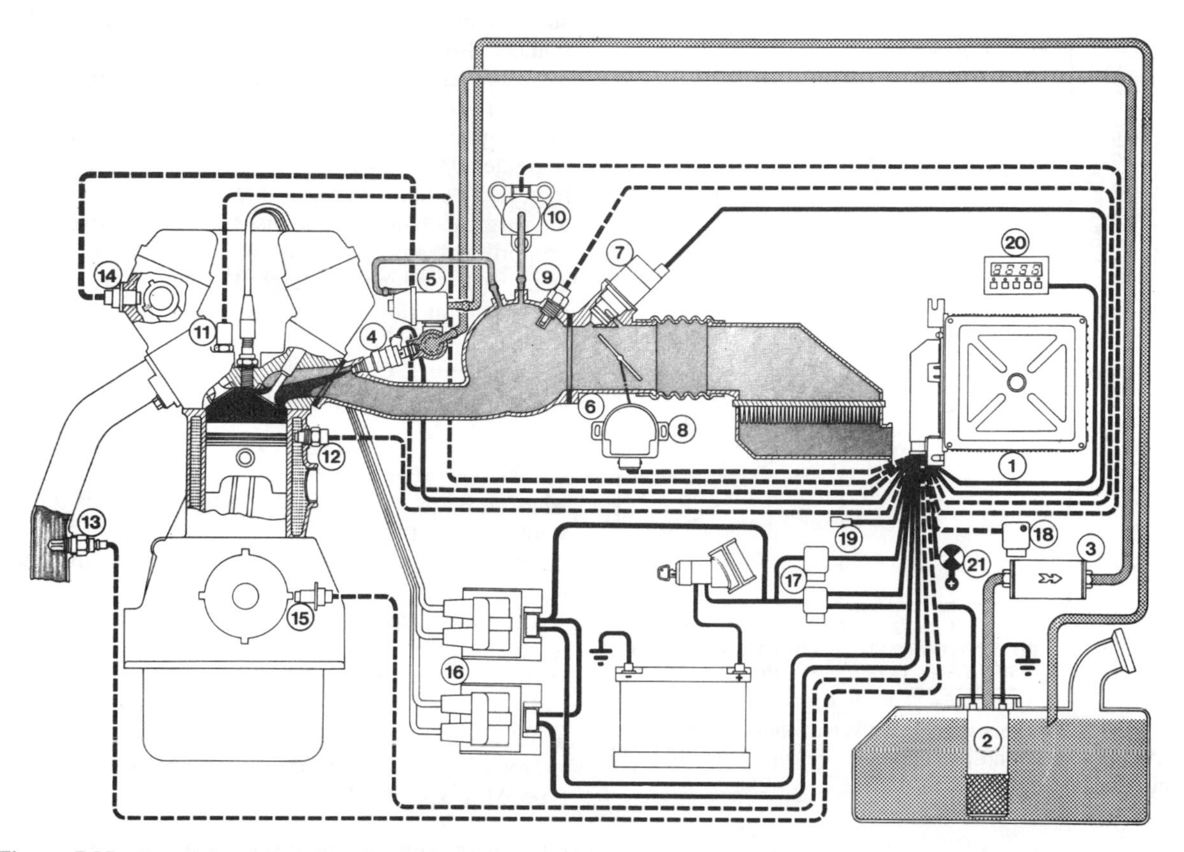

Figure 5.35 Speed-density engine management system (*Magneti Marelli*): [1] microcomputer-based ECU; [2] fuel pump; [3] fuel filter; [4] injectors; [5] fuel pressure regulator; [6] throttle body; [7] idle air control valve; [8] throttle position potentiometer; [9] inlet air temperature sensor; [10] MAP sensor; [11] knock sensor; [12] coolant temperature sensor; [13] exhaust gas oxygen sensor; [14] timing sensor; [15] engine rpm sensor; [16] double-ended ignition coils; [17] system relays; [18] trimmer; [19] diagnostic connector; [20] trip computer; [21] malfunction indicator light

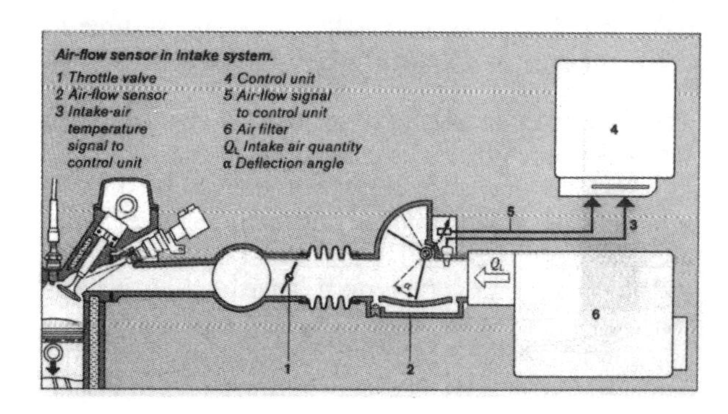

Figure 5.36 Flap-type air-flow sensor (*Bosch*)

To ensure long-term reliability the potentiometer track is made from a ceramic-metallic ('cermet') composite, fired directly onto an alumina substrate. The range of measurable air-flow variation is about 80:1 with such a construction.

Hot-wire air-flow sensor

If a heated platinum wire ('hot-wire') is exposed to the intake air flow its temperature will drop as a result of the air removing heat from the wire and reducing its resistance. A larger air flow results in greater heat removal and lower resistance, thereby providing the basis for air-flow sensing. A typical hot-wire AFS circuit is illustrated in Figure 5.38. Two sensing wires are placed in the intake air stream; R_H is the hot wire and is heated to 100–200°C above ambient air temperature by the portion I_2 of the op-amp's output current, I_1. R_K is the cold-wire and is used to sense the ambient air temperature. The two wires, together with resistors R_1, R_2 and R_3, form a bridge circuit, but the resistance of $R_K + R_1 + R_2$ is much greater than $R_H + R_3$ and therefore I_3 is much less than I_2. With a steady air flow, the bridge is balanced and so the two input voltages to the op-amp, V_1 and V_2, are equal. If the throttle is opened, more air flows past the hot-wire

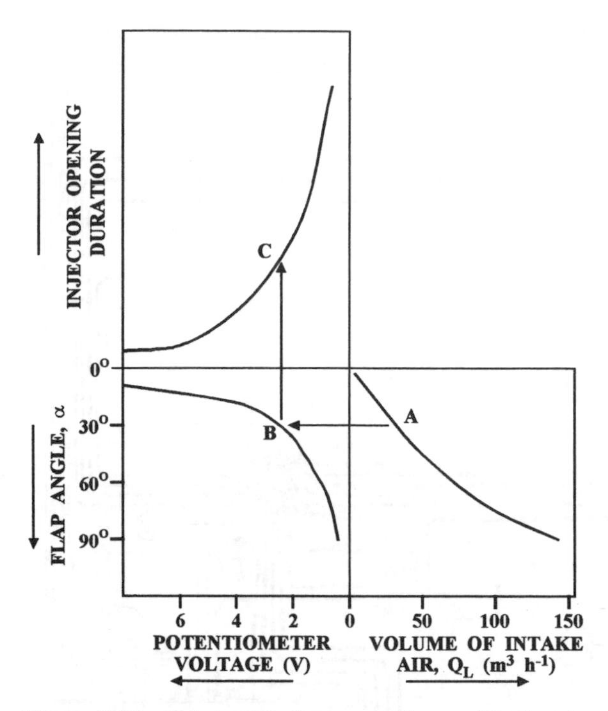

Figure 5.37 Relationship between volume of intake air [A], flap angle and potentiometer output voltage [B] and injector opening duration [C]

Table 5.2 Available types of air-flow sensor

Sensor type	Measured quantity	Output signal	Comments
Movable flap	Volume flow	DC voltage	Cheap and simple. Poor response
Turbine	Flow velocity	Frequency	Slow to follow sudden flow changes
Hot wire and Hot film	Mass flow	DC voltage	Fast response. Can be over-sensitive
Karman vortex	Flow velocity	Frequency	Slow to follow sudden flow changes
Ultrasonic	Mass flow	DC voltage	Limited measurement range
Ion drift	Mass flow	DC voltage	Needs high voltage

and it cools down, reducing its resistance, R_H. The voltage dropped across R_H therefore diminishes, V_1 increases and the inputs to the op-amp become unbalanced, resulting in a rise in the output current I_1 due to the op-amps' gain. The higher value of I_1 provides a greater heating current, I_2, to the hot-wire and so increases its temperature and restores its resistance to the original value. I_3 also rises and so V_2 rises by a proportional amount. V_1 and V_2 thus become equal again, but at a higher voltage. V_1 is therefore a measure of the air flow past the hot-wire and it increases with increasing flow.

The quantity, Q, of intake air is related to the heating current, I_2, by the equation,

$$Q = K \times I_2^n$$

where n takes a value in the range 5–8, K is a constant for the particular sensor construction and the current I_2 lies in the range 0.5–1.5 A, depending upon throttle opening.

The output voltage, V_1, therefore has a non-linear relationship to Q and this must be taken into account by the ECU when processing the AFS signal. Typically, V_1 measures about two volts at idle, rising to about seven volts at maximum air flow.

Hot-wire AFS offers several advantages over flap-type AFS, specifically:

(i) It provides a signal related to the *mass* of the inducted air and therefore requires no corrections for air temperature or altitude.
(ii) It offers minimal obstruction to the intake air flow.
(iii) There are no moving parts.
(iv) It offers an exceptionally fast response time to changes in air flow (a few milliseconds).

The principal disadvantage is that it can be over-sensitive. Since there are pulsations in the intake air flow, which the hot-wire sensor inevitably detects, some means of electronically 'smoothing' the signal must be employed.

The construction of a typical hot-wire AFS is illustrated in Figure 5.39. It consists of a cast aluminium housing fitted with a hybrid IC module. Air enters the housing and divides between the main housing and the bypass housing which contains the hot-wire and cold (temperature compensating) wire. This arrangement minimizes signal offset due to the build-up of dirt on the wires and reduces the risk of wire damage from inlet-manifold back-firing.

The wires are approximately 70 μm diameter platinum, which may be wound on a glass support or suspended from insulated supports. Some recent designs have dispensed with wires and instead use heated

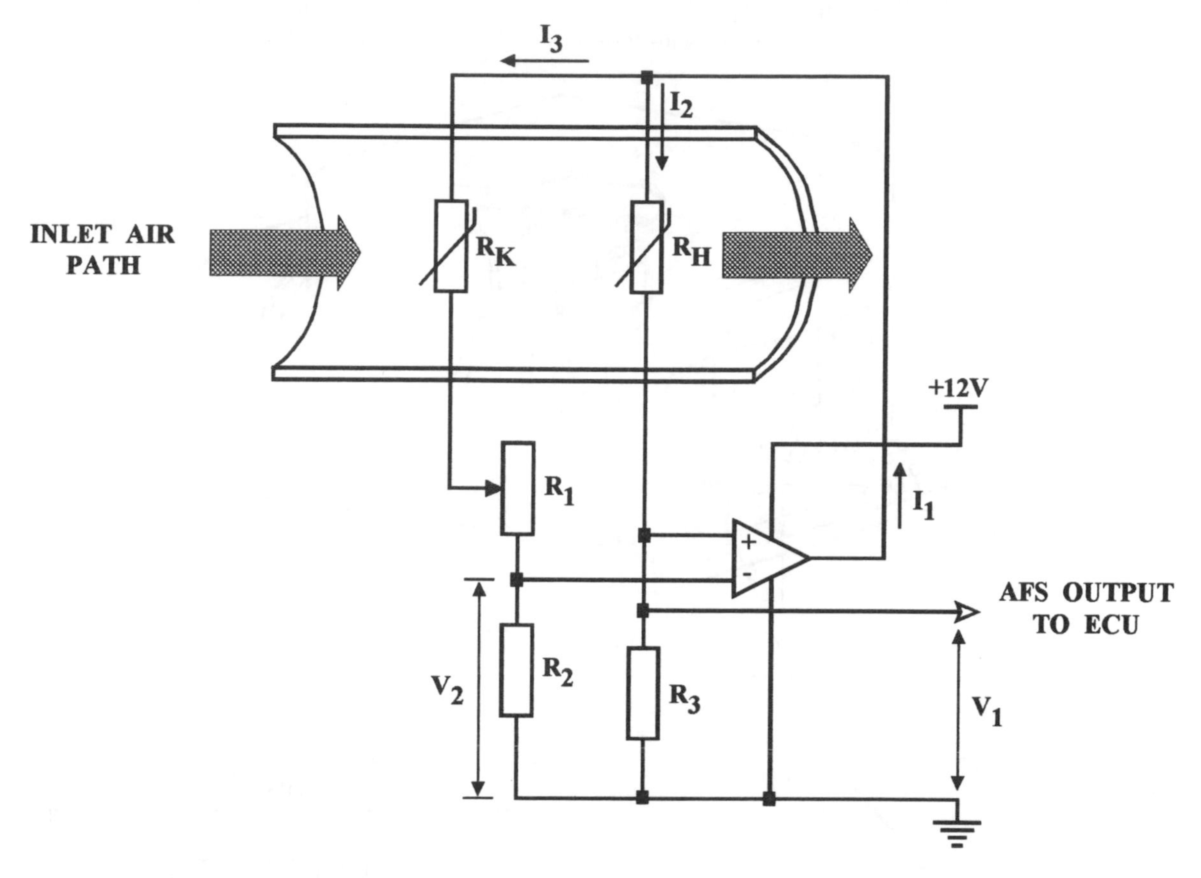

Figure 5.38 Circuit diagram of hot-wire air-flow sensor

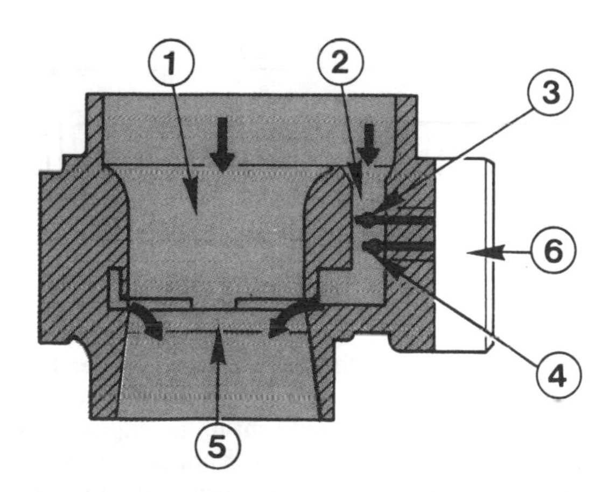

Figure 5.39 Section through a hot-wire air-flow sensor: [1] main air channel; [2] bypass channel; [3] hot-wire; [4] cold wire; [5] venturi; [6] electronic circuit (hybrid IC)

metal-film resistors deposited on an insulating substrate. This approach is claimed to give a more robust and reliable sensor.

The hybrid IC module contains the components illustrated in Figure 5.38, as well as voltage regulating and signal filtering circuits.

A further feature provided by some manufacturers is a 'self-cleaning' function whereby the module heats the wire to about 1 000°C for one second each time the ignition is switched off, thus burning away any dirt deposits which may influence calibration.

Karman vortex air-flow sensor

If a narrow column is placed across the intake air path, regular vortices are generated as the air flows around it. The Karman vortex AFS works by counting the number of vortices generated by the column each second. The air-flow velocity, V, is then found according to the equation,

$$V = f_a \times (d/St)$$

where d is a constant relating to the width of the vortex-generating column, St is a parameter called the Strawhal number (approximately 0.23 for a typical AFS design) and f_a is the vortex generation frequency. Using this equation it is possible to obtain

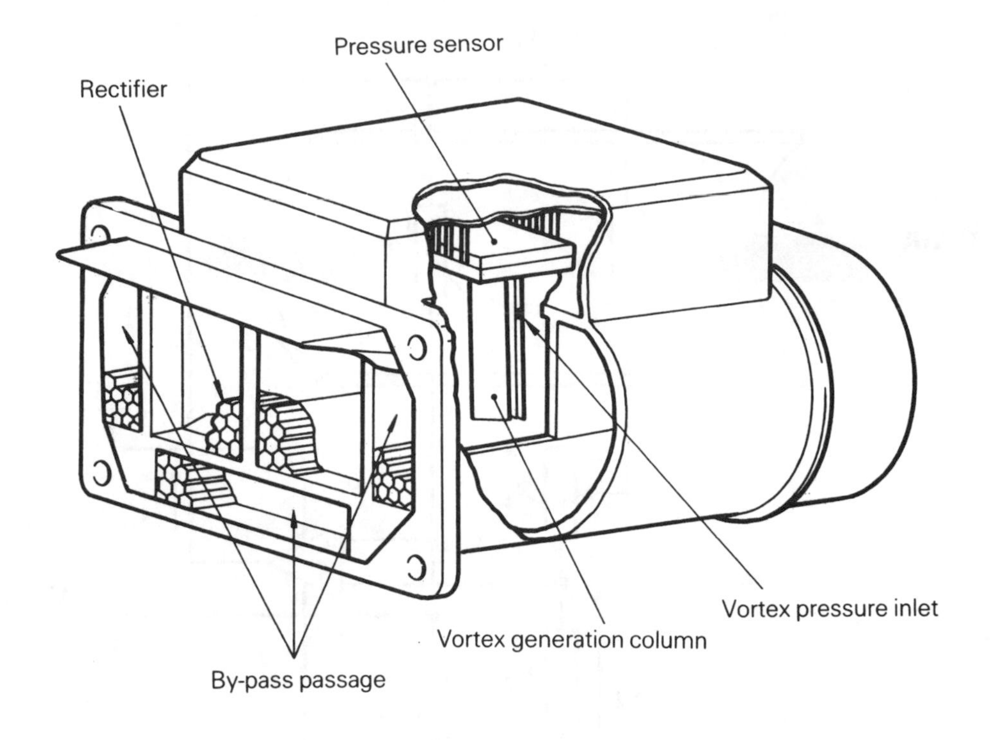

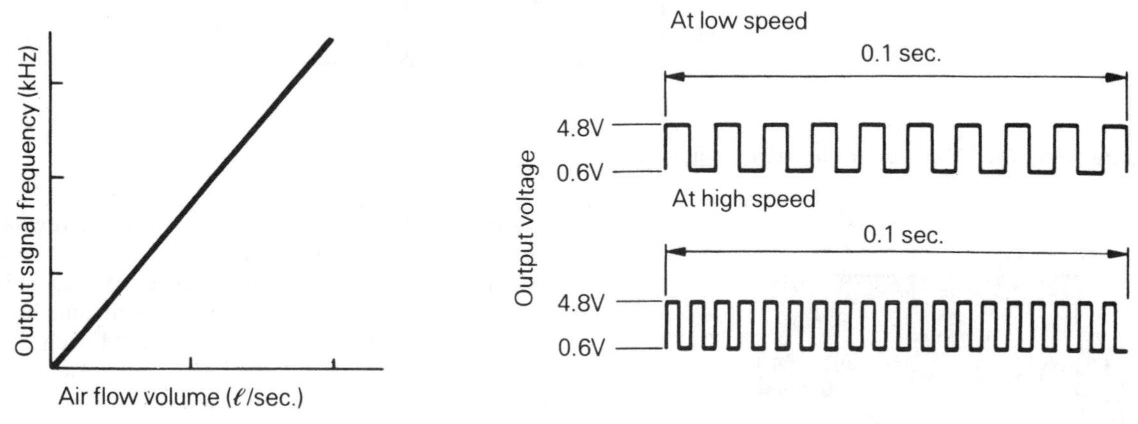

Figure 5.40 Construction of a Karman vortex air-flow sensor (*Mitsubishi*)

V by counting f_a; the air volume flow rate is then calculated by multiplying V by the air-flow sensor cross-sectional area.

There are two common methods for measuring the vortex frequency, f_a; ultrasonic and pressure-sensing.

Ultrasonic sensing relies on the vortices slightly altering the frequency of a 50 kHz ultrasonic signal which is transmitted across the air-flow path. f_a is obtained by counting the frequency of these disturbances.

Pressure sensing is a much cheaper alternative and relies on the detection of the vortices by using a semiconductor pressure sensor placed just downstream of the vortex column. Each time a vortex is generated it causes a small pressure variation which is converted to a square-wave signal and sent to the engine ECU. The construction of this type of Karman vortex AFS is illustrated in Figure 5.40. It consists of an air-flow straightener ('air rectifier'), triangular-section vortex generation column and pressure sensor. Sensors to measure air temperature and barometric pressure are also incorporated in the housing.

The output response is as shown, with f_a in the region of 100 Hz at idle, rising linearly to about 2 kHz at high engine speed.

5.7 FUEL INJECTION VOLUME CONTROL

The engine ECU controls the injector opening current pulse-width in accordance with pre-programmed data held in its memory. The exact way in which this data is used depends upon the strategy adopted by the ECU designer, however, the operation of a typical system will be similar to that described here. For normal engine operation the basic injector opening time is stored as data in a map and is read-out in accordance with the instantaneous values of MAP and engine speed (speed-density EFI) or intake air-flow and engine speed (mass air-flow EFI). The engine ECU's microprocessor then modifies this data to take account of various compensation factors, examples of which are presented in Table 5.3 for the Honda PGM-FI speed-density system. Additional sets of data are held to provide information on fuelling for special circumstances, e.g. starting, idling and wide-open throttle.

Engine starting

When the ignition key is first turned the ECU calculates an injection duration using the formula,

Injection Duration = Cranking Duration
+ Temperature Correction
+ Battery Voltage Correction

The ECU thus holds data for the injection duration appropriate to an engine operating at cranking speed. This duration is modified by information from the intake air and coolant temperature sensors. Correction must also be made for variations in battery voltage since a low voltage will lead to a slower opening response from the injectors. Once the engine fires, its rpm signal rises rapidly, the ECU detects this and progressively reduces the injection duration to remove the starting enrichment.

Engine idling

Since modern traffic conditions dictate that vehicles spend a large proportion of city journeys in queues it is important that a steady idle speed can be maintained with great precision, even for sudden changes in engine load. When the ECU detects that the engine is idling (via the throttle potentiometer and engine speed signals) it selects an idle fuelling map and implements feedback control of idle speed. As a general guide, approximately 25% of the software in a modern ECU is dedicated to idle speed control, it detects loads on the engine from a variety of sensors, specifically:

(i) Air-conditioner operation via operation of the air-con compressor's magnetic clutch.
(ii) Automatic transmission engagement via opening of the park/neutral inhibitor switch contacts.
(iii) Increasing electrical load via a rising alternator output voltage.
(iv) Operation of the power steering via a power steering pressure switch which closes when the hydraulic pressure in the steering system rises above a pre-set limit.

Control of the idle speed is achieved by varying the amount of intake air passing through a bypass passage around the closed throttle valve. Three common types of actuator are used to control the quantity of idle air,

(i) Stepper-motor idle air control (IAC) valve (Figure 5.41). Adjustment of the idle air flow is achieved by rotation of a stepper-motor under ECU control. The air-valve shaft engages in the motor via a screw thread so that as the motor turns, the clearance between the valve and its seat is varied.
(ii) Linear solenoid IAC valve. This is mounted in the bypass passage in a similar manner to the stepper-motor IAC but uses a duty-cycle controlled solenoid to vary the valve opening. The solenoid is typically energized at frequencies in the range 30–200 Hz, depending on manufacturer.
(iii) Rotary solenoid IAC valve. This type of valve features a solenoid armature which rotates against a return spring to provide a variable valve opening cross-section. The drive coils are duty-cycle controlled by the ECU at a frequency of around 100 Hz to provide the required air flow.

At low coolant temperatures the IAC valve is held open to give a fast idle and aid smooth running. Once the engine is warm, the ECU actuates the IAC to provide the target idle speed, irrespective of auxiliary loads. Additionally, a higher idle speed can be commanded for special circumstances; e.g. to increase the alternator output when the battery voltage is low, or to increase the air-conditioner cooling capacity when 'COLD' is selected.

Normal operation

During normal driving the fuel injection volume is controlled according to the formula,

Table 5.3 Compensation factors and sensors used in the Honda PGM-FI speed-density EFI system

Compensation	Sensor and input	Description
Compensation during cranking	Cranking signal	Increase injector pulse width to improve starting
Battery voltage	Battery voltage signal	Increase injector pulse width to compensate for injector lag caused by low battery voltage
Engine coolant temperature	Engine coolant temperature sensor	Increase injector pulse width according to engine temperature
Intake air temperature	Intake air temperature sensor	Adjust injector pulse width according to intake air temperature
Compensation immediately after starting	Cranking signal	Gradually decrease injector pulse width after starting
Compensation during acceleration	Throttle position sensor	Increase injector pulse width according to rate of increase in sensor output voltage during acceleration
Compensation during deceleration	Throttle position sensor	Reduce injector pulse width according to rate of decrease in sensor output voltage during deceleration
Fuel cut-off during deceleration	Throttle position sensor, crank position sensor	When the throttle angle sensor indicates that the throttle is closed and the engine speed is above 1500 revs/min fuel is cut off to improve fuel economy
Maximum engine fuel cut-off	Crank position sensor	Fuel is cut off restricting engine speed to 6600 rev/min
Compensation after fuel cut-off	Crank position sensor	Increase injector pulse width after fuel cut-off at low speeds
Compensation under heavy load	Manifold absolute pressure sensor	Increase injector pulse width when inlet manifold absolute pressure exceeds prescribed value

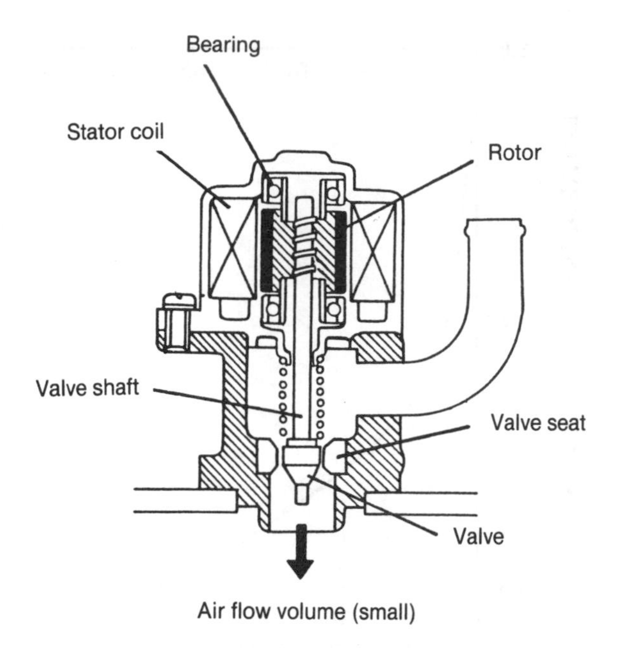

Figure 5.41 Construction of stepper-motor idle air control valve

$$\text{Injection Duration} = \text{Basic Injection Duration} \times \text{Correction Coefficients} + \text{Battery Voltage Correction}$$

The basic injection duration is read directly from the normal running map and then modified by various correction coefficients. The nature and value of each correction coefficient is obtained from extensive tests performed during engine development work and is therefore unique to a particular engine variant. Typical corrections are:

(a) *After-start enrichment.* Once the engine has been started the ECU implements after-start enrichment to keep the engine running smoothly. The amount of enrichment is dependent on coolant temperature at starting and is progressively reduced over a short time (Figure 5.42).

(b) *Warm-up enrichment.* A cold engine needs a richer than normal mixture to prevent it hesitating and stalling. Figure 5.43 shows how the enrichment coefficient for one engine varies with coolant temperature.

(c) *Intake air temperature correction.* The density of the intake air varies according to its temperature, being more dense when cold and less dense when hot. Accordingly, the injection volume is slightly increased when the air temperature is low.

(d) *Acceleration enrichment.* When the throttle is opened the ECU provides a momentary additional enrichment to prevent hesitation ('flat-spotting'). The amount of enrichment depends upon the throttle opening rate and angle.

(e) *Deceleration fuel cut-off.* When the ECU detects that the vehicle is being decelerated, and the engine is above idle speed, it implements fuel cut-off to reduce emissions and improve economy. Fuelling is progressively re-introduced as the engine speed falls to idle.

(f) *Air-fuel ratio feedback correction.* In order to operate the three-way catalytic converter at maximum efficiency, the air-fuel ratio must be precisely maintained at 14.7:1 ($\lambda = 1$). The exhaust gas oxygen sensor continuously monitors oxygen concentration and returns a 'rich' or 'lean' signal to the ECU. When the EGO sensor indicates a rich mixture, the ECU applies a correction coefficient in the range 0.8–1.0 and so reduces the injection duration, making the mixture leaner. Conversely, if a lean mixture is detected the ECU applies a correction coefficient in the range 1.0–1.2 and so makes the mixture richer.

Air-fuel ratio correction is only applied during steady-state engine operation and is not implemented under certain conditions, specifically:

(i) on engine starting;
(ii) during after-start enrichment;
(iii) during warm-up enrichment;
(iv) during deceleration fuel cut-off;
(v) when operating with a wide-open throttle (maximum power).

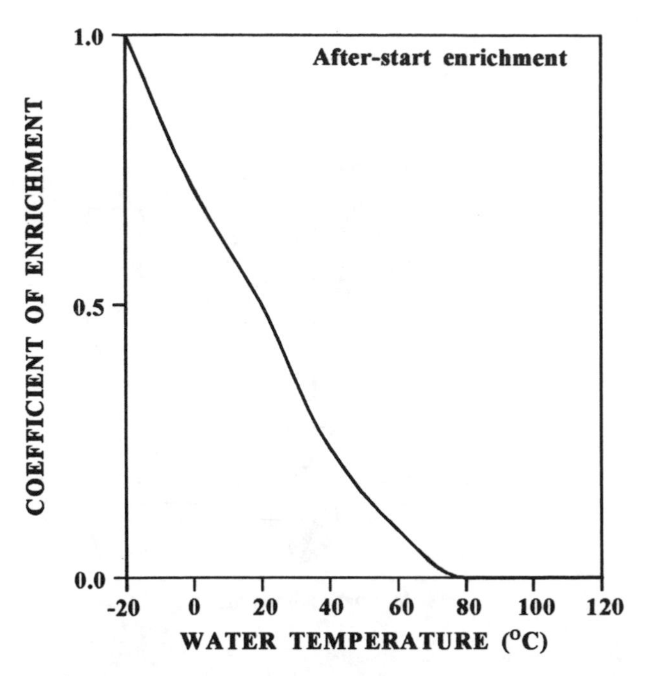

Figure 5.42 After-start enrichment. The amount of enrichment is progressively reduced with each rotation of the engine

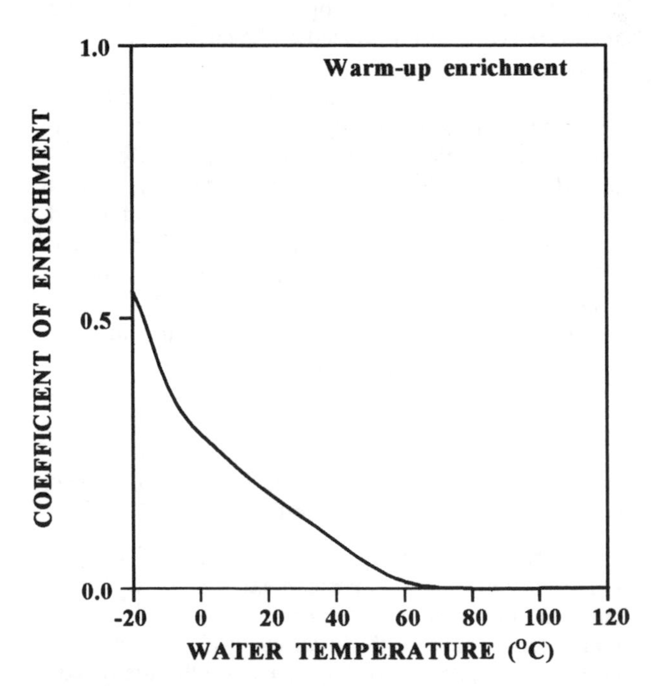

Figure 5.43 Warm-up enrichment

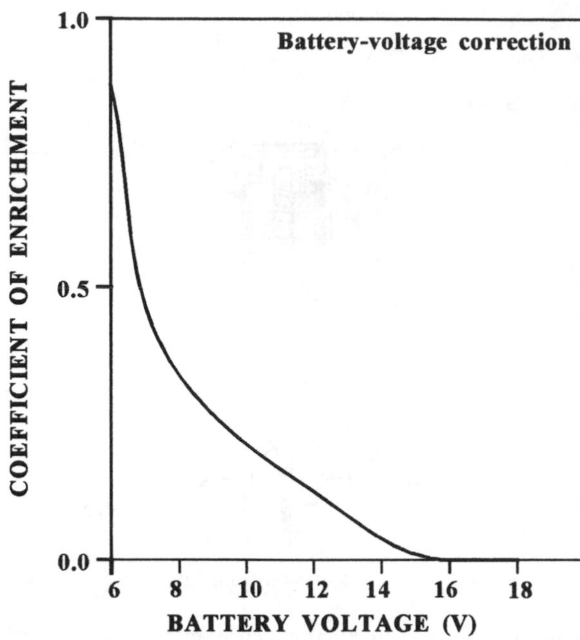

Figure 5.44 Battery voltage correction

The details of air-fuel ratio correction are discussed in greater detail in the following section.

(g) *Battery voltage correction.* Since the injector opening delay increases with decreasing battery voltage, a correction duration must be added to the injection duration to provide additional time for the injector to open when the battery is voltage low (Figure 5.44).

5.8 STOICHIOMETRIC MIXTURE CONTROL

Oxygen sensor operation

When an engine is fitted with a three-way catalyst accurate control of the A/F ratio close to $\lambda = 1$ is essential for effective pollution reduction. This is achieved by using an exhaust-gas oxygen sensor, fitted in the exhaust system, which returns a signal dependent on the oxygen concentration in the exhaust gas. This signal is then used as a correction factor to vary the fuel injection duration and maintain stoichiometry.

Two types of oxygen sensor are in common use; the Zirconia sensor and the Titania sensor. Both types utilize the fact that the amount of free oxygen in the exhaust gas increases sharply as λ changes from 0.99 to 1.01.

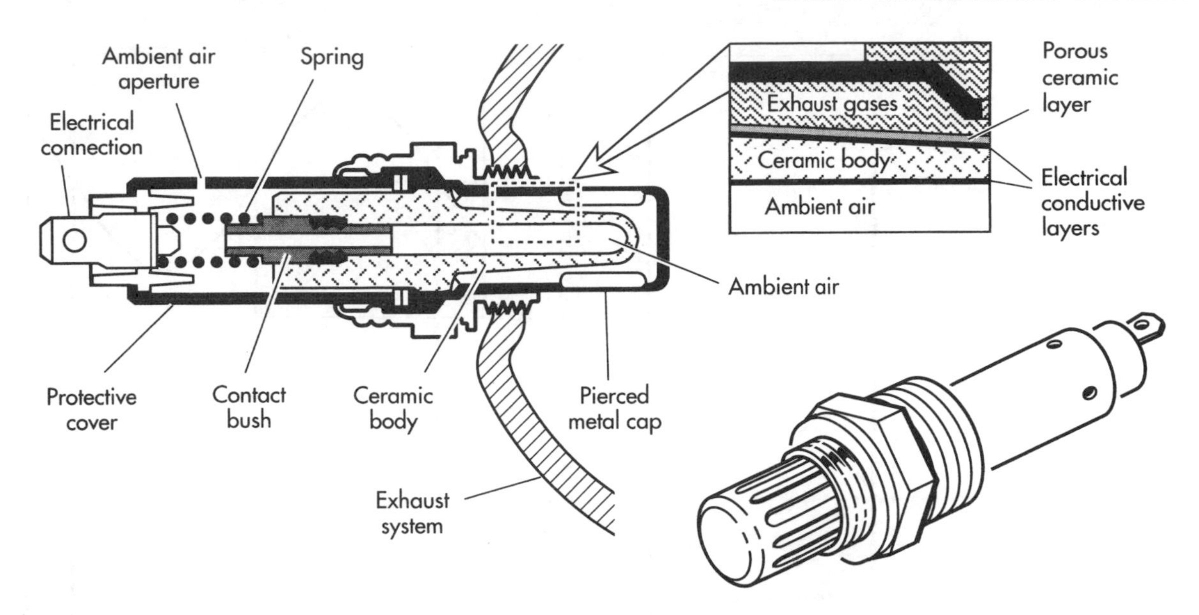

Figure 5.45 Cross-section of the zirconia exhaust gas oxygen sensor (*Lucas*)

Zirconia EGO Sensor

Figure 5.45 shows a cross-section of a zirconia EGO sensor. It consists of a yttria (Y_2O_3) stabilized zirconia (ZrO_2) ceramic tube which functions rather like a small battery in that it generates a voltage according to the ratio of oxygen in the exhaust gas to that in the atmosphere. The inner and outer surfaces of the zirconia ceramic tube are each coated with a 10 μm thick gas-permeable platinum layer which acts both as an electrode and oxidizing catalyst. The outer layer is exposed to the exhaust gas and so is protected from erosion by a 100 μm layer of porous ceramic and a perforated metal cap. The inner layer is exposed to the ambient air and is enclosed by a metal cover.

During operation at high temperatures (300–900°C) the zirconia ceramic allows the passage of oxygen ions between the two platinum electrode layers; a characteristic that can be exploited to produce a small voltage.

When the outer electrode layer is exposed to the emissions of a rich A/F mixture combustion, there is very little free oxygen in the exhaust gas around the sensor. Furthermore any oxygen present at the exhaust electrode layer is immedately combined with CO (to give CO_2) by the catalytic action of the platinum. The oxygen concentration at this layer is therefore extremely low, in contrast with that at the inner (atmospheric) layer where it is very high. Oxygen atoms contacting the atmospheric electrode gain four electrons and travel through the zirconia ceramic to the exhaust electrode where they then shed the extra electrons, thus leaving positive charge on the atmospheric electrode and negative charge on the exhaust electrode. Through this mechanism a small voltage of about 0.8 V is generated by the sensor.

Conversely, when the outer platinum electrode is subjected to the emissions of the combustion of a lean mixture the concentration of free oxygen in the exhaust gas is large. Even the oxidizing action of the platinum cannot significantly diminish it and so there is little difference in oxygen ion concentration across the zirconia tube. This situation produces very little electromotive force and so the output from the sensor is close to 0 V. The output of the Zirconia EGO sensor is therefore a step voltage change from about 0.8 V to about 0 V, at the stoichiometric mixture point (Figure 5.46). At the ideal operating temperature of 600°C the sensor will react to mixture changes in less than 50 ms.

Titania EGO sensor

Titania (TiO_2) is a high-temperature semiconductor which exhibits a resistance which is dependent on oxygen concentration. A titania EGO sensor uses a small bead of n-type TiO_2 supported in the exhaust pipe on a ceramic former. When the exhaust-gas oxygen concentration is low (rich mixture) the resistance of the titania bead is also low (a few ohms). If the oxygen concentration is sharply increased (lean mixture) the resistance of the sensor also sharply

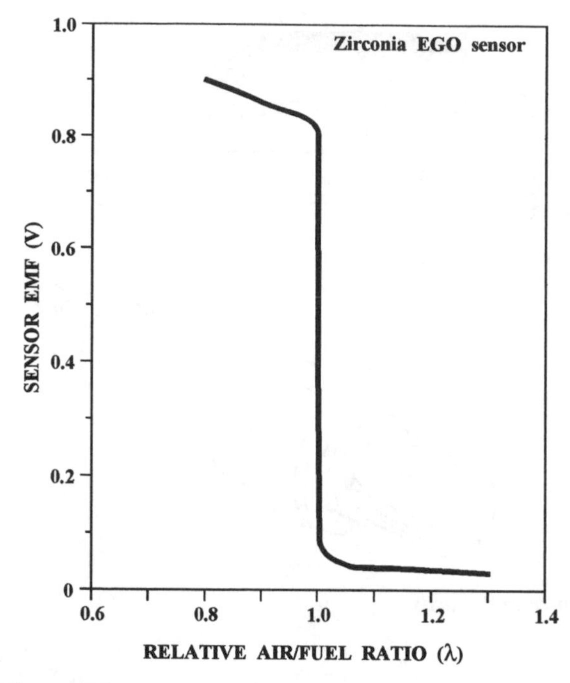

Figure 5.46 Output emf as a function of relative air–fuel ratio for the zirconia EGO sensor

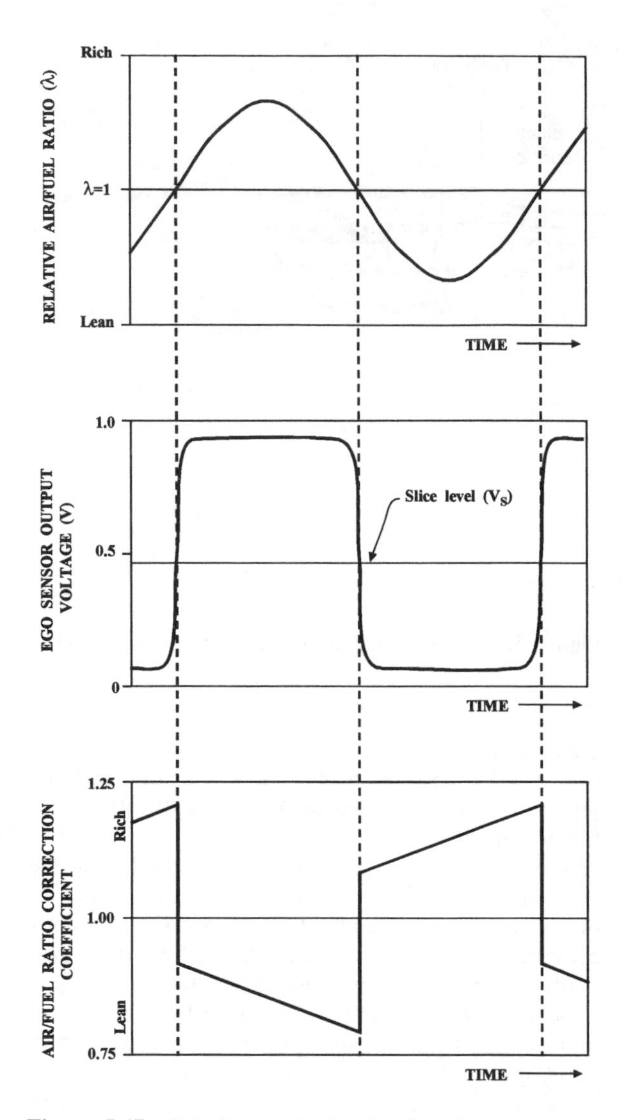

Figure 5.47 Relative air–fuel ratio of engine, sensor output voltage and ECU air–fuel correction coefficient

increases (to in excess of 10 kΩ). The sensor is therefore connected in series with a 50 kΩ resistor to form a potential divider which gives a step change in voltage at the stoichiometric point.

Since both zirconia and titania sensors operate only at temperatures above about 300°C it is necessary to either mount them in the exhaust manifold where the gases are still very hot, or to fit an electrical heating element into the sensor body. This quickly brings the sensor up to working temperature and enables stoichiometric mixture control to commence within a few minutes of starting a cold engine.

Stoichiometric mixture control software

The signal from the EGO sensor is routed to the engine management ECU where it is compared with a fixed reference voltage called the 'slice level', V_s, which is set mid-way between the 'rich' and 'lean' voltage levels; i.e. about 0.45 V for a zirconia sensor.

When the EGO sensor output is above the slice level the ECU's software judges that the mixture is rich and so provides a step reduction in injection duration correction coefficient to correct for this. The sensor signal does not respond at once to this change and remains higher than the slice level for some time, during which the mixture is gradually weakened yet further. The longer the sensor indicates a rich mixture, the greater the reduction in injection duration. As a result of this action the sensor output eventually falls below the slice level, indicating to the ECU that the mixture is now weaker than stoichiometric. The mixture control software thus applies a step increase to the injection duration correction coefficient to enrich the mixture and then progressively adds further enrichment until the sensor output switches once again to indicate an excessively rich mixture. Thus the mixture continuously oscillates 'rich' and 'lean', even under steady driving conditions (Figure 5.47). This behaviour is called limit cycle operation, with the frequency, f_{LC}, of the rich–lean switching cycle being given by the equation:

$$f_{LC} = 1/4t_L$$

where t_L is the time lag for the fuel to travel from the injector through the engine to the EGO sensor. For most engines f_{LC} lies in the range 0.5 Hz to 2 Hz at idle speed. It is important to appreciate that although the mixture oscillates between 'rich' and 'lean' it actually *averages* to stoichiometric and since the catalytic converter has considerable exhaust gas capacity it converts at this *average* A/F ratio.

Over a short period of time the A/F ratio correction coefficient should average to 1.0, thus indicating that the data stored in the engine's map is valid for stoichiometry. If, over a period of time, the average A/F ratio correction coefficient is larger or smaller than 1.0 then the ECU's software may 'learn' that there needs to be an offset applied to the map data to take account of changed engine conditions. This offset is stored in the ECU's memory and is retained even when the ignition is switched off.

5.9 ADDITIONAL ENGINE MANAGEMENT FUNCTIONS

Pulse air injection

Pulse air injection systems are fitted to some vehicles to provide a supply of fresh air to the catalyst just after engine starting when the engine is operating with an enriched mixture. This helps oxidize unburned hydrocarbons within the catalyst, enabling it to quickly reach operating temperature and thereby reducing pollutant emissions from a cold engine. The pulse air injection system operates by having the engine management ECU actuate a solenoid valve to open a passage between the inlet and exhaust manifolds. Pulsations of the gas in the exhaust manifold then cause fresh air to be drawn into the exhaust system and react in the catalyst. As soon as the engine warms up the air injection valve is closed.

Electrically heated catalysts

Electrically heated catalysts (EHCs) have been developed to solve the problem of high pollutant emissions that prevail in the first 30–60 seconds of operation after a cold start. Such devices will increasingly be required as emission regulations become more severe and strategies such as pulse air injection are no longer feasible.

Early prototype EHCs consisted of normal catalyst structures that were heated by a current of 300 A for the first 30 seconds of engine operation. By thermally isolating the catalytic element from its container with an air-gap and by mounting it on tiny 'pin' supports, this heating requirement has been reduced to 50–100 A for 5 seconds.

Newly-developed EHCs have a small volume and are mounted directly up-stream of the main three-way catalyst, which operates when the engine has warmed up.

The EHC current is switched on by the ECU, just as the engine fires, and heats up the catalyst element which soon gains further warmth as a result of the heat-releasing chemical reactions that take place. An operating temperature of about 400°C is obtained 15 seconds after engine start, with the catalyst reaching 90% efficiency after about 40 seconds at which time the heating current can be switched off.

Variable intake air-flow control

During the normal operation of an engine the continuous motion of the intake air flow gives a 'ram' effect and so helps achieve purging of exhaust gas from the cylinder and therefore optimum cylinder charging. At low engine speeds the intake air velocity must be high to overcome exhaust gas pressure in the cylinder, this can be achieved by having a long and narrow intake tract. Conversely, at high engine speeds the intake air already possesses a high velocity and so it is desirable to have a short, large diameter, intake tract to reduce flow resistance. These conflicting requirements can be satisfied by using two intake tracts for each cylinder. A main intake tract and a second ('bypass') tract, fitted with a disc valve whose position is electronically controlled by the engine ECU according to engine speed, as illustrated in Figure 5.48. At low speed the bypass valve is kept shut, forcing all of the intake air to flow at high velocity through the main tract. At high speed the valve is held fully open to provide least resistance to intake air flow. At medium engine speed the valve is slightly opened to prevent a dip in output torque at the transition between the two modes of operation. The bypass valve can be actuated either by a solenoid-controlled vacuum diaphragm or by a servo-motor, the latter giving a more precise control of valve position.

Evaporative emission control system

The evaporative emission control system is used to draw-off and burn the HC emissions from the vehicle's fuel tank. A typical arrangement is illustrated in Figure 5.49.

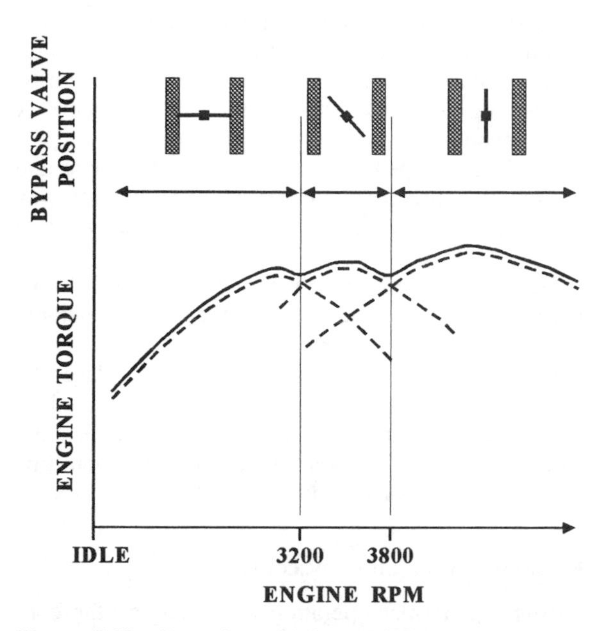

Figure 5.48 Operation of inlet manifold tract bypass valves used to improve engine torque characteristics

Evaporative fumes are drawn from the fuel tank, pass through a check valve and are temporarily stored in a charcoal canister. The quantity of fumes drawn into the engine and burned is regulated by a duty-cycle controlled solenoid valve in accordance with pre-programmed values stored in the ECU's memory which correspond to a variety of engine operating conditions.

The solenoid valve is opened when the following conditions are met:

(i) warm engine;
(ii) driving in a gear;
(iii) engine above idle speed;
(iv) oxygen sensor functioning normally.

5.10 LEAN-BURN ENGINE CONTROL

Pressure to reduce CO_2 emissions (which are not presently regulated) may increase the demand for more fuel-efficient engines. This could provide stimulus to lean-burn research, which potentially offers greater fuel efficiency. Operating an engine with lean-of-stoichiometric mixtures (A/F ratios of 16:1–25:1) means that the fuel burns in an excess of oxygen and this can provide fuel consumption improvements of at least 10%, together with drastic reductions in CO and NO_X emissions. Since a lean-burn engine is operating close to the limits of combustion its full exploitation requires good mixture preparation, a very high energy spark to ignite the weak mixture and very good monitoring of combustion quality and mixture strength via a closed-loop system. Many of these concepts are incorporated into recent Japanese-designed engines, such as Honda's VTEC-E and Toyota's Carina-E. These engines can typically run at A/F ratios of 22:1, meet European and US emission regulations and offer fuel consumption improvements of up to 25% under cruising conditions. A vital component of their control systems is a type of EGO sensor (called a universal exhaust gas oxygen sensor, or UEGO sensor) which gives an output current that varies smoothly in proportion to exhaust oxygen content.

Operation of the UEGO sensor

The UEGO sensor is a development of the conventional heated zirconia EGO sensor, however in addition to detecting the stoichiometric point it can measure a wide range of A/F ratios from very rich (10:1) through to very lean (35:1).

The construction of one type of sensor is shown in Figure 5.50. It consists of two oxygen transfer cells; an oxygen pumping cell (I_P cell) and an oxygen detecting cell (V_S cell). The V_S cell is supplied with a small constant current which moves a tiny amount of oxygen to the right and keeps the O_2 cavity filled with oxygen. This oxygen acts as a 'reference gas' for the sensor.

Exhaust gas enters the detecting cavity and a voltage is developed across the V_S cell, depending upon the oxygen concentration in the exhaust gas. The I_P cell then controls the oxygen concentration in the detecting cavity by pumping oxygen to, or from, the atmosphere to keep the V_S voltage at a steady 0.45 V. The pumping current, I_P, is therefore a measure of the air–fuel ratio of the exhaust gas.

Combustion monitoring

In-cylinder sensors can provide valuable data on the timing and quality of the combustion process, which is particularly important when engines are operated with lean mixtures or heavy EGR. Techniques available include direct pressure measurement with a pressure sensor (used on some Japanese-market vehicles) and indirect monitoring via spark-plug ionization current. The 'spark-plug voltage ion method' detects the ion density within the combustion chamber by measuring the decay time of voltage after a spark (Figure 5.51). Using this technique, the limits of lean-burn or EGR can easily be detected.

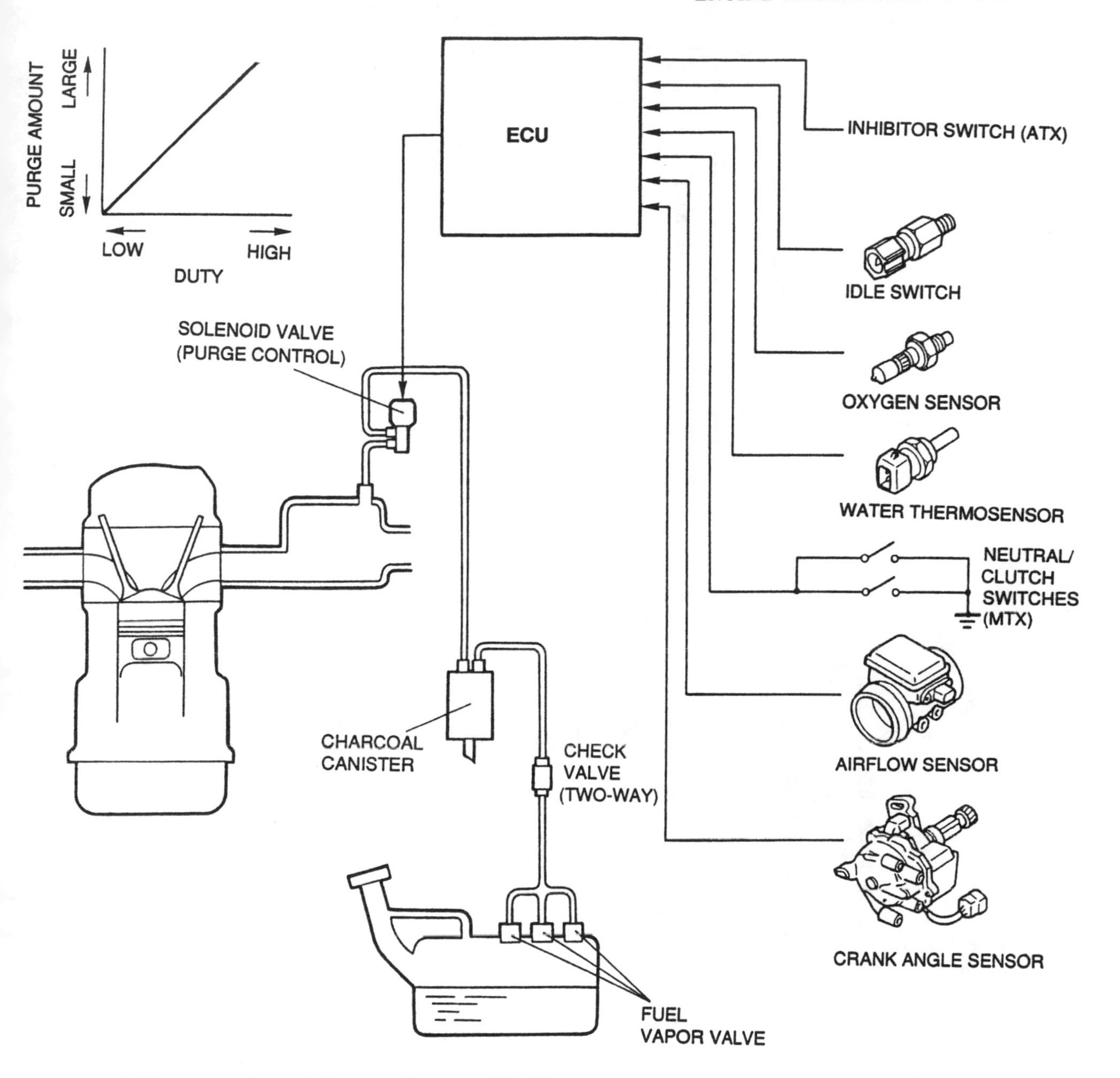

Figure 5.49 Evaporative emission control system (*Mazda*)

Another (much cheaper) option is to derive a combustion quality value from piston acceleration. This can be obtained from the measurement of crankshaft speed fluctuations using the crankspeed sensor.

Toyota lean-burn engine

Figure 5.52 illustrates a lean-burn engine control system developed by Toyota. It features in-cylinder sensors to monitor combustion pressure, and a sequential injection system which can control fuelling on a cylinder-by-cylinder basis. The A/F ratio for the engine is monitored using a wide-range oxygen sensor mounted in the exhaust down-pipe.

A novel design feature is the use of a *swirl control valve* (SCV) fitted in the intake system. The intake tract for each cylinder is divided into two; one passage is smooth to allow maximum gas flow and good cylinder

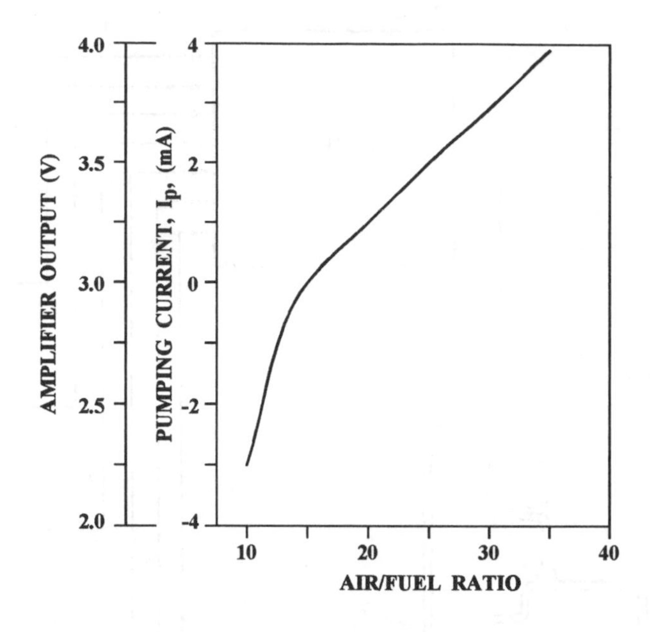

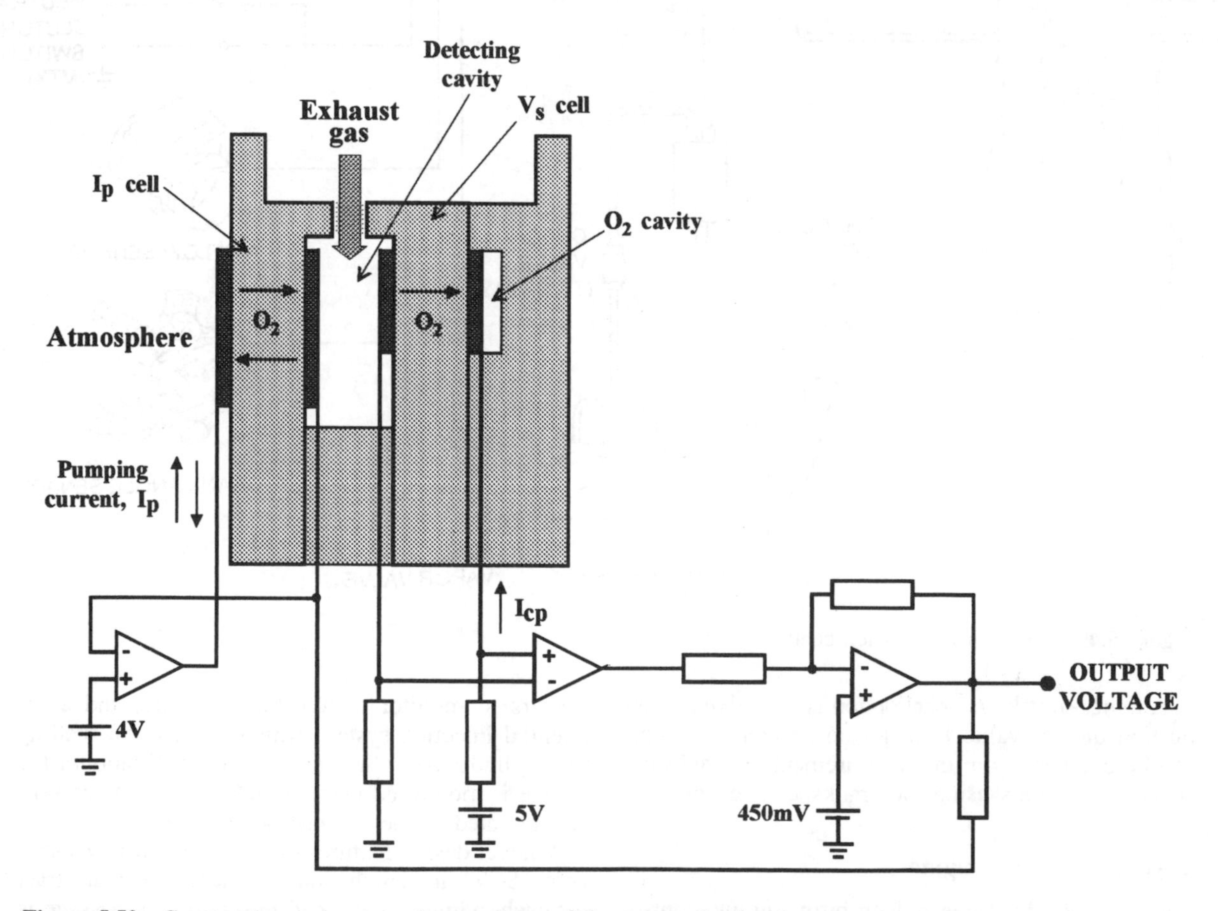

Figure 5.50 Construction and output characteristics of the universal exhaust gas oxygen sensor

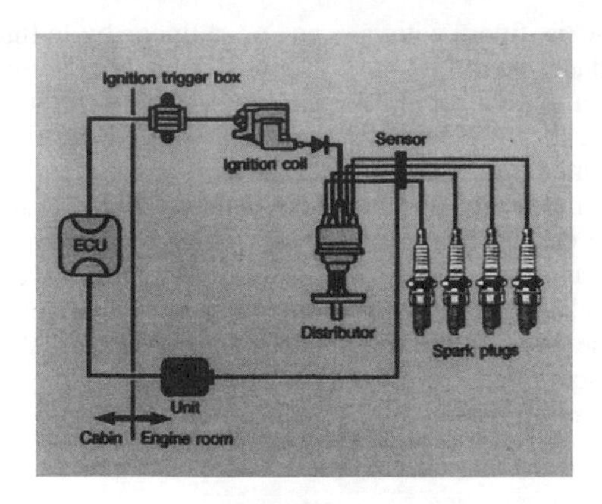

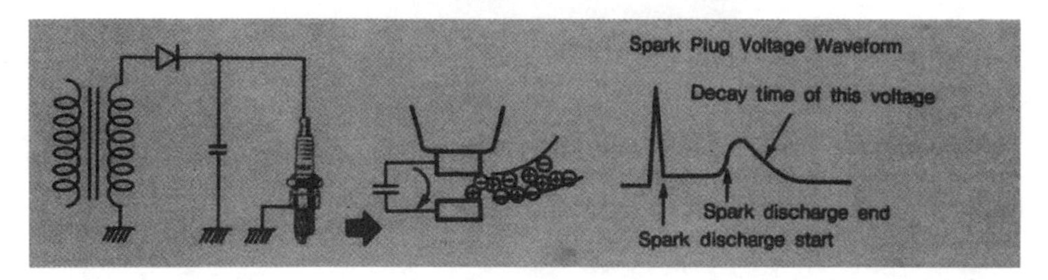

Figure 5.51 Combustion quality measurement using spark-plug ionization current method (*NGK*)

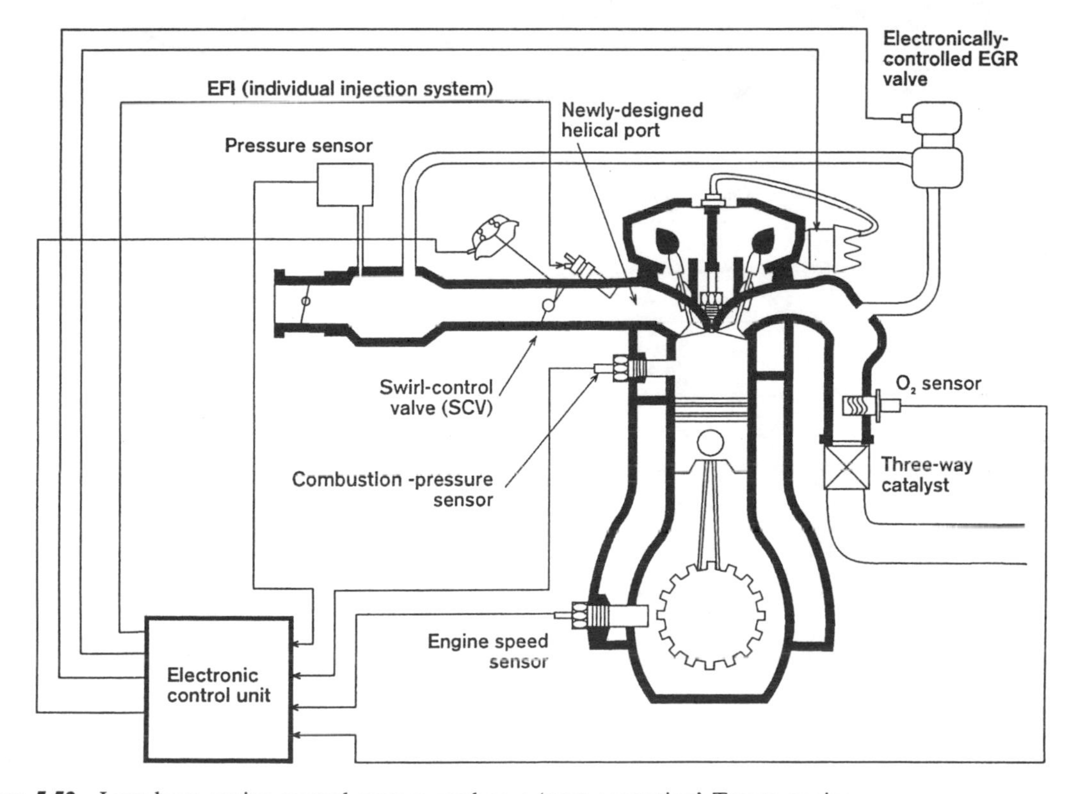

Figure 5.52 Lean-burn engine control system used on a 'next generation' Toyota engine

charging, whilst the other passage is fitted with a corkscrew-shaped flange that introduces swirl into the incoming air. The engine ECU can switch air flow between the two tracts by actuating the SCV. When the engine is running under light to medium load conditions it operates in a lean-burn mode. The ECU commands the SCV to open-up the curved inlet passage so that the incoming vapour has a high level of turbulence and is therefore very well mixed. During combustion, the ECU monitors the pressure sensor signals for any signs of slow or incomplete burning and then modifies the A/F ratio or ignition timing on a cylinder-by-cylinder basis, as appropriate. Coupled with continuous monitoring of the exhaust gas oxygen concentration, these facilities allow the engine to work with high efficiency at A/F ratios up to 25:1.

When the engine is required to work under conditions of high load, perhaps during overtaking or when hill-climbing, the ECU switches to stoichiometric operation for maximum power. This is achieved by commanding the SCV to open the smooth intake passage and by using the oxygen sensor signal to maintain an A/F ratio of 14.7:1.

6

Electronic transmission control

6.1 INTRODUCTION

The idea of using automatic and semi-automatic gear-changing systems to lessen driver fatigue is almost as old as the car itself. Early motorists were confounded by heavy and obstructive crash gearboxes, leading many car designers to investigate ways of providing a more satisfactory means of changing gear.

German engineers were particularly active in the area of transmission design, and during the period up to World War I a variety of gearless drive systems were developed. The 1901 Manurer-Union car used two friction discs, mounted at a steep angle to each other, to create a stepless transmission system. An idea that is known today as continuously variable transmission (CVT). The driver still had to effect ratio changes by manually moving the edge of the driven disc radially across the drive disc, but was relieved of the fatiguing task of manipulating gears, clutch and throttle.

As the power output of engines increased, crude friction drives proved unreliable and so a variety of semi-automatic transmissions were developed, mostly based around conventional stepped-ratio gearboxes, but with automated operation of the clutch. Unfortunately most of these experiments were short-lived, due largely to poor reliability and inadequate materials.

American car manufacturers, in particular, were eager to eliminate the need for gear changing on their luxury vehicles, and it was with this in mind that General Motors Corporation established a new division to develop a type of fully-automatic transmission which they called Hydramatic. Hydramatic utilized variations in hydraulic fluid pressure to control an epicyclic gear train that was driven from the engine through a fluid coupling. The transmission used a low gear to start the vehicle moving, before shifting to a higher gear for cruising. The project was spectacularly successful and the world's first truly automatic transmission, with two self-selecting forward gears, was available on the 1939 Oldsmobile saloon.

Competition between manufacturers resulted in rapid improvements in transmission design, and three-speed units were soon developed. Since these transmissions were ideally suited to American motoring conditions the market for automatics grew strongly, so that today only one in ten cars sold in the US is equipped with a manual gearbox.

In Europe automatic transmissions have never been particularly popular and currently only account for about 10% of the market. This is partly because automatic gearboxes are less efficient than their manual counterparts, giving rise to a noticeable deterioration in vehicle performance when coupled with the small capacity engines that are common in Europe. Another factor is that European roads tend to be less well graded than those of North America, and so the additional driver control afforded by a manual gearbox is often to be valued. Finally, automatic transmissions are generally only available as an extra-cost option, with a price premium of around 10% of the total vehicle cost, and this often deters potential purchasers.

The impact of electronics on transmission control

The potential of electronics to offer enhanced control of semi- and fully automatic transmissions has long been realized. Many transmission manufacturers, including AP (in the UK), ZF (in Germany), Renault (in France) and, naturally, the US car makers, were experimenting with electrohydraulic transmission controls as early as the mid-1960s. These early analogue systems had no processing power but simply used transistors to switch current to solenoid valves and actuators on the transmission casing. The first such system to enter series production was available on a 1968 model Renault. Toyota and Nissan soon followed with their own versions in 1970. Since they were expensive and offered few advantages to the driver they failed to gain popularity.

The arrival of microcomputers allowed much more sophisticated control of automatic transmissions and enabled Toyota to introduce one of the first computer controlled automatic gearboxes on their 1982 model year cars. In 1983 the evolution of powertrain control electronics was continued in Europe when Bosch announced their Motronic system, which incorporated both engine and transmission control in one unit.

Today, sophisticated microprocessor control is a standard feature of almost all passenger car automatic transmissions, including those using CVTs; additionally the first truly driver-friendly semi-automatic transmissions have now been made possible.

6.2 ELECTRONICALLY CONTROLLED SEMI-AUTOMATIC TRANSMISSION

The rationale behind the development of electronically controlled semi-automatic transmissions is to produce a gearbox with the efficiency and controllability of a manual transmission, but the ease of driving offered by a clutchless automatic. In all cases, semi-automatic transmissions have been based around layshaft or epicyclic-type gearboxes, with automated actuation of clutch engagement and disengagement.

Early designs of the 1960s and 1970s, such as the Manumatic gearbox (developed by AP and subsequently installed in some Hillman cars) and the system developed by NSU for their Wankel-engined Ro80, used a microswitch under the gear lever knob to detect the driver's hand-pressure and hence his intention to change gear. When the switch was operated, a solenoid valve was energized to allow either vacuum or hydraulic pressure to actuate the clutch mechanism. Since clutch control was of a rather 'on–off' nature, these transmissions were not particularly smooth and therefore fluid couplings were sometimes used to cushion the drive. Deficiencies in the control electronics remained, however, and these systems made little impact on the market.

Electronic clutch control

During the late 1980s renewed interest in semi-automatic transmissions, coupled with the ready availability of cheap and powerful microprocessors, led to the development of low-cost semi-automatic gearboxes with good performance. In Europe, the Valeo company in France, Fichtel and Sachs in Germany, Magneti Marelli in Italy and AP Borg and Beck in Britain have all developed fully commercialized systems that enable semi-automatic operation of a conventional layshaft gearbox. At the heart of all of these systems is a microcomputer-based electronic control unit that monitors engine and transmission speeds, as well as detecting the positions of the accelerator pedal and gear lever. Commands issued by the control unit then energize an electric, vacuum or hydraulic actuator to engage or disengage a conventional dry-plate clutch. In this way the driver can have all the advantages of a manual gearbox without the fatigue of manipulating the clutch. This basic idea has been modified to give a variety of semi-automatic control strategies, as detailed below.

Electronic clutch control

This is the basic semi-automatic system described above. With the vehicle at rest, the driver simply engages a gear and depresses the accelerator pedal to move away. Once moving, gear changes are effected simply by moving the gear lever, permitting very fast gear changes. Clutch operation is fully automated, but during the gear-change interval the driver must still control engine speed with the accelerator pedal. When the vehicle is slowed, the clutch is not disengaged until the vehicle is almost at a standstill. This maximizes engine braking but avoids a stall. Among the first vehicle applications of electronic clutch control are the Ferrari Modial (introduced in 1993 with the Valeo TEE2000 system), Renault Twingo Easy (introduced into Europe during 1994 with an AP Borg and Beck system) and Volkswagen's Golf diesel Ecomatic, also introduced in 1994.

Electronic clutch and engine speed control

This is an enhancement to the system described above. The accelerator cable is replaced by an electrical connection between a position sensor mounted on the accelerator pedal and a servomotor operating the engine's throttle disc. Normally, the throttle disc is actuated directly at the command of the driver, but during gear changing the control unit intervenes and takes command of engine speed to synchronize it with transmission speed for the newly selected gear. This strategy ensures smooth and comfortable clutch re-engagement and minimizes clutch wear.

Remote gear selection

With this system the features described above are enhanced by the addition of a remote gear selector. The usual gear selection linkage is replaced by an electrical connection between the gear change switch (usually a small lever or a push-button) and a transmission-mounted electrohydraulic actuator. Gear changes are requested simply by flicking the small lever forwards or backwards to shift down or up a gear. Since the gear lever is simply an electrical switch it can be sited in a location best suited to the driver and requires only a small operating pressure.

Automatically shifted manual transmission

Having designed a transmission that is equipped with electronically controlled clutch and remotely selected gears it is comparatively straightforward to entirely automate the gear-selection process. The result is a small, light and highly efficient automatic transmission,

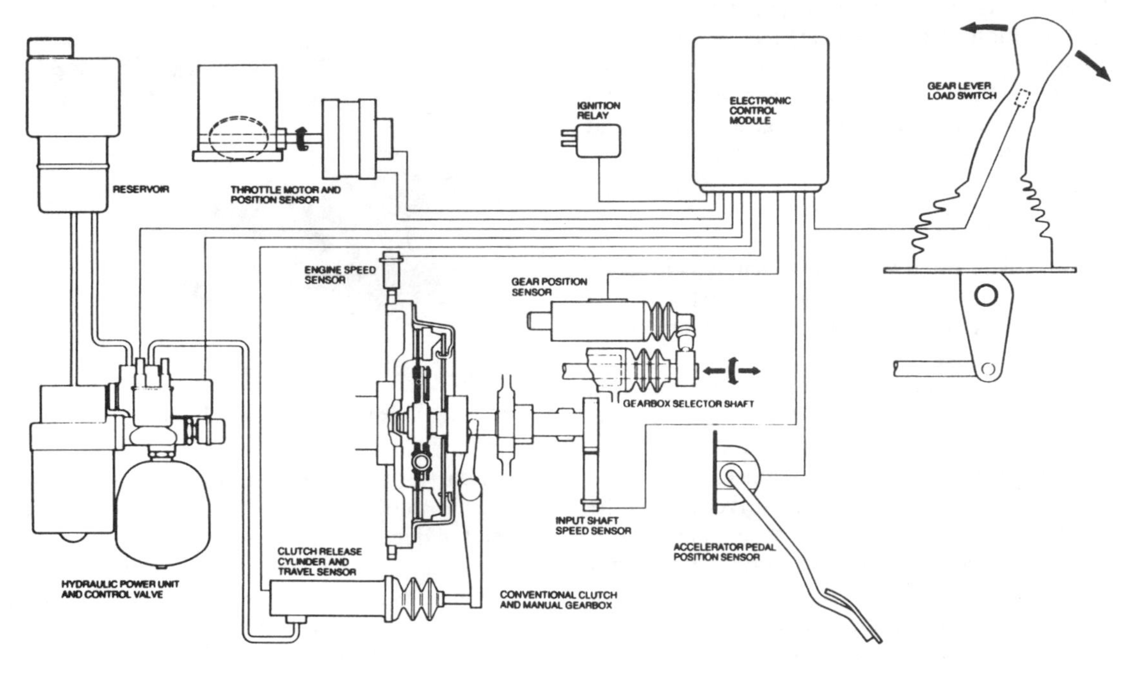

Figure 6.1 Functional diagram of Automatic Clutch and Throttle System (ACTS) semi-automatic transmission (*AP Borg & Beck*)

well suited to small-capacity cars for which the provision of a conventional automatic transmission (with torque-converter and epicyclic gears) would cause an excessive performance degradation.

The AP Borg and Beck ACTS semi-automatic transmission

To illustrate the operation of a semi-automatic transmission the Automatic Clutch and Throttle System (ACTS) developed by AP Borg and Beck will be examined in detail (Figure 6.1). ACTS is a system that allows the driver to change gear using a conventional 'H'-gate lever, but with automated control of clutch and engine-speed during the gear-changing process.

ACTS is comparatively simple and low-cost in that it uses a conventional manual transmission, but with the addition of a sophisticated electronic control unit and a variety of sensors and actuators, specifically:

(i) An accelerator pedal position sensor.
(ii) A throttle disc actuator with a built-in position sensor.
(iii) A gear lever load switch to detect the driver's hand pressure on the knob.
(iv) A gear position sensor.
(v) A magnetic pickup to sense engine speed.
(vi) A magnetic pickup to sense gearbox input shaft speed.
(vii) A controllable source of hydraulic energy consisting of a fluid reservoir, electrically driven pump, pressure accumulator and electro-hydraulic control valve.
(viii) A clutch release cylinder with a built-in position sensor.

The hydraulic system is used to operate a clutch release cylinder that engages and disengages the clutch via a conventional release lever. Pressure to the release cylinder is controlled by a solenoid valve mounted on the hydraulic power unit. This valve is operated by the controller in response to signals from the sensors. A servo loop based on the release lever travel sensor and the engine and transmission speed sensors provides fine control of the clutch friction plate speed gradient, ensuring very smooth engagement.

The throttle motor is fitted with a feedback potentiometer to report throttle opening to the control unit. During normal driving the motor's position is controlled by the driver via another potentiometer, mounted on the accelerator pedal. During gear changing, however, the microcomputer intervenes to temporarily modify the engine speed. A switch in the gear lever detects the driver's hand pressure (and therefore his intention to change gear) and so signals the controller to disengage the clutch. Once a gear change has been made the gear position sensor sends a digital code to the controller to report the gear selected, and a decision is then made to re-engage the clutch.

Operation of ACTS

To start a vehicle fitted with ACTS safety considerations demand that the transmission must first be placed in neutral to close the inhibitor switch, permitting operation of the starter motor.

When the engine starts, the clutch cylinder position sensor finds the current release lever position (thus compensating for clutch wear) and the controller establishes an engine idle-speed reference.

With the vehicle still at rest, the driver moves the gear lever to engage first gear, so causing the controller to disengage the clutch. It remains disengaged until the driver depresses the accelerator, causing the clutch to re-engage under microcomputer control. During re-engagement the controller continuously modifies the throttle disc and release lever positions to ensure a smooth take-up whilst simultaneously maintaining the engine speed requested by the driver. In this manner it is possible to perform starts with a high engine speed (for hill-starts or rapid acceleration) or with a low engine speed (for leisurely acceleration). If the driver initially selects an inappropriate gear then this is detected by the microcomputer; a warning is issued to the driver (via a buzzer) and clutch engagement is inhibited until a suitable gear is engaged.

With the vehicle in motion, the driver performs gearshifts simply by moving the lever to the desired gear position. The microcomputer then momentarily takes command of clutch position and engine speed. Engine speed is modified depending on whether the gear position sensor detects an upward gearshift or a downward gearshift.

When an upshift is detected, and the microcomputer determines that the engine speed is greater than that of the transmission input shaft, the clutch can be engaged immediately and engine-speed control remains with the driver. When a downshift is detected, the engine speed is usually below that of the transmission input shaft and so the microcomputer delays clutch re-engagement and commands the throttle motor to momentarily increase engine speed, matching it with that of the transmission input shaft. Once the clutch has been re-engaged engine-speed control reverts to the driver.

If the vehicle is operated down to a very low speed, the engine revs fall and the microcomputer disengages the clutch to prevent a stall. The clutch is then not re-engaged until a suitable gear is selected and the accelerator depressed.

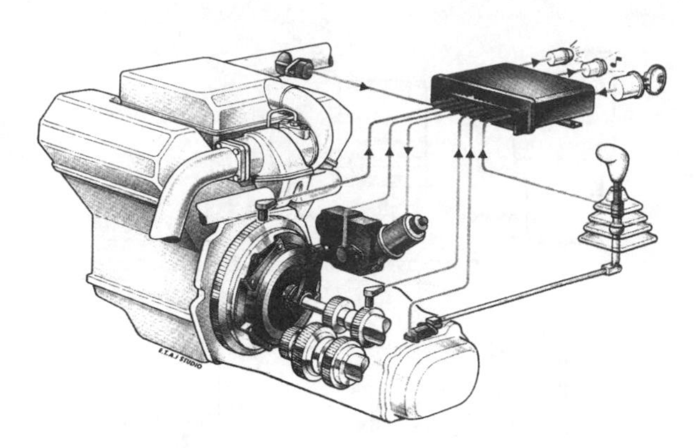

Figure 6.2 Valeo electronic clutch transmission (*Valeo*)

As a means of reducing drive-line shunt in 'stop–go' driving it is possible for ACTS to allow an amount of controlled clutch slip. Any large discrepancy between the actual engine speed and the driver-demanded engine speed is detected by the microcomputer and a limited amount of slip introduced accordingly.

The Valeo electronic clutch transmission TEE 2000

Valeo's TEE 2000 system (Figure 6.2) has many elements in common with AP Borg and Beck ACTS. It is somewhat less complex in that it does not intervene to adjust engine speed during gear changes and so requires a little more skill from the driver. A noteworthy feature is the use of an electromechanical (rather than electrohydraulic) actuator system for clutch operation. The actuator is composed of an electric motor that drives the release arm through a worm reduction gear. To reduce the power demand on the motor the actuator incorporates a strong spring, working against the clutch diaphragm spring, to counterbalance the disengagement load. Even with a relatively small motor, clutch actuation takes less than one-fifth of a second.

Volkswagen Ecomatic system

An interesting development of electronic clutch control, the Ecomatic system, has been commercialized by Volkswagen for use on certain diesel-engined cars. In Volkswagen's case, the clutch is operated by a vacuum servo unit controlled by the Ecomatic ECU via a solenoid valve.

Ecomatic is a semi-automatic transmission that

improves fuel economy and reduces exhaust emissions by disengaging the engine from the transmission when the driver eases back on the accelerator to slow down. When the driver entirely releases the accelerator pedal for more than about two seconds, for example in a traffic queue, the engine is switched off and the clutch disengages.

Whilst the engine is not running a 50% oversize 92 Ah battery, charged by a 90 A alternator, is used to provide power for normal electrical loads plus the electrically driven vacuum, coolant and power-steering pumps.

When the driver requires engine power again, simply engaging first gear and depressing the accelerator automatically operates the starter motor and engages the clutch to allow the vehicle to move off. Diesel engines are particulary well-suited to this type of operation since when warm they generally start on the first turn of the crankshaft – petrol engines normally require several turns.

The Ecomatic adds about 40 kg to the weight of a standard vehicle, but with a 1.9 litre diesel engine it offers a claimed 22% lower urban fuel consumption, 36% reduction in CO emissions, 22% reduction in CO_2 and 25% in HC and NO.

6.3 ELECTRONICALLY CONTROLLED AUTOMATIC TRANSMISSIONS

An automatic transmission is a type of gearbox that selects the most appropriate gear ratio for the prevailing engine speed, powertrain load and vehicle speed conditions, without any intervention by the driver. All gear-shifting is undertaken by the transmission system itself, the driver merely selects the desired operating mode with the selector lever. The Society of Automotive Engineers (SAE) recommends that the selector be sequenced **P R N D 3 2 1** (in the case of a four-speed transmission) where the letters have the following meanings:

P (Park) In this mode the transmission is in neutral and the transmission output shaft is locked by means of a 'parking pawl'.

R (Reverse) A single-speed reverse gear is selected and held. Engine braking is effective.

N (Neutral) The same as Park, but the output shaft is not locked.

D (Drive) This is the normal gear selection for forward motion. The vehicle may be operated from a standstill up to its maximum speed, with automatic upshifts and downshifts being made by the gearbox depending upon its assessment of vehicle speed and engine load. When rapid acceleration is required for overtaking, the driver can push the throttle pedal to its full travel to invoke a speedy 'kick-down' downshift into a lower gear.

3 (Third) Operation varies between manufacturers; in general the transmission operates as in **D** range, but is prevented from upshifting into fourth gear.

2 (Second) Again, operation varies between manufacturers, but normally the transmission can only operate in first and second gear. **2** is usually selected to provide engine braking when driving in hilly country or when towing.

1 (First) Locks the transmission in first gear to provide powerful engine braking. Used when driving on steep hills or when towing.

To prevent the vehicle being inadvertently started in a gear, a gearbox 'inhibitor switch' (sometimes called a 'neutral switch') is wired in series with the starter motor solenoid supply. The inhibitor switch contacts are closed only when the selector lever is in Park or Neutral and so the engine may be started only in these positions. Additional safety is provided by having the selector lever fitted with a mechanical interlock that prevents, for example, the lever being moved out of Park unless a spring-loaded release button is pressed.

Although automatic transmissions are usually less fuel-efficient than their manual counterparts, they do offer many driving advantages, especially in urban road conditions:

(i) Driver fatigue is reduced since there is no clutch or gear lever to manipulate. This is very significant when driving in dense traffic.
(ii) Both hands can remain on the steering wheel at all times, so increasing safety.
(iii) Since the transmission always engages the correct gear for the prevailing driving conditions, the possibility of labouring or over-revving the engine is eliminated.

The use of a microcomputer control system enhances the performance of the automatic transmission by precisely controlling the hydraulic system, offering:

(i) Crisp and smooth gear shifts with consistent quality.
(ii) Perfectly timed gear shifts.
(iii) Elimination of 'hunting shifts'.

(iv) Protection of the transmission by constant monitoring of engine and transmission speed, temperature and so on.
(v) Driver-selectable shift pattern options for extra performance or economy, or for icy road conditions.
(vi) A simplified hydraulic control system.

An additional advantage of electronic control is the ability of the microcomputer to store diagnostic trouble codes, greatly assisting in the speedy repair of faulty transmission units.

Component parts of an automatic transmission

Figure 6.3 shows a sectional view of an electronically controlled automatic transmission designed for front-wheel drive cars. Although transmission designs vary, the major components shown are common to all. In particular the transmission comprises:

(i) *The torque converter*. The transmission bell-housing is bolted to the rear of the engine block and encloses the torque converter, which is secured to the engine flex-plate (a lightweight flywheel) by several small bolts. The torque converter is a virtually wear-free fluid coupling which multiplies and transfers engine torque to the transmission's geartrain via the input shaft.

(ii) *The friction elements*. To obtain different gear ratios, various hydraulically operated brake-bands, multiplate clutches and multiplate brakes are engaged to couple or lock the appropriate sets of planetary elements.

(iii) *The geartrain*. The geartrain is usually of the compound epicyclic form, designed to give several different ratios in a compact configuration. Commonly used gearsets include the Simpson, Ravigneaux and Wilson types. Variable-reluctance type sensors are located in the transmission housing to monitor the input (turbine) and output shaft speeds.

(iv) *The oil pump*. The hydraulic pressure required to operate the various clutches and bands is produced by an oil pump, located just behind the torque converter, driven by the engine via the torque converter housing. Oil pump output pressure (known as line pressure) is regulated by an electronically controlled solenoid valve and directed to the appropriate clutches and bands by shift solenoids. The solenoid valve precisely modulates line pressure during gearshifts to ensure smooth and rapid ratio changes.

(v) *The electrohydraulic control unit*. Often known simply as the 'valve body', the electrohydraulic control unit is generally located beneath the geartrain and houses a number of mechanical and solenoid valves which direct and modify the hydraulic fluid flow to the various clutches and brake-bands. In an electronically controlled transmission it is the solenoid valves that execute gearchange commands issued by the microcomputer in the transmission control unit (transmission ECU). The transmission ECU thus controls the sequence and timing of ratio changes, and also the quality of the changes.

(vi) *The transmission housing*. The transmission housing is a lightweight aluminium casting that locates all of the transmission components as well as various sensors that measure shaft speeds, oil temperature and so on. It is usually designed so that the valves and sensors can be replaced while the transmission is still installed in the vehicle. At the base of the housing a sump pan is kept filled with transmission oil (more commonly known as ATF – automatic transmission fluid), basically an SAE 20 grade mineral oil with various additives to improve its frictional and low-temperature properties. Various pressure taps are provided on the side of the housing to enable service technicians to connect a pressure gauge and undertake basic diagnostic tests.

Torque converter operation

Crankshaft rotation is transferred to the automatic transmission via a torque converter. A torque converter permits the engine to smoothly drive the transmission from a standstill up to maximum speed, absorbing the shocks of gear changing and damping-out vibrations. The exterior of the torque converter is shaped like a large metal donut, with a sealing weld around its outer edge. Internally, it incorporates the four elements illustrated sectionally in Figure 6.4, namely an impeller, a stator (sometimes called a reactor) and a turbine. Most converters are also fitted with an electrohydraulically controlled lockup clutch that locks the turbine to the impeller under certain circumstances to improve the power transmission efficiency of the system.

The impeller is fixed to the torque converter shell and so is directly driven by the engine, which also drives the transmission pump, so that with the engine running the torque converter is filled with transmission fluid under pressure. As the impeller is rotated, centrifugal force throws fluid from the centre outwards and onto the turbine blades. Fluid striking the turbine blades causes the turbine to rotate, turning the gearbox input shaft. Fluid leaving the turbine blades is then redirected by the specially curved stator blades back onto the impeller blades at such an angle that it helps the engine in driving the impeller (Figure 6.5). It is this redirection of fluid energy that makes the

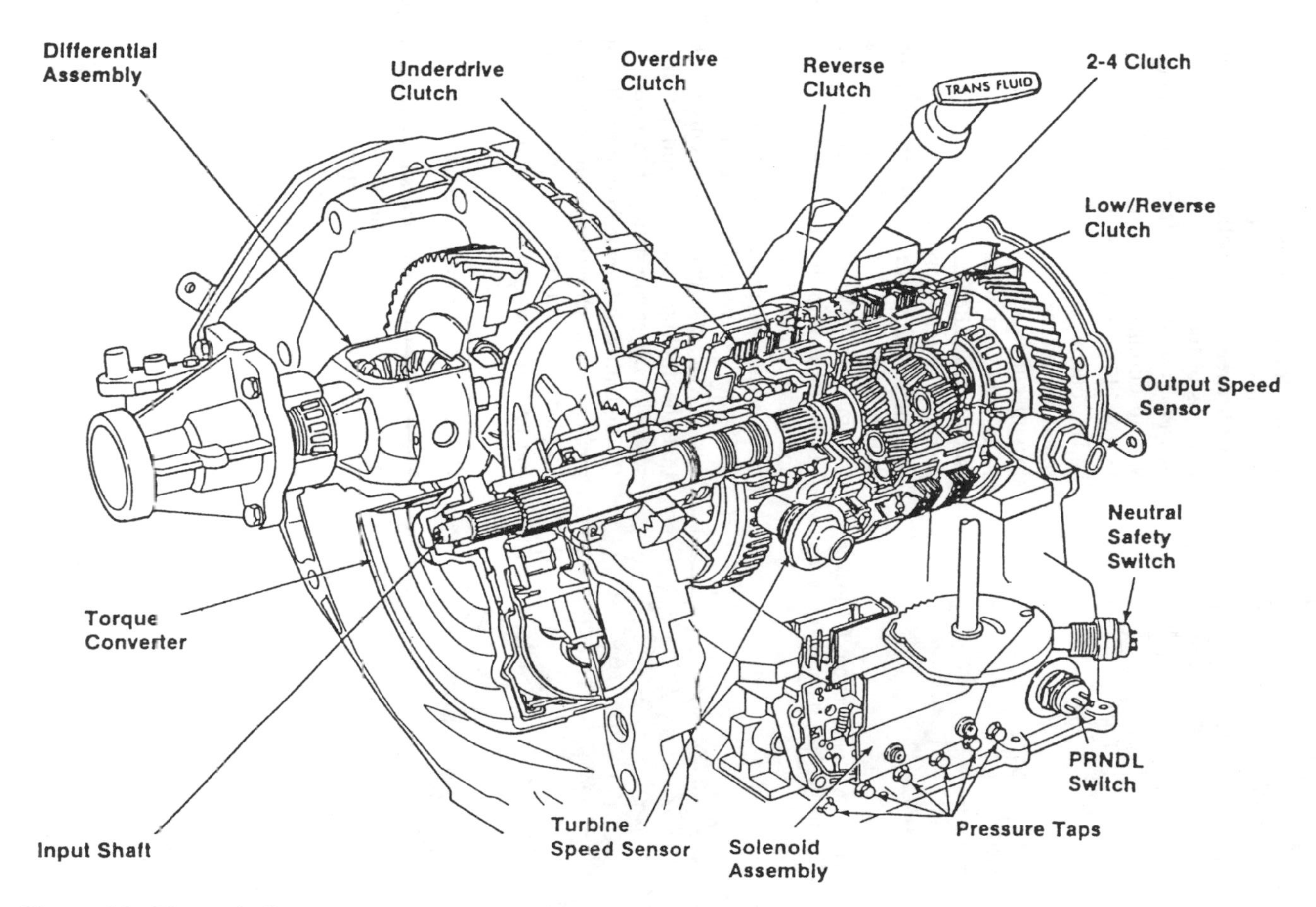

Figure 6.3 Electronically controlled automatic transmission for front-wheel drive vehicles

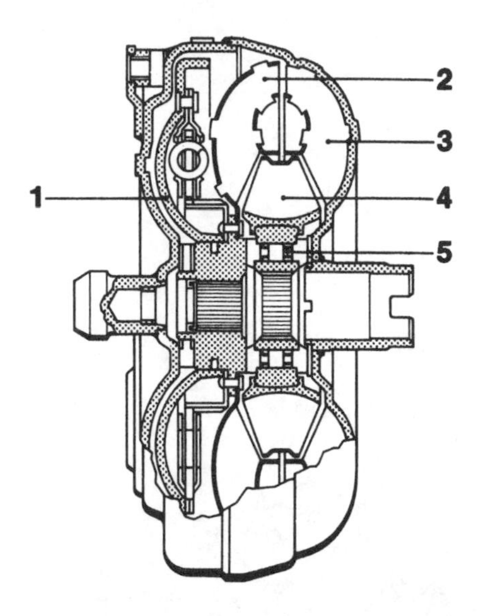

Figure 6.4 Torque converter construction: [1] lockup clutch; [2] turbine; [3] impeller; [4] stator; [5] one-way clutch (*Bosch*)

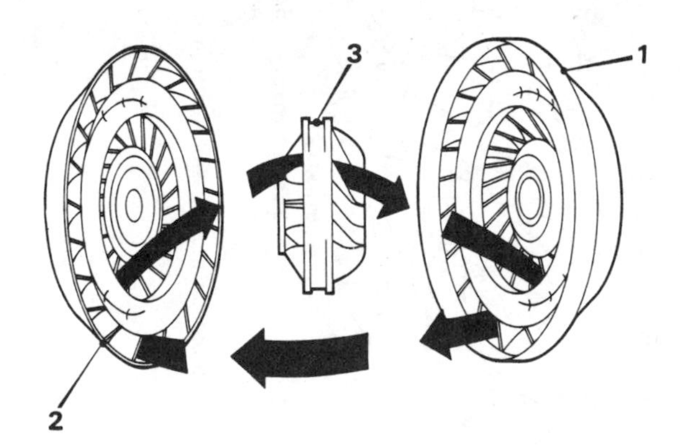

Figure 6.5 Exploded diagram of torque converter showing fluid path: [1] Impeller (driven by engine); [2] turbine (drives transmission); [3] stator (*Rover*)

torque converter capable of multiplying engine torque by a factor of up to two, providing good 'drive-away' performance.

When the impeller and turbine are revolving at a higher speed, with only a light load on the transmission, the fluid centrifugal force is almost the same in both units and so the impeller, turbine and stator all revolve at about the same speed. This is known as the torque converter 'coupling point' and the torque conversion ratio is 1:1, with a coupling efficiency of better than 90%. During normal driving the torque conversion ratio will continuously vary between about 2:1 and 1:1, depending upon the load on the engine.

Lockup clutch operation

When a torque converter is operated in the coupling phase, a speed difference ('slip') of up to 10% can exist between the impeller and the turbine. This leads to a loss of fuel economy during high-speed cruising, as well as wasteful heating of the transmission oil. It was to eliminate this slip that the torque converter lockup clutch was introduced in the 1970s.

The operation of the lockup mechanism is illustrated by the sectional drawing of Figure 6.6. The lockup clutch consists simply of a narrow friction lining, 2–3 cm wide, bonded to a thin metal disc (sometimes called a piston) which is attached to the turbine through a torsional damper spring. Fluid flow into the torque converter chamber is controlled by the transmission ECU via solenoid valves. When the transmission ECU decides that the torque converter can be locked-up, fluid is directed into the port 'C', and allowed to exit via ports 'D' and 'E'. The lockup piston thus engages against the converter cover and the torque converter is placed in direct-drive.

Conversely, when the transmission ECU decides that the lockup clutch should be disengaged, the solenoid valves are actuated to direct fluid into port 'E' and allow it to exit from ports 'C' and 'D'. This pushes the clutch piston away from the impeller and so places the converter in hydraulic drive, enabling torque-multiplication to take place.

Damper lockup torque converter

The damper (or controlled-slip) torque converter was introduced by Mitsubishi in 1982 and has since been adopted by several manufacturers as a way of dealing with the main drawbacks of conventional lockup torque converters; engagement shock and the transfer of engine vibrations to the passenger compartment, especially at low speeds.

In the damper torque converter the lockup clutch engagement is carefully controlled by having the transmission ECU constantly monitor the slip in the converter (i.e. the impeller input-speed minus the turbine output-speed). The measured slip is then compared to a setpoint value stored in the transmission ECU's computer memory. Based on the result of this comparison the transmission ECU operates the lockup clutch fluid-control solenoid valve in a duty cycle mode at a high frequency (about 30 Hz) to vary the fluid flow to the converter and so obtain just the right clutch contact pressure for the required slip. In this

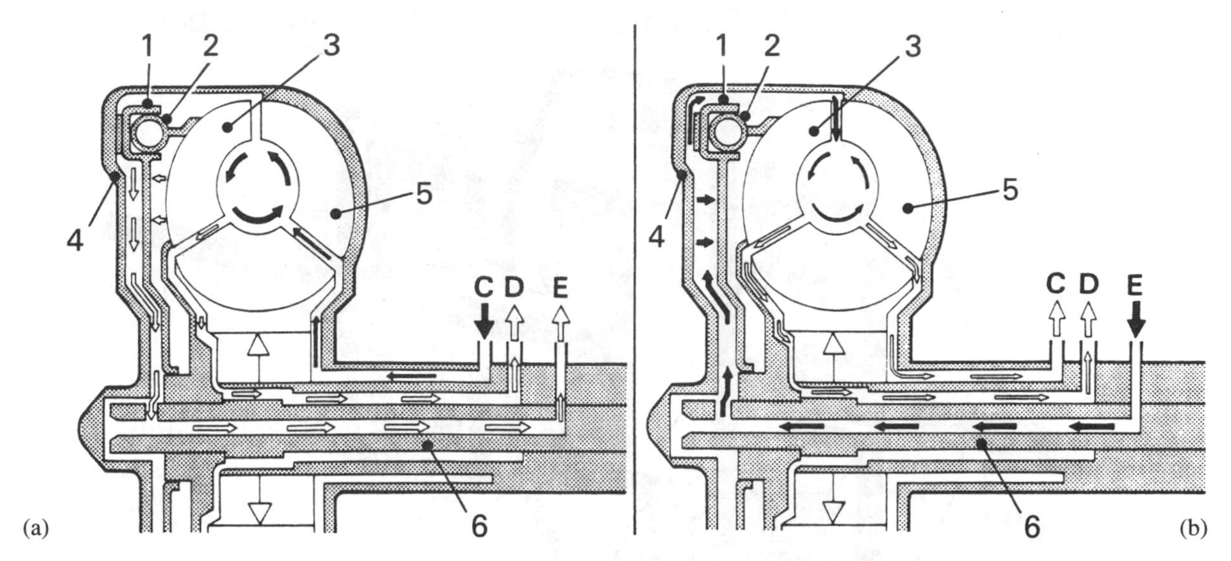

Figure 6.6 Operation of torque converter lockup clutch showing lockup mechanism engaged (a) and disengaged (b). [1] Lockup piston (with friction lining); [2] damper spring; [3] turbine; [4] torque converter cover; [5] impeller; [6] mainshaft

manner clutch slip is regulated in the range 1–10% of input rpm, resulting in smooth and vibration-free operation.

Damper torque converters are particularly efficient in that they allow partial lockup at low speeds in low gears, something not possible with conventional lockup converters due to engagement shock and vibration.

Provision of gear ratios

In a conventional automatic transmission a selection of gear ratios is provided by an epicyclic geartrain. By way of a simple example, Figure 6.7 shows a Simpson geartrain capable of providing three forward gear ratios (termed 'low', 'intermediate' and 'high') and reverse. The selection of the appropriate ratio depends on the application of hydraulic pressure to clutches and brake-bands operating in a set pattern.

Control of friction element engagement is accomplished by the electrohydraulic control valves which respond to signals from the transmission ECU and direct line pressure accordingly.

Low gear is selected by engaging the forward clutch, which directs engine power to the ring-gear of the first gearset, thus rotating the planet wheels clockwise to drive the common sun gear anticlockwise. The one-way sprag clutch prevents the planet carrier of the second gearset from rotating and so the planet wheels rotate to drive the output shaft in compound reduction. Engine braking can be provided by having the low-reverse brake-band applied to disable the free-wheeling effect of the sprag clutch.

Intermediate gear is selected by applying the intermediate brake-band to lock the common sun gear against rotation. With the forward clutch engaged, power is applied to the first ring-gear, so driving the planet wheels around the sun gear and rotating the output shaft in simple reduction drive.

High gear is selected by engaging both the forward and the reverse-high clutches. This action locks the sun gear to the first ring-gear and the whole geartrain revolves at the same speed as the input shaft to give direct drive.

To obtain reverse gear the reverse-high clutch and the low-reverse brake-band are applied. This action transfers drive through the sun gear to the planet wheels of the second gearset, turning the ring-gear in the opposite direction to the input shaft. The low-reverse brake-band holds the planet-carrier of the second gearset against rotation, causing the output shaft to turn.

Transmission control system structure

The operation of a typical transmission control system will be illustrated by considering the GF4A-EL electronically controlled automatic transmission manufactured by Jatco (Japanese Automatic Transmission Company). Transmissions of this type are installed

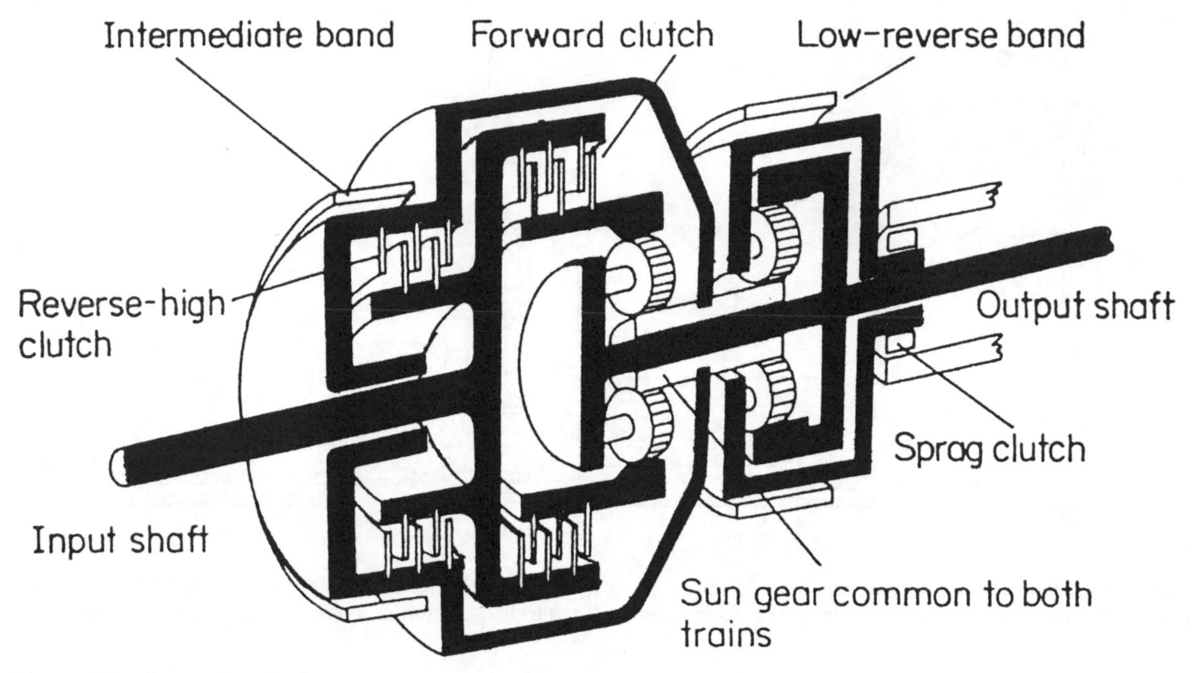

Figure 6.7 Gear ratio selection

by a number of vehicle manufacturers, including Mazda, Nissan and Rover.

The GF4A-EL is typical of modern automatic gearbox design in that it offers four forward gears, controlled by a microcomputer-based transmission control unit (transmission ECU) which communicates with the engine control unit (engine ECU) to provide total powertrain management. The transmission ECU makes decisions based on electrical signals received from various sensors located on the engine and gearbox (Figure 6.8). Its microcomputer contains data relating to the ideal gear for every speed and load condition, along with correction factors for engine and transmission temperatures, brake pedal depression and so on. Using this corrected data the transmission ECU energizes solenoid valves to engage the most suitable gear ratio for the prevailing driving conditions. The transmission ECU also provides a self-diagnosis function on some of the sensors and operates a fail-safe mode if a fault is found. Some features of the GF4A-EL that are typical of current electronically controlled automatic transmissions include:

(i) A duty-cycle solenoid valve to vary line pressure and so optimize clutch and brake-band engagement force during each gearshift.
(ii) Total control of engine and transmission by reducing the engine torque during gearshifts, thereby smoothing the ratio change.
(iii) A duty-cycle lockup solenoid enabling controlled-slip lockup at low speeds and giving less shock when engaging full lockup at cruising speeds.
(iv) Optional control programs, selected by the driver, allowing POWER or HOLD shift patterns to be engaged. The POWER program delays each upshift by a few hundred rpm to permit a sporty driving style. HOLD causes the transmission ECU to hold a selected gear; useful when driving in hilly country or on difficult road surfaces.

The transmission ECU input system

The GF4A-EL transmission control system is illustrated in Figure 6.9. At its heart is the transmission ECU, which accepts inputs from various sensors and communicates with the engine ECU.

The *pulse generator* is a magnetic pickup that is located in the upper part of the transaxle casing and detects the geartrain's reverse-forward drum speed (i.e. the torque converter output speed).

The *speedometer sensor* is a magnetic pickup sensor that detects the gearbox's output shaft speed (i.e. the vehicle speed).

The *throttle sensor and idle switch* is a combination sensor composed of a potentiometer and also a

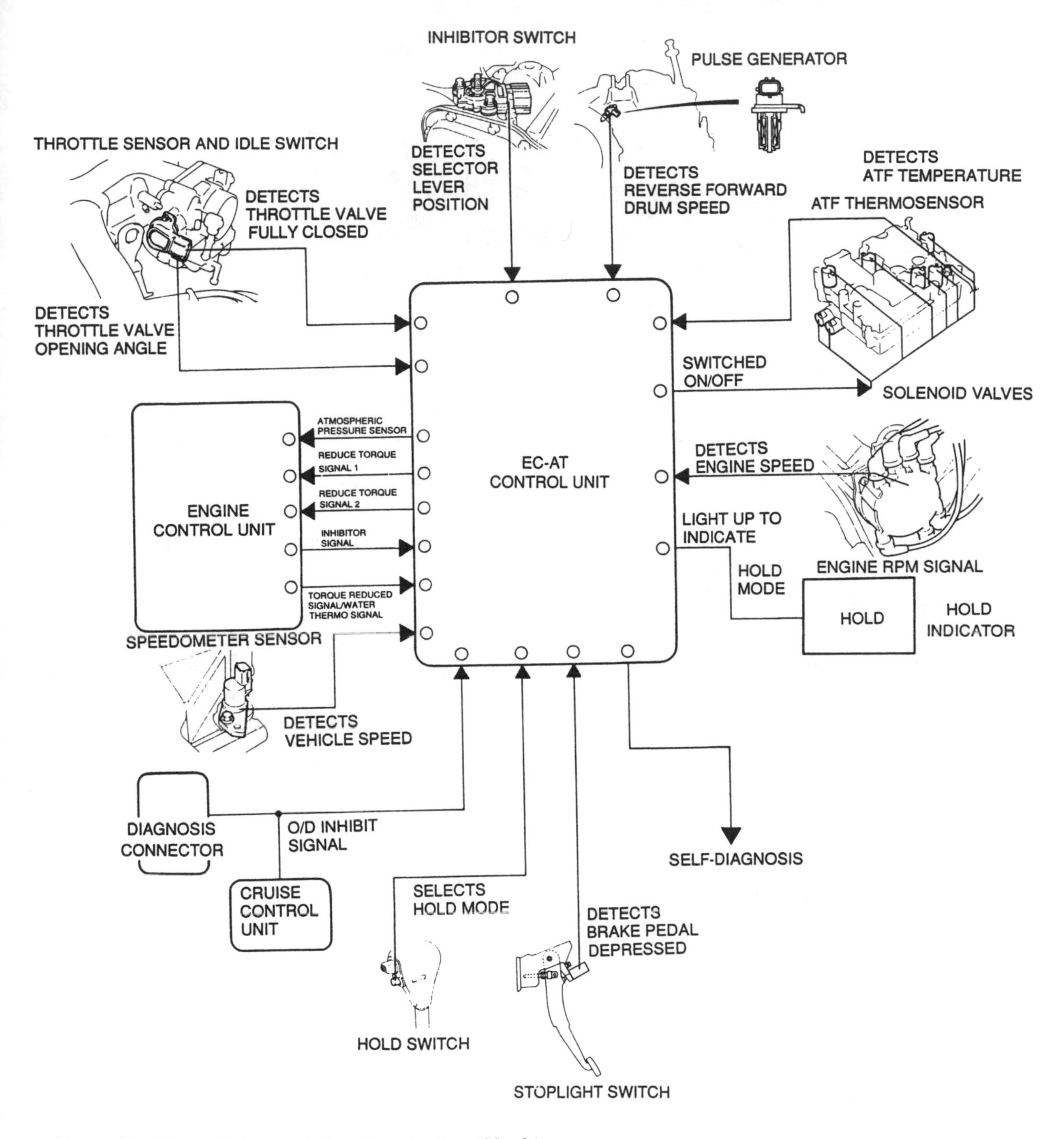

Figure 6.8 Transmission control system structure (*Mazda*)

pair of contacts that close when the engine is at idle. It detects throttle angle and hence engine load. When idle is detected, and the vehicle is stationary, the transmission ECU engages second gear. This minimizes torque loads on the transaxle and reduces 'creep' when stationary in traffic. First gear is engaged as soon as the ECU detects movement of the accelerator pedal.

The *inhibitor switch* reports the position of the selector lever to the transmission ECU. It also prevents

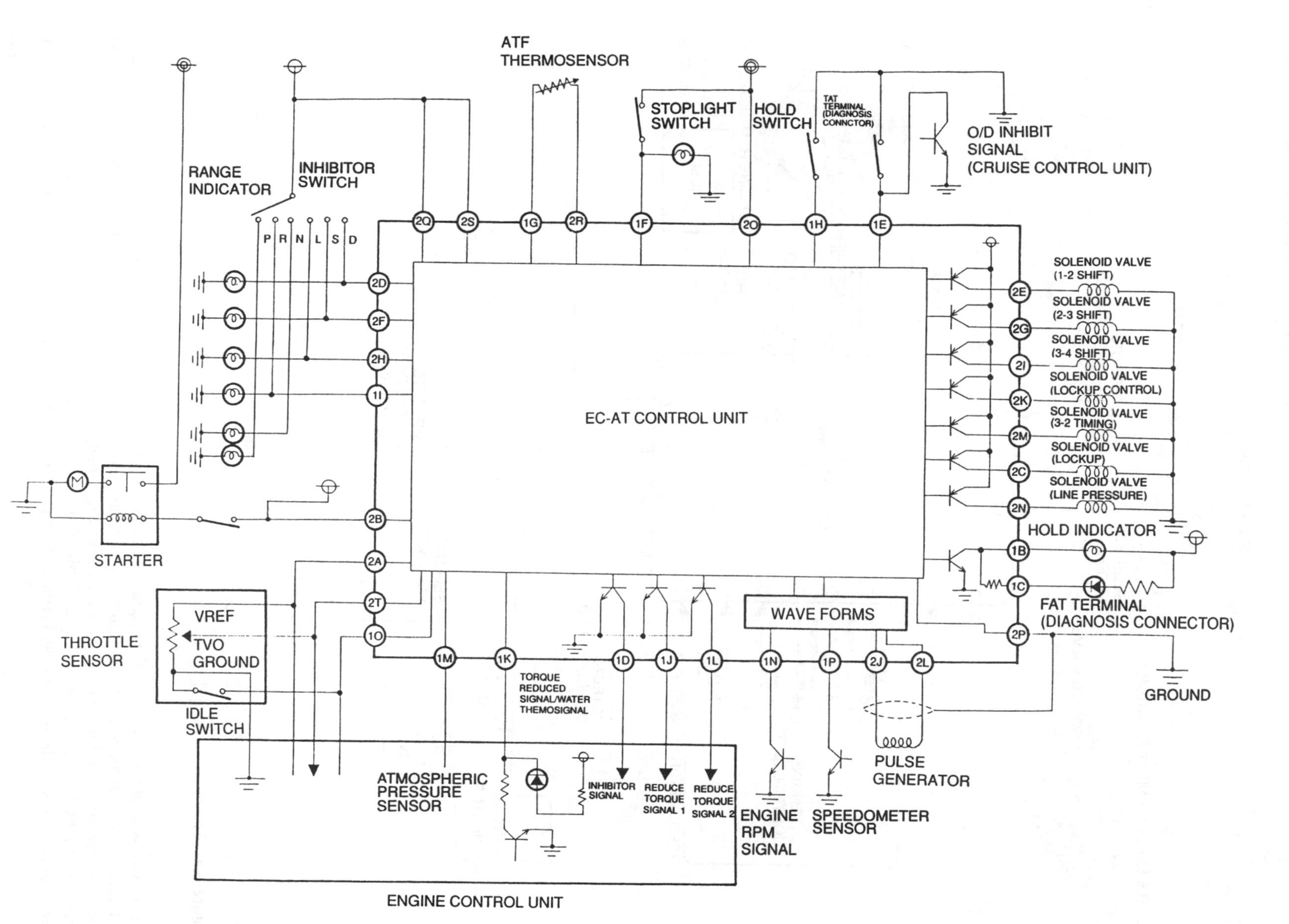

Figure 6.9 Details of control system used with Mazda GF4A-EL transmission (*Mazda*)

operation of the starter motor if it is not in either Park or Neutral.

The *HOLD switch* is a press-button switch located on the selector lever. It is used by the driver to instruct the ECU to hold a particular gear ratio, for example when descending a hill. When HOLD is in use the HOLD indicator lights up.

The *stoplight switch* detects that the brakes have been applied. If the brakes are used while the torque converter is in lockup, the transmission ECU releases lockup to provide a smooth deceleration.

The *O/D inhibit signal* is used together with the cruise control. It prevents the transmission from shifting into overdrive (fourth gear) when the cruise control is operational and vehicle speed is more than 8 kph below the set cruising speed.

The *ATF thermosensor* is a thermistor that registers the temperature of the transmission fluid. The ECU uses this information to modify line pressure at extremes of temperature, thus taking account of the fluid's higher viscosity at low temperatures and the danger of overheating at high temperatures.

An *engine rpm signal* is taken from the ignition coil primary winding.

An *atmospheric pressure sensor* sends a signal to the transmission ECU when the measured atmospheric pressure indicates that the vehicle is at a height of 1 500 m or more. The engine develops less power at high altitudes and so the ECU must modify the gearshift points.

The output system

To change gear, the transmission ECU sends energizing signals to seven solenoid valves located in the valve body of the transmission. The solenoid valves are either of the ON/OFF type or, where a variable pressure is required, the duty-cycle type. The three *shift solenoids* (1–2, 2–3, 3–4 shift) are either ON or OFF and direct line pressure to the shift valves. The transmission ECU selects a programmed shift pattern based on selector position and then measures vehicle speed and throttle opening angle. Based on the results of calculations performed on these signal values the ECU energizes the appropriate solenoid valve, setting the gear ratio (Figure 6.10).

The *3–2 timing solenoid* controls the precise timing of clutch and brake-band engagement by directing pressure to the 3–2 timing valve.

The ***lockup control solenoid*** is used to direct line pressure to the torque converter lockup system when the transmission ECU judges it appropriate. It is either ON or OFF.

Engagement slip in the lockup clutch is controlled by the *lockup pressure reducing solenoid*. This solenoid is operated on a duty-cycle basis (at about 30 Hz) and modifies the engagement force on the clutch piston. The duty-cycle is continuously modified to maintain the target slip value.

The engagement force applied to the brake-bands and multiplate clutches is controlled by the *line pressure solenoid*. This is a 30 Hz duty-cycle controlled valve which modifies the line pressure based on signals sent from the ECU. Line pressure is increased in proportion to throttle opening angle, providing increased clamping force on the friction members to cope with greater engine torque. During gearshifts the line pressure is momentarily reduced to minimize shift shock.

Total engine and transmission control

Shift comfort is a key feature of modern automatic transmissions, especially those with four or five gears where shift frequency is likely to be greater than in older three-speed designs. To reduce shift shock the transmission and engine ECUs exchange digital data to temporarily lower the engine output torque during gearshifting. This feature is known as 'total control' or 'engine intervention' and is a key function of most automatic transmission controllers.

Lowering engine torque during gearshifting reduces torque fluctuations at the transmission output shaft and offers several additional advantages:

(i) A lower line pressure can be used to clamp the friction elements, reducing engagement shock.
(ii) Reduced slippage of friction elements during engagement results in less lining wear.
(iii) Less slippage of the friction elements means less heating of the transmission fluid, extending its life and increasing the transmission's efficiency.

Engine intervention is usually achieved by either cutting off fuel injection or retarding the ignition timing (or a combination of both). In the Mazda GF4A-EL transaxle, fuel injection is cut off during upshifting and the ignition timing is retarded during downshifting (Figure 6.11).

Fuel cut control

When the transmission ECU decides that an upshift (1–2 or 2–3 shift) is appropriate it sends a 'Reduce Torque Signal 1' (logic 1) to the engine ECU. If engine conditions allow a fuel cut, the engine ECU confirms its action by sending a 'Torque Reduced Signal' back to the transmission ECU. The transmission ECU then

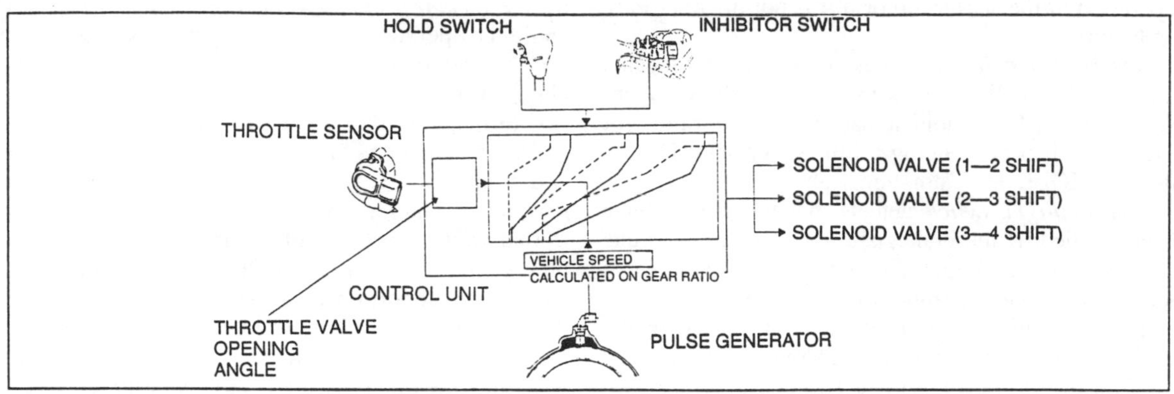

Solenoid Operation and Gear Position

Range	Mode	Gear position		Solenoid valve 1-2 shift	Solenoid valve 2-3 shift	Solenoid valve 3-4 shift
P	—	—				○
R	—	Reverse	Below approx. 4 km/h {2.5 MPH}		○	○
R	—	Reverse	Above approx. 5 km/h {3 MPH}			
R	—	—	Above approx. 30 km/h {19 MPH}	○		
N	—	—	Below approx. 4 km/h {2.5 MPH}			○
N	—	—	Above approx. 5 km/h {3 MPH}	○		
D	★ POWER NORMAL (Except HOLD)	1st			○	○
D	★ POWER NORMAL (Except HOLD)	2nd		○	○	○
D	★ POWER NORMAL (Except HOLD)	3rd		○		
D	★ POWER NORMAL (Except HOLD)	O/D		○		○
D	HOLD	2nd	Below approx. 15 km/h {9.3 MPH} 14 km/h {8.7 MPH}	○	○	
D	HOLD	2nd	Above approx. 18 km/h {11.2 MPH} 17 km/h {10.5 MPH}	○	○	○
D	HOLD	3rd		○		
D	HOLD	❈O/D		○		○
S	POWER (Except HOLD)	1st			○	○
S	POWER (Except HOLD)	2nd		○	○	○
S	POWER (Except HOLD)	3rd		○		
S	POWER (Except HOLD)	❈O/D		○		○
S	HOLD	2nd		○	○	○
S	HOLD	❈3rd		○		
S	HOLD	❈O/D		○		○
L	POWER (Except HOLD)	1st			○	○
L	POWER (Except HOLD)	2nd		○	○	
L	HOLD	1st			○	
L	HOLD	❈2nd		○	○	

○ : Solenoid valve is ON.

❈ : Engine overspeed protection.

★ : The EC-AT control unit automatically switches between POWER and NORMAL modes corresponding to the speed at which the accelerator pedal depression.

16G0KX-529

Figure 6.10 Solenoid operation and gear selection (*Mazda*)

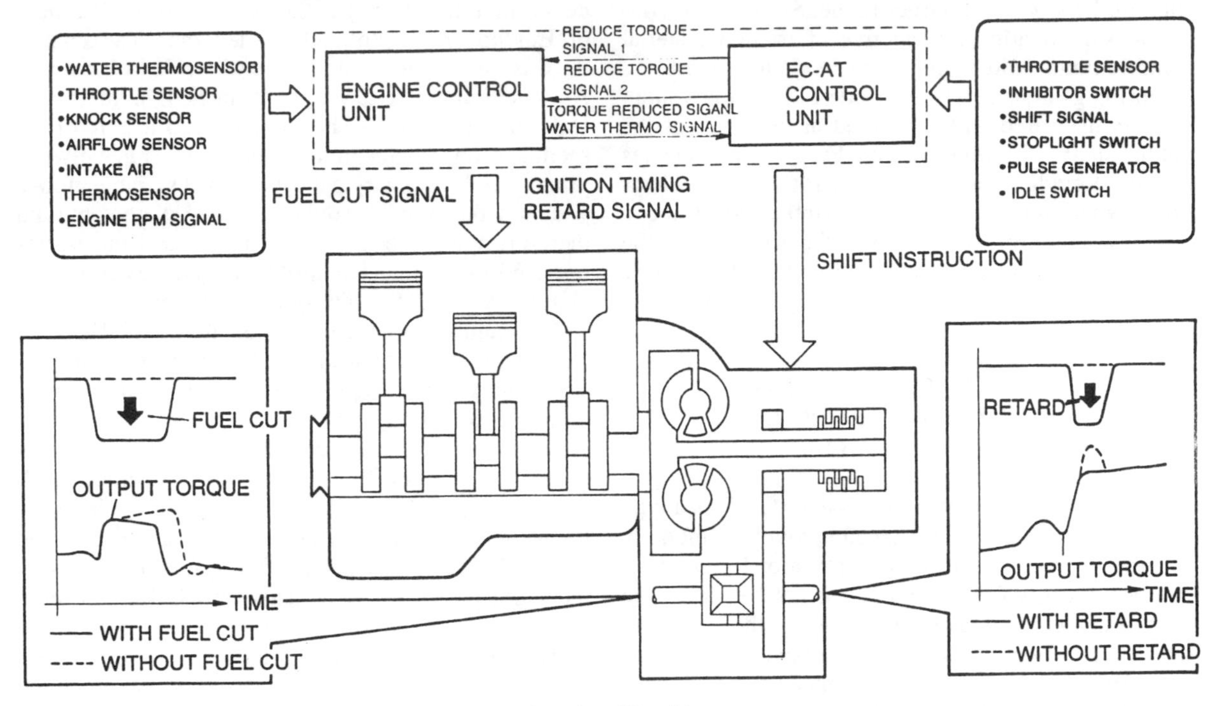

Figure 6.11 Engine torque reduction during gearchanging (*Mazda*)

performs the gearshift and the 'Reduce Torque Signal 1' is removed, allowing fuel injection to be reinstated.

Ignition timing retardation control

When the transmission ECU judges that a downshift (3–2 or 2–1 shift) is required, it transmits a 'Reduce Torque Signal 2' (logic 1) to the engine ECU. The engine ECU then reduces the spark timing by several degrees to lower the engine torque and sends a 'Torque Reduced Signal' as confirmation. Once the transmission ECU has completed the downshift, the 'Reduce Torque Signal 2' is removed and the ignition timing is returned to normal.

Feedback control

The ultimate method of giving smoother gear changes is to provide feedback control of the friction element engagement force. This can be achieved either by using sensors in the transmission to directly monitor various fluid pressures, a technique used on some Chrysler transmissions, or by using speed sensor signals to monitor the instantaneous slippage in the transmission (Figure 6.12).

The latter technique relies on precise control of the line pressure during gear engagement to ensure that the input shaft speed variations, and hence band/clutch

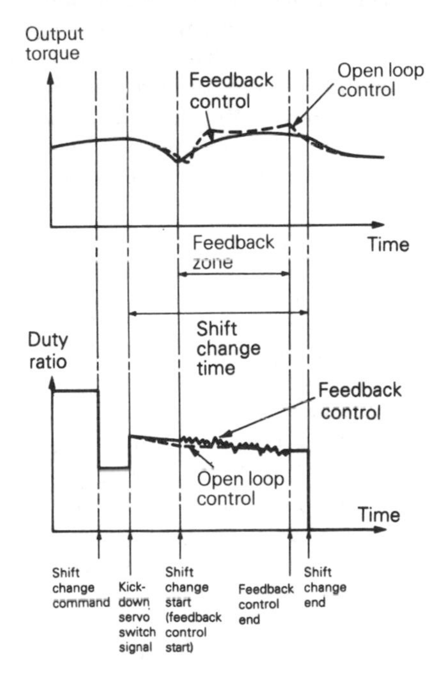

Figure 6.12 Feedback control of clutch engagement pressure by monitoring clutch slip and controlling drive duty ratio of pressure control solenoid valve (*Mitsubishi*)

slip, are kept within a target range. Smooth and consistent shift quality, irrespective of friction material condition and fluid temperature, is then assured.

During gearchanges, the speed of the gearbox input shaft is measured and compared against that of the output shaft. The transmission ECU then calculates the speed gradient of the input shaft (i.e. its deceleration). Whenever this calculated quantity moves above or below a preprogrammed value the drive duty to the line pressure solenoid is lowered or raised to restore it, ensuring a smooth and steady engagement.

6.4 ELECTRONICALLY CONTROLLED CONTINUOUSLY VARIABLE TRANSMISSION (ECVT)

A continuously variable transmission (CVT) is a type of automatic transmission that can provide a smoothly varying gear ratio. Unlike a conventional automatic gearbox the CVT has no fixed gears. It varies the drive ratio continuously by changing the operating diameters of two pulleys that are linked by a steel V-belt. The transmission can thus alter its ratio imperceptibly, with no interruption of drive.

Potentially, the CVT offers the ease of driving associated with an automatic gearbox together with the efficiency of a manual transmission, without any loss of vehicle performance. Unlike a conventional automatic, no torque converter is used and so there are no hydraulic slippage losses. The CVT can respond instantly to throttle pressure, giving smooth and rapid acceleration. Unfortunately the difficulties of manufacturing a reliable and durable system have meant that few car makers offer such a transmission.

The history of CVT

The idea of using a belt drive system with variable diameter pulleys dates from about 1908, when the Rudge company developed their 'Multi' motorcycle. The Rudge Multi employed a CVT system consisting of a variable-diameter drive pulley fitted to the engine, coupled by a leather belt to a fixed-diameter pulley on the rear wheel. By rotating a small handwheel fitted next to the fuel tank the rider could vary the running diameter of the drive pulley to alter the gear ratio. Simultaneously, the position of the rear wheel was also slightly altered to maintain the drive-belt tension. So successful was this system that the Rudge Multi had to be banned from the Isle of Man TT race in order to give other motorcycles some chance of winning.

The first practical CVT system for cars was developed in 1955 by the Van Doorne brothers of Eindhoven in Holland. Called the 'Variomatic' it was manufactured by DAF (Van Doorne's Automobielfabriek NV) and launched in their tiny Daffodil saloon car. The system used a pair of V-section rubber belts running under tension between primary and secondary pulleys that had variable groove widths. Control of the pulley running diameters (and therefore the drive ratio) was achieved using servos operated by an inlet manifold vacuum. A centrifugal clutch was used for setting off from rest.

Although it remained in production until 1992 the Variomatic suffered from several drawbacks, in particular its torque capacity was limited and it occupied a large installation volume.

DAF engineers, meanwhile, established a new company called VDT (Van Doorne's Transmissie BV) to continue work on CVT in an attempt to improve the technology. Their efforts were rewarded in 1979, when VDT launched a new CVT system using a segmented steel thrust belt. This new system offered increased torque capacity and was much lighter and more compact than the Variomatic system. It has subsequently been used in small cars manufactured by Rover, Ford, Fiat, Subaru and Nissan.

The Subaru ECVT

The Subaru ECVT was co-developed by Fuji Heavy Industries in Japan (Subaru's parent company) and VDT in Holland. It was the world's first practical electronically controlled CVT. The system uses an electromagnetic powder clutch and electronic control unit developed by Subaru, together with VDT's steel thrust belt and pulleys (Figure 6.13).

ECVT was introduced in Japan in February 1987 on the Subaru Justy and has since been used on several other small vehicles (most notably the Nissan Micra) with considerable success. The microcomputer-based transmission controller ensures that the engine is always operating in its most efficient speed range, reducing emissions and improving fuel economy.

Operation of the VDT drive

Engine torque is transferred through an electromagnetic powder clutch to the primary pulley which, in turn, drives the steel thrust belt to rotate the secondary pulley. Each pulley has two tapered sides, called sheaves. One sheave is fixed to the appropriate transmission shaft and the other is movable under hydraulic pressure. The pressures are manipulated so that the groove widths between the sheaves of each pulley widen or narrow in inverse proportion to each other

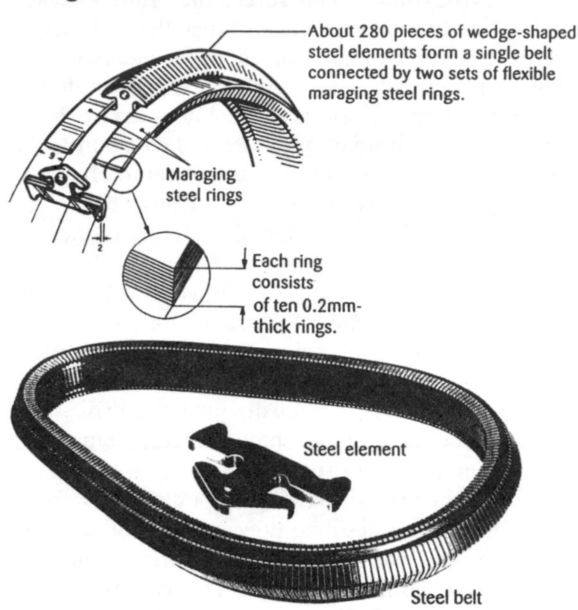

Figure 6.13 Construction of the Van Doorne steel thrust belt (*Subaru*)

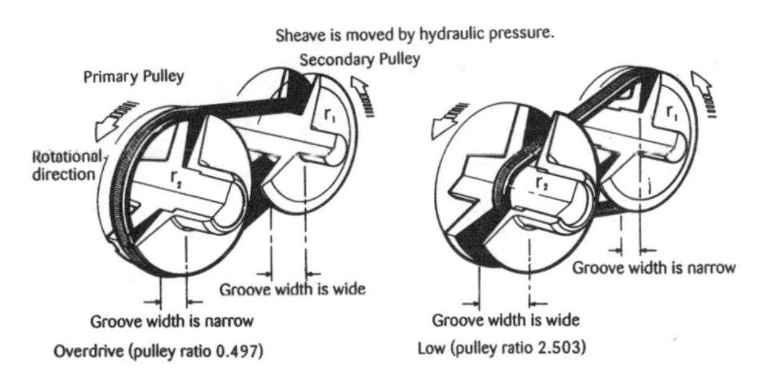

Figure 6.14 Use of sheave clutches to effect ratio changes (*Subaru*)

(Figure 6.14), clamping the belt and giving a stepless variation in ratio from 2.503:1 to 0.497:1.

The drive belt is composed of about 280 wedge-shaped blocks, each precision-ground from high-friction steel. The blocks transmit thrust by pressing against each other, and are guided between the pulleys by two thin steel rings.

Operation of the electromagnetic clutch

Important components of the electromagnetic powder clutch are illustrated in Figure 6.15. The clutch operates by having the ECVT microcomputer direct an energizing current to the exciting coil, magnetizing

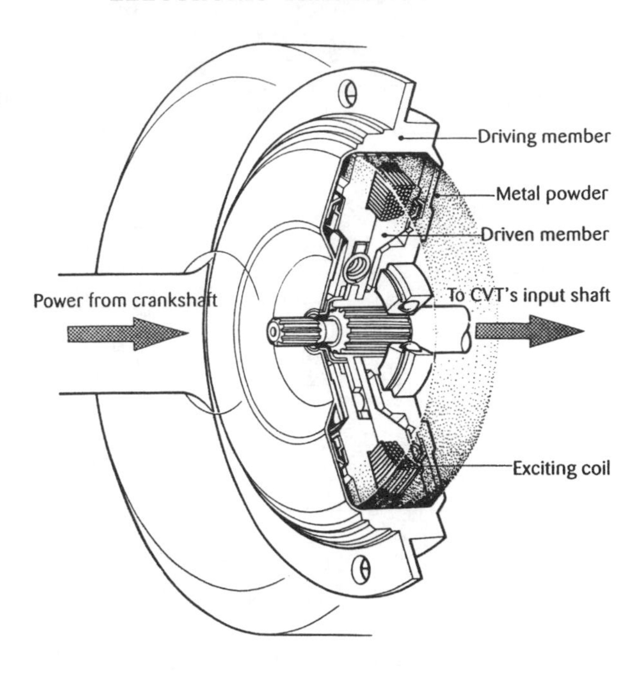

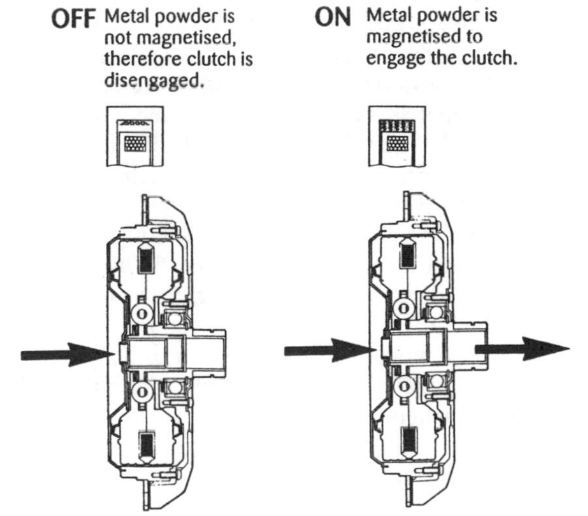

Figure 6.15 Operation of the electromagnetic powder clutch (*Subaru*)

the metal powder. The magnetized powder progressively binds together and so locks the external driving member to the internal driven member, smoothly transferring engine torque to the primary pulley.

The control system

The electrohydraulic control system is illustrated in Figure 6.16. An 8-bit microcomputer controls both

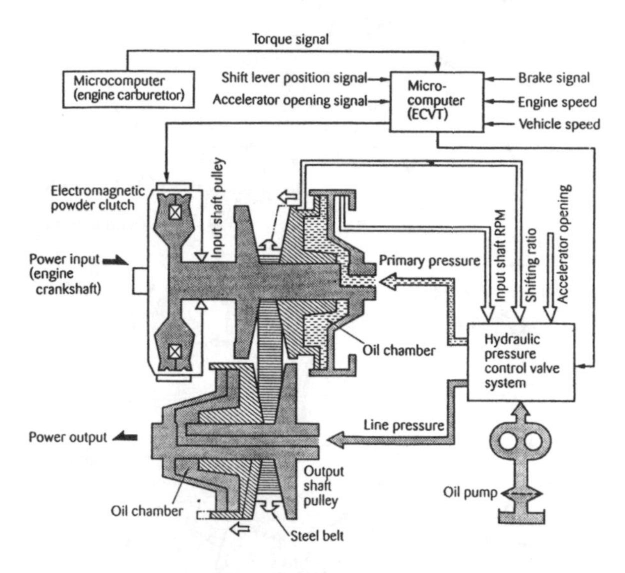

Figure 6.16 The ECVT control system (*Subaru*)

the clutch and the hydraulic system, using inputs from various sensors to indicate the vehicle operating mode.

A shift lever position switch is used to prevent engagement of the clutch when the transmission is in the 'P' or 'N' position. Similarly, a brake pedal switch is used to signal to the controller that the vehicle is slowing down. When the controller receives this signal it disengages the electromagnetic clutch, preventing a stall.

An accelerator pedal position switch informs the control unit that the driver has pressed the pedal and so wishes to drive away. An energizing current is then supplied to the clutch electromagnet, causing gradual engagement of drive. The controller judges vehicle speed, maximizing this current as normal road speeds are reached.

Besides the essential requirements of stopping, starting and smooth ratio changing, the microcomputer provides other features, including:

(i) Prevention of fierce clutch engagement when the engine is running at a fast idle during cold weather starting.
(ii) A self-diagnosis and back-up system designed to prevent damage to the transmission in the event of a control failure.
(iii) Provision of a small clutch energizing current to prevent the vehicle from rolling backwards during a hill-start.

The pressure control valve system

The electrohydraulic valve unit is located in the body of the transmission. Control of the gear ratio is set by the primary pressure which is applied to the primary pulley servo cylinder. This forces the primary pulley to the desired running diameter. Since the steel belt is of a fixed length it, in turn, forces the secondary pulley, against line pressure, to a running diameter that is inverse to that of the primary. The line pressure therefore opposes the primary pressure and so controls the clamping force on the belt. It must be set to eliminate damaging slip but avoid excessive loads.

To improve drivability the ECVT incorporates a solenoid valve to switch the line pressure between two values. When the engine torque signal informs the ECVT microcomputer that output torque is below 60% of maximum, the solenoid is switched to give LOW line pressure. This reduces the pulley clamping force applied to the belt, cushioning the drive. The transmission operates more smoothly, eliminating jolts and jerks in stop–go driving.

Conversely, when the engine output torque is above 60% of maximum, the line pressure control solenoid is switched to give HIGH line pressure (about 50% increase over the LOW setting). This causes the pulleys to clamp the belt very firmly, eliminating any possibility of slip and ensuring maximum power transfer.

The ZF Ecotronic ECVT

In contrast to the Subaru ECVT, which is designed for use on small cars, the ZF company of Germany have designed an ECVT suitable for use on medium-size cars. The ZF Ecotronic operates on the same principle as the Subaru ECVT, but uses a wider 30 mm steel thrust belt to give a higher torque capacity of 210 Nm, making the transmission suitable for use with engines of up to 2.5 litres displacement. The transmission ratio varies from 2.44:1 to 0.46:1 and a lockup torque converter (rather than electromagnetic clutch) is used to transfer power from the engine. An interesting feature of the electronic control system is that it 'learns' driver behaviour and so optimizes the interaction of the engine and transmission to give best performance and economy. This results in a fuel saving of about 10% when compared to a conventional four-speed automatic transmission.

Driving with ECVT

The ECVT is equipped with a selector lever very similar to that of a conventional automatic transmission. When 'D' is engaged the driver can accelerate away, with the transmission operating at the optimum ratio for the desired engine rpm. In comparison with their manual counterparts, ECVT cars are generally much quicker from a standing start.

Once underway the ECVT is characterized by a very wide ratio span, roughly equivalent to that of a six-speed manual transmission. Theoretically, this should enable exceptional fuel economy to be obtained by keeping the engine at the optimum speed and load point for a given road speed. Unfortunately, frictional losses in the transmission waste a considerable amount of energy and so ECVT cars are about 5% less fuel efficient than their equivalent manual transmission counterparts.

7

Electronic control of chassis systems

Chassis systems are those elements of the automobile that are associated with controlling the motion of the vehicle; accelerating, braking, turning and vertical movements over bumps.

Driving a car represents the operation of a closed-loop control system. The controller is the driver, who interacts with the chassis systems to control the motion of the vehicle within its surroundings. The vital link between the vehicle and its surroundings is the frictional contact between tyres and road surface. All control forces are fed through the tyres, and so any loss of grip results in a loss of control of the vehicle. Good chassis control is therefore vital to the promotion of active safety – the avoidance of accidents through driver action. A key aspect of this is the precision of the vehicle's response to a driver input and its relationship to the road surface.

Under normal driving conditions the chassis behaviour remains entirely predictable and 'comfort' features such as power-assisted steering and semi-active suspension enable the driver to operate the vehicle with minimal stress and effort.

Occasionally, however, most drivers are faced with the need to unexpectedly make emergency manoeuvres. Under these circumstances an electronically controlled braking or traction control system can assist the driver by intervening more rapidly and precisely than human reactions permit. Such systems enable the vehicle to maintain stability in situations where even an experienced driver would be unable to cope.

7.1 ANTI-LOCK BRAKING SYSTEMS (ABS)

During heavy braking on wet or icy roads it is very easy for the driver to inadvertently lock the wheels. If the rear wheels lock, the car tends to yaw, becoming very difficult to control and possibly spinning. Conversely, if the front wheels lock, all steering action is lost. Experienced drivers avoid these situations by rapidly pumping the brake pedal, repeatedly taking the wheels to the point of locking and then allowing them to roll again. This technique, called cadence braking, requires a high level of skill and concentration under panic conditions.

An anti-lock braking system (ABS) allows even an unskilled driver to retain control of a vehicle during emergency braking. By rapidly increasing and decreasing the braking pressure, the rate of wheel deceleration is maintained at a desired value to prevent the wheels from locking. This allows the vehicle to be stopped quickly and, most important, steering control and stability are maintained.

Development of ABS

During the 1950s design engineers working on aircraft braking systems found that aeroplane tyres gave their greatest grip just before locking. Using entirely mechanical techniques, they managed to design braking systems that released the braking force the moment locking was detected. This allowed the tyre to be continuously operated in the region of maximum grip, close to locking, reducing braking distances. These early anti-lock braking systems therefore allowed much safer aircraft landings, especially on slippery surfaces.

Automotive engineers soon realized the potential of ABS to maintain directional stability and steering control during emergency braking. One of the first vehicles to use ABS was the 1965 model Jensen FF. This was fitted with the Dunlop Maxaret mechanical ABS system, adapted from aircraft use. Although it provided comparatively good performance, the Maxaret was too expensive and bulky for use on anything other than very high-priced cars and so automobile ABS was shelved until technology improved.

During the late 1960s many vehicle manufacturers started to experiment with anti-lock braking systems that were controlled by analogue electronic circuits. Ford demonstrated such a system in 1968 and Chrysler showed a pioneering 4-wheel ABS controller in 1971. In Japan, both Nissan and Toyota announced electronic ABS and in Germany a joint-venture company formed by Telefunken and Bendix attempted to market an ABS system called Teldix. Unfortunately the electronic design and manufacturing techniques of that time were insufficiently developed for a safe and reliable system to be developed and so none of these systems was commercially successful.

Development work continued however, and in 1978 Bosch announced the availability of their pioneering *Anti-Blockier System* (German for anti-skid system and from which the abbreviation ABS actually derives).

Although initially available only as optional equipment on German prestige vehicles, the good performance and reliability of Bosch ABS made it the first commercially successful system and it was subsequently fitted to many cars. Although Bosch still commands a large share of the European ABS market, many other companies also supply systems. Today, most drivers understand the benefits of ABS and it is a standard fitment on most large and medium size cars, and a low-cost option on most small cars. In 1995, approximately 50% of automobiles manufactured in Europe were fitted with ABS.

Limitations of ABS

Despite the widely accepted advantages of ABS, it does have limitations. In particular, the laws of nature still apply to an ABS-equipped vehicle. If the driver takes a corner at too great a speed the car will still slide off the road.

Moreover, there are some circumstances in which ABS is of only limited effectiveness, these include:

(i) Deep snow or loose chippings which may build-up to form a wedge in front of the wheel.
(ii) Driving at speed on a very wet road, when aquaplaning may occur.
(iii) Frost or shallow snow, where a locked wheel may usefully 'bite' down to the road surface.

Tyre dynamics

To appreciate the operation of an anti-lock braking system it is essential to understand tyre dynamics. Tyres are involved with the transference of accelerating and decelerating forces between the road surface and the vehicle. The grip exerted by the tyre on the road surface is measured in terms of the coefficient of friction, μ (pronounced 'mew'). The value of μ depends on the nature of the two surfaces; for example, it reaches a maximum of about 1.0 on a dry tarmac road, falls to about 0.7 on wet tarmac, but is only about 0.2 on a snow-covered road. Thus a greater value of μ means more braking grip and a shorter stopping distance.

Tyre grip also depends on the slip ratio; the ratio of the tyre speed to the road speed. When a car is travelling at a steady speed there is no slippage between the road and the tyre, and so the slip ratio is zero. On the other hand, the application of a powerful braking (or accelerative force) may cause the wheel to lock (or spin) and the tyre then has a slip ratio of one.

During gentle braking the tyres are slowed to slightly less than the vehicle speed and so some slip occurs. The value of μ rises proportionately with this slip, reaching a maximum at a slip ratio of about 0.15–0.30. Heavier braking causes increasing slip, resulting in μ falling sharply and rapid locking of the wheel. Maximum braking force is thus achieved with a slip ratio of about 0.15–0.30. Lateral grip (the resistance to sideways forces on the tyres) also varies with the slip ratio, falling from a maximum at zero slip to almost nil at a slip ratio of one. These characteristics are illustrated in Figure 7.1, which shows tyre grip as a function of slip ratio for both negative slip (braking) and positive slip (acceleration). As may be seen, the maximum values of frictional coefficient and lateral grip do not occur at the same slip ratio. However, by setting a target braking slip ratio of about 0.20 the vehicle will be stopped very swiftly, and with adequate lateral grip. Under dry road conditions stopping distances are typically reduced by about 15% when using ABS operating at this slip ratio. This advantage increases to about 40% when braking on a wet road surface.

The function of the ABS control system is therefore to continuously monitor the speed of each wheel and then modulate the hydraulic braking pressures to keep the slip ratio in the range 0.15–0.30. The ABS control module does this not by directly calculating slip but rather by calculating wheel deceleration, which increases sharply just after the coefficient of friction, μ, reaches a maximum. By controlling the braking force to limit wheel deceleration, wheel locking is avoided.

Configuration of the ABS system

A typical ABS configuration is illustrated in Figure 7.2 with the components shown in more detail in Figure 7.3. The speed of each wheel is continuously detected by a magnetic reluctance type wheel speed sensor located in each hub. The sensor consists of a small permanent magnet surrounded by a pickup coil, and aimed at a toothed reluctor ring that turns with the wheel (Figure 7.4). As each tooth moves past the sensor the magnetic flux within the pickup coil alternately rises and falls, so generating an alternating current in the coil. Typically the ring has 45 teeth, which means that a current alternation occurs for every

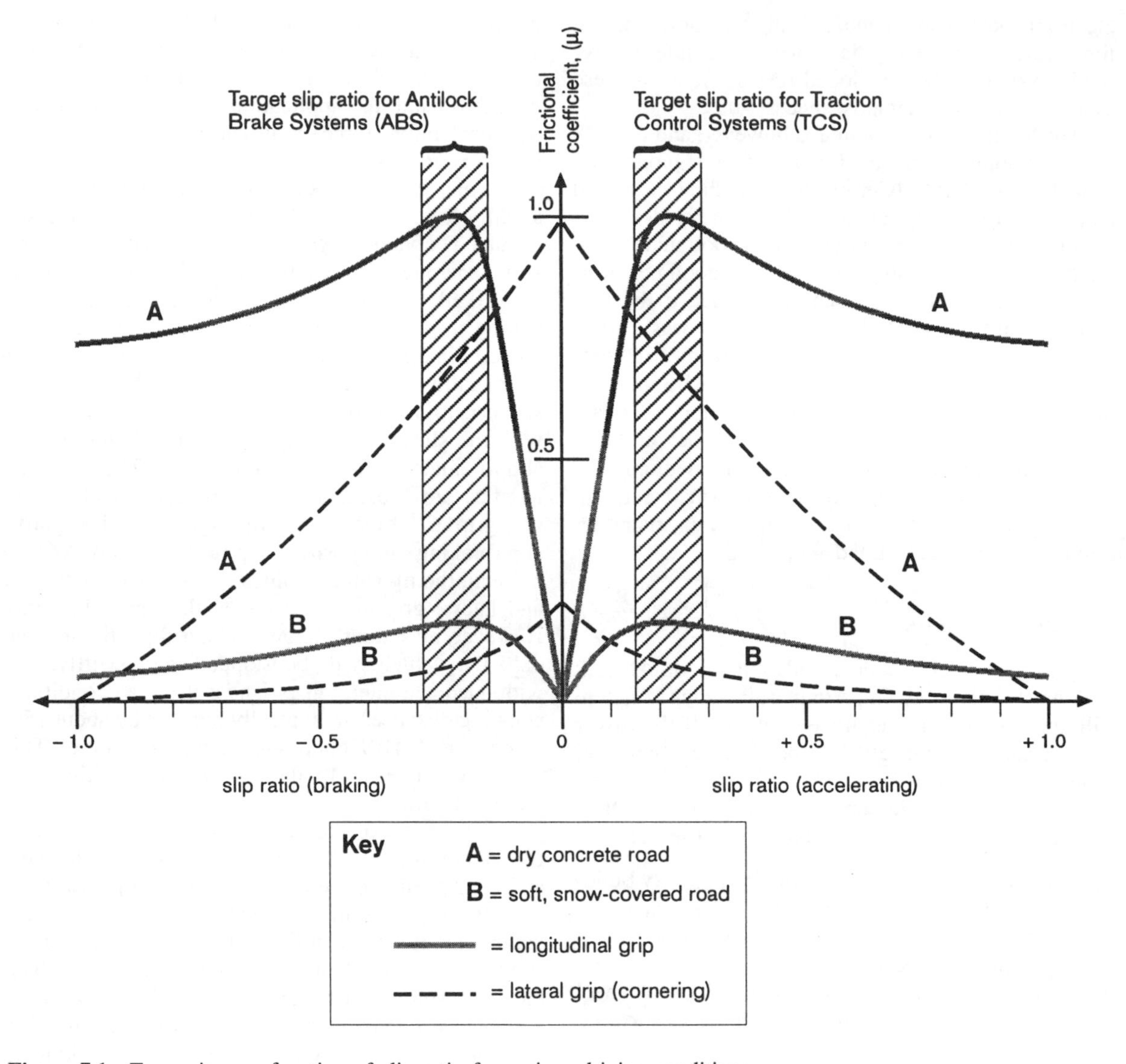

Figure 7.1 Tyre grip as a function of slip ratio for various driving conditions

8° of wheel rotation. The frequency of the generated ac is thus proportional to the speed of rotation of the wheel. After filtering, to remove interference, this speed signal is converted into clean 5 V pulses, suitable for use by a microprocessor which forms part of the ECU. The microprocessor uses this frequency information to calculate the wheel speed and rate of deceleration during braking. If a set limit is exceeded, the ECU can issue commands to solenoid valves in the hydraulic modulator unit to either prevent any increase in braking force (HOLD braking pressure), or if necessary, to reduce the braking force (DECREASE braking pressure). Once the target slip ratio is re-established, the ECU de-energizes the solenoid valves and so returns brake control to the driver. If the driver is still applying too much brake pressure the ECU will again HOLD or DECREASE the braking pressure to maintain the required slip. This control cycle is repeated rapidly (usually between 5 and 15 times per second) to keep the tyres in the region of maximum grip and prevent locking.

The Bosch hydraulic modulator unit

Although there are now many different ABS systems on the market, they all operate on the same basic

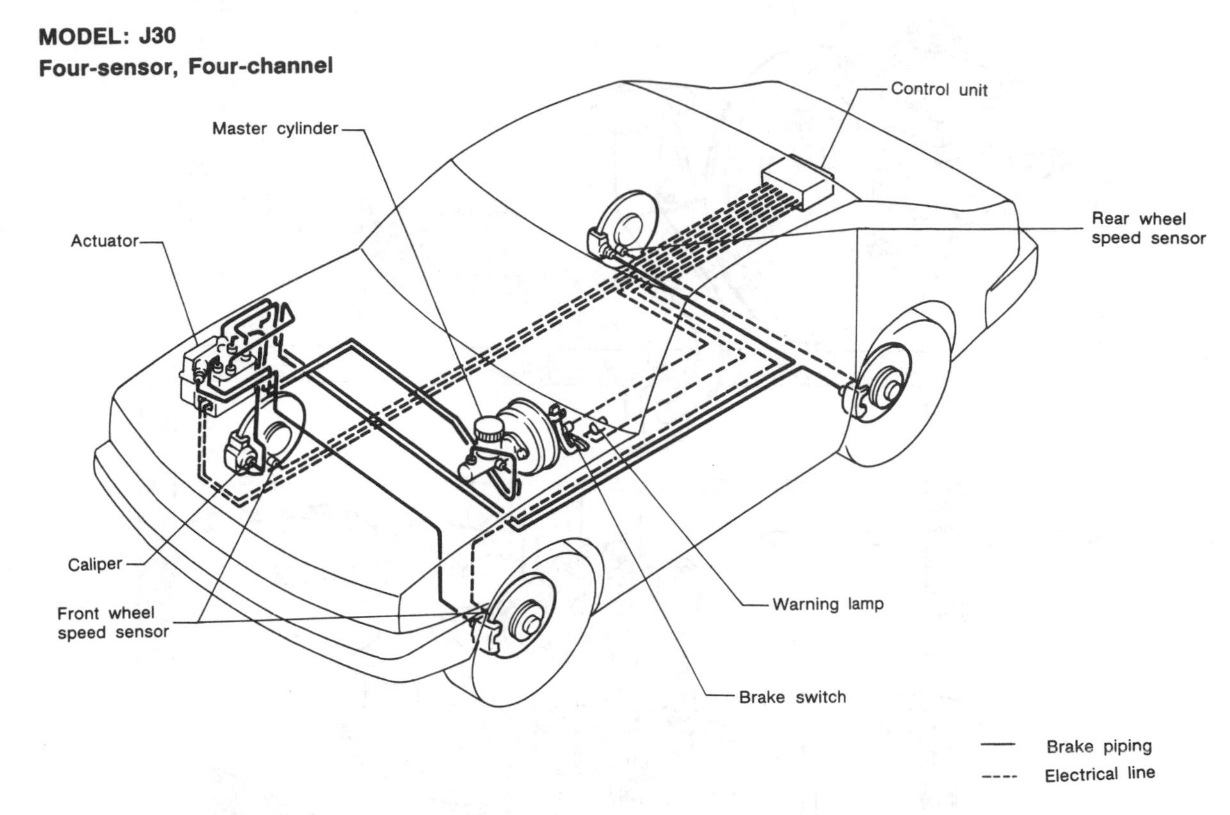

Figure 7.2 Configuration of ABS system (*Nissan*)

principle. Since Bosch ABS is still a market leader, the Bosch hydraulic modulator system will be considered here in detail.

The hydraulic modulator unit is a key component of the Bosch ABS system. It is a sealed unit consisting of solenoid valves, hydraulic accumulators and a fluid return pump. The solenoid valves and return pump are energized by the ECU via their respective relays.

The operation of the hydraulic control circuit for a single wheel cylinder is illustrated in Figure 7.5 (a,b,c). When no energizing current is applied to the solenoid valve (normal driving) it connects the brake wheel cylinder directly with the master cylinder, allowing conventional brake operation as in Figure 7.5(a). Since the driver may readily increase the braking force via the brake pedal this is termed the pressure INCREASE position.

If the ECU detects excessive wheel slip during braking, a low-level energizing current (about 2 A) is applied to the solenoid valve [2] to move it to the pressure HOLD position, as shown in Figure 7.5(b). This isolates the wheel cylinder from the master cylinder and so maintains the fluid pressure in the wheel cylinder (and therefore the braking force) at its previous value. Any increase in braking force due to an increase in the driver's pedal pressure is thus prevented.

If the wheel slip continues at an excessive rate then the ECU must reduce the braking force to avoid wheel locking. A higher energizing current (about 5 A) is applied to the solenoid valve, moving it to the pressure DECREASE position which is illustrated in Figure 7.5(c). The wheel cylinder is still isolated from the master cylinder, but is now connected to the return pump circuit. The pump is switched on by the ECU and draws fluid away from the wheel cylinder, pumping it back, via the accumulators, to the appropriate master cylinder circuit. In this way the braking force is reduced and fluid is returned to the master cylinder against pedal pressure. The driver feels the pedal rise slightly ('kickback') beneath his foot.

The function of the accumulators is to momentarily hold the returning high pressure fluid before it flows into the master cylinder, thereby minimizing kickback and preventing heating and foaming of the fluid.

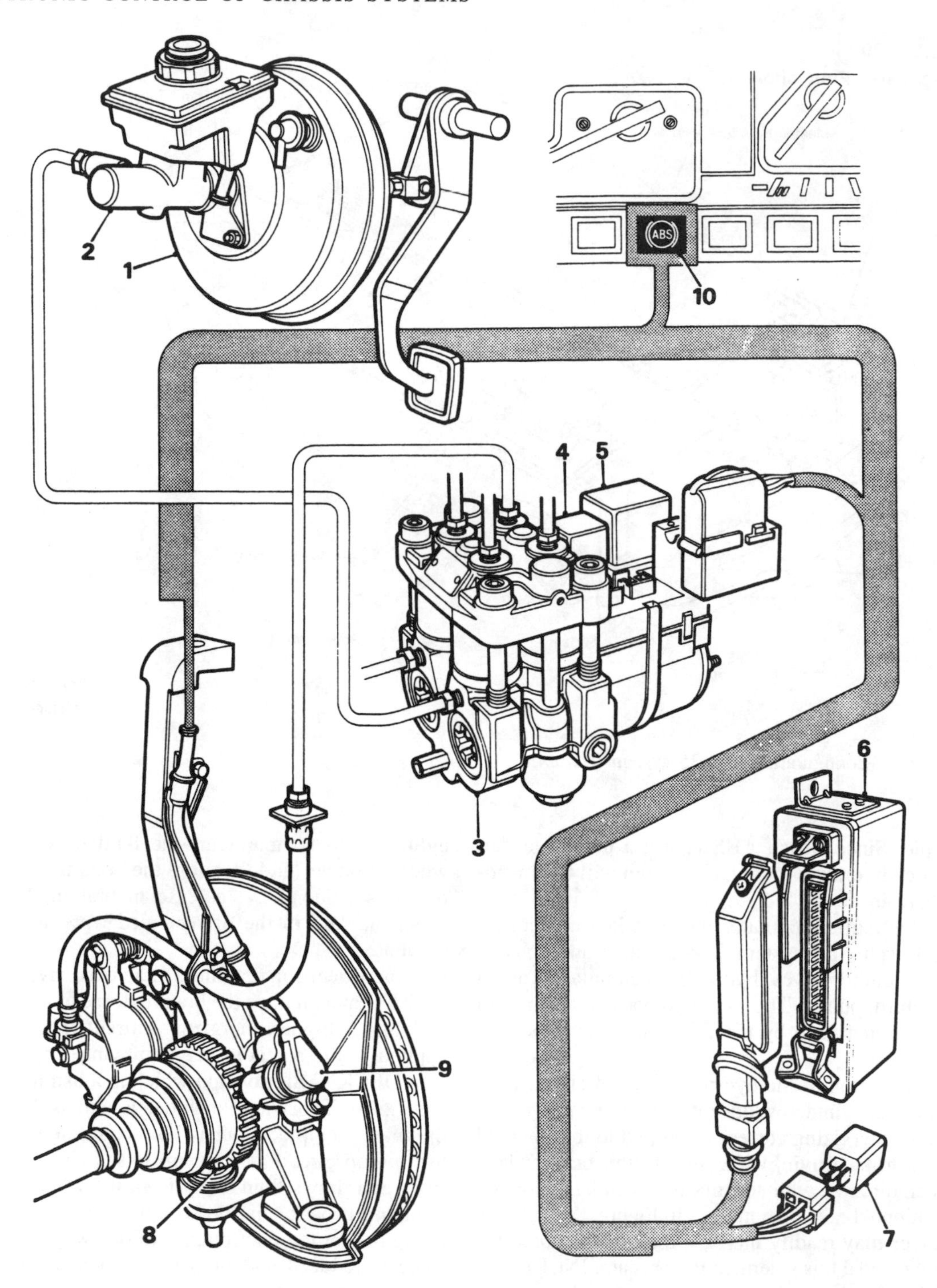

Figure 7.3 ABS system components: [1] brake servo; [2] master cylinder; [3] hydraulic modulator; [4] solenoid valves relay; [5] return pump relay; [6] ECU; [7] relay; [8] reluctor ring; [9] wheel speed sensor; [10] warning lamp (*Rover*)

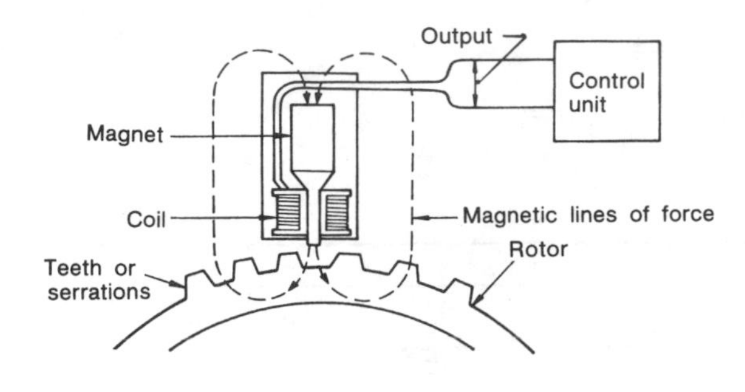

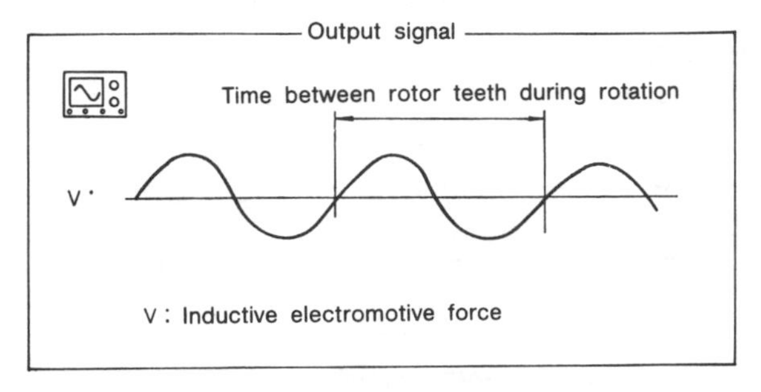

Figure 7.4 Magnetic reluctance (MR) ABS wheel speed sensor (*Nissan*)

Hydraulic actuator unit

The hydraulic actuator unit is used as an alternative to the modulator unit described above. There are many variants, including systems manufactured by Teves, WABCO and Honda, but all operate on the principle of using hydraulic energy stored in an accumulator. An example of actuator unit operation is given here.

The actuator unit itself consists of a motor-driven pump, hydraulic accumulator, pressure switch, ECU-controlled solenoid valves and expander piston assemblies.

The accumulator is a robust steel flask, separated into two chambers by a thick rubber diaphragm. One chamber is factory-filled with highly compressed nitrogen gas that maintains pressure on the brake fluid which fills the other chamber. The pump produces the high pressure brake fluid (typically at 100–200 bar) that is stored in the accumulator and used as a power source for the operation of the ABS system. A pressure switch controls the motor, ensuring that the accumulator is always fully charged with fluid at working pressure.

The operation of the actuator unit is illustrated in Figure 7.6 for a single wheel cylinder. The pressure applied to the wheel cylinder is controlled by the inlet and exhaust solenoid valves, which govern the position of the expander piston and cut-off valves. Four control modes are possible; normal braking (ABS not in operation), REDUCED pressure mode, HELD pressure mode and INCREASE pressure mode.

(i) *ABS not in operation.* The ABS ECU opens the inlet solenoid valve but closes the exhaust valve. The expander piston is pushed down the bore to open the cut-off valves. Master cylinder pressure is directed to the wheel cylinders.

(ii) *REDUCE pressure mode.* The inlet valve is closed and the exhaust valve is pulsed open, allowing fluid above the expander piston to return to the reservoir. The expander piston moves up the bore, allowing the cut off valves to isolate the wheel cylinder from the master cylinder and creating an expansion volume for the wheel cylinder fluid. This reduces the pressure in the wheel cylinder.

(iii) *HELD pressure mode.* Both solenoid valves are closed, keeping the expander piston at a fixed position in its bore. This isolates the wheel cylinder from the master cylinder, with no change in fluid volume or pressure.

(iv) *INCREASE pressure mode.* In this mode the exhaust valve is closed and the inlet valve is pulsed open. High pressure fluid from the accumulator forces the expander piston downwards. The cut-off valves remain closed but since the wheel cylinder fluid volume is decreased, fluid pressure rises.

Control channels

An ABS hydraulic control unit is described by the number of braking circuits (or 'channels') that it regulates. Two-, three- and four-channel systems are available.

In a two-channel system, the hydraulic unit controls two diagonally split circuits, each circuit consisting of one front wheel with its diagonally opposite rear wheel. The system is cheap and simple since speed sensors need be fitted to the front wheels only. Whatever pressure can be applied to each front wheel without causing it to lock is also applied to its opposing rear wheel. This is not a particularly good compromise since the rear wheels could be under or over-braked and some directional instability may then arise, requiring quick steering inputs from the driver to keep the vehicle straight. Nevertheless several manufacturers, including AP Lockheed and ATE (Teves), have developed such systems for use on small cars. Despite their disadvantages they still offer a useful gain in performance over conventional braking systems.

Three-channel systems have become the basic ABS configuration for medium and large cars. The ECU controls each front wheel separately and the rear wheels as a pair. Four speed sensors are used, but the braking force simultaneously applied to both rear wheels is governed by the wheel with least grip (this

(a)

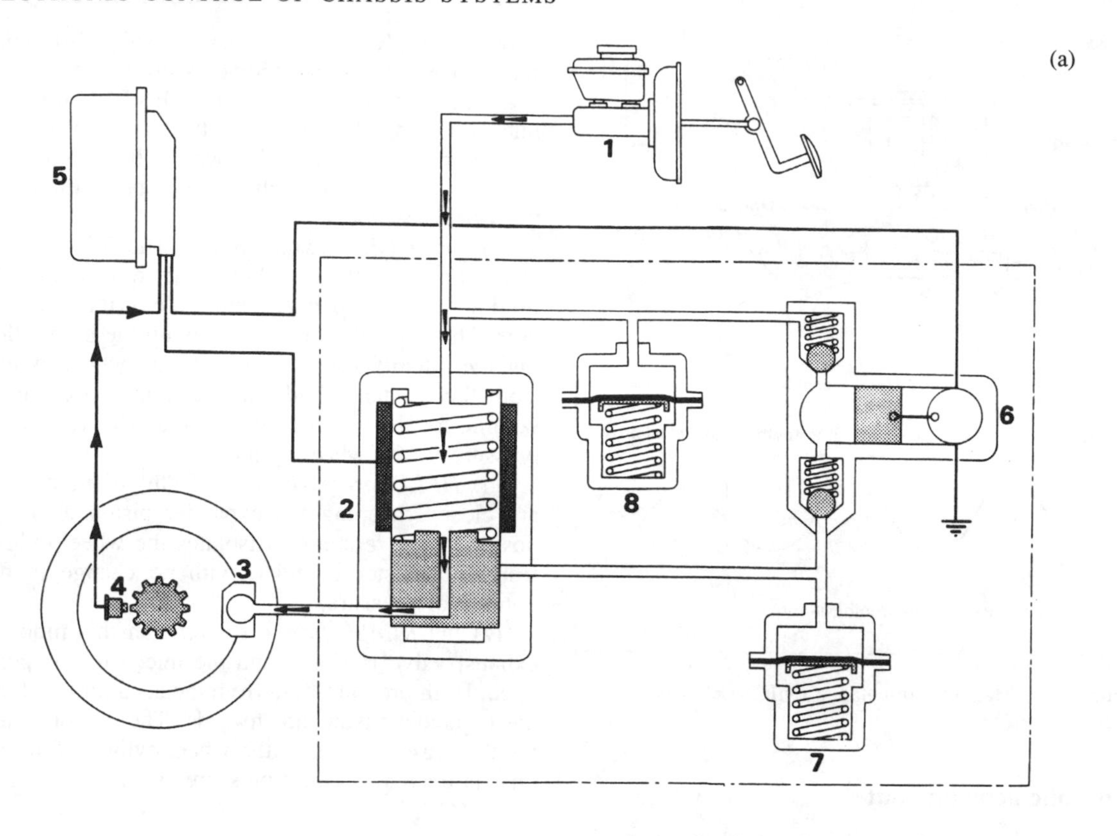

(b)

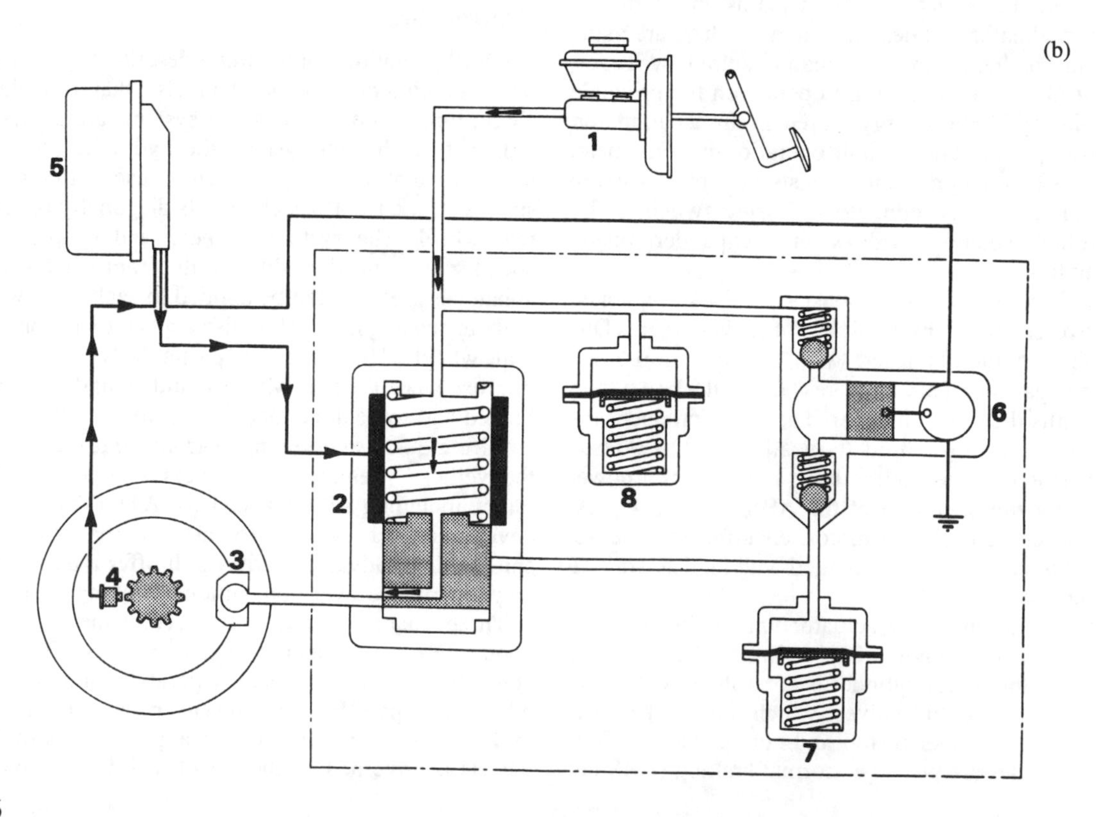

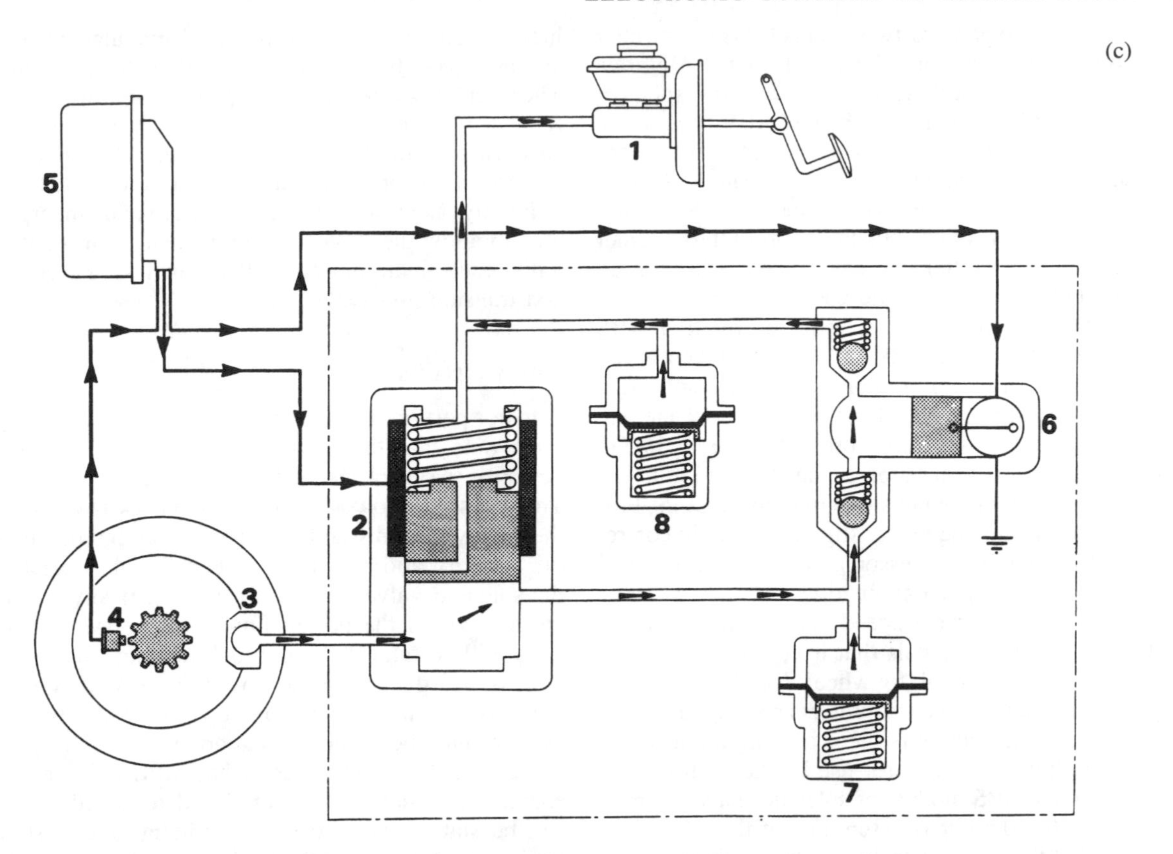

Figure 7.5 Operation of ABS hydraulic control circuit: (a) normal braking and ABS INCREASE pressure position; (b) pressure HOLD position; (c) pressure DECREASE position. Key: [1] master cylinder; [2] solenoid valve; [3] brake wheel cylinder; [4] wheel speed sensor; [5] ECU; [6] return pump; [7] accumulator

is termed 'select low' control). Three-channel control has certain advantages over two-channel control. Since the rear wheels are braked as a pair it provides good yaw stability where conditions are difficult, as on a winter road that is dry near to the centre, but covered in ice and slush near to the kerb (a so-called 'split-μ' surface).

Four-channel ABS provides individual monitoring and control of each wheel, resulting in optimum stability and the shortest possible stopping distance. To maximize stability on split-μ surfaces many four-channel systems incorporate additional signal processing to compare wheel deceleration rates on opposite sides of the vehicle. Large differences in deceleration suggest that the braking forces are severely imbalanced, possibly leading to yaw, and making the vehicle difficult to control. The ABS ECU detects this situation and takes corrective action by limiting the rate of brake pressure rise applied to the high-grip wheels. This means that the braking imbalance grows slowly and so the yawing force rises only gradually, making driver correction much easier.

Unfortunately this delay can cause problems when braking while taking a corner on a high-μ surface. In these circumstances some braking imbalance is essential to promote understeer and ensure steering stability. To avoid this problem the four-channel ABS system can be fitted with a fifth sensor that detects lateral acceleration, causing the delay to be switched off when lateral grip is high during cornering.

Electronic control unit

Most ABS ECUs are microcomputer-based controllers, incorporating ASICs and at least one microprocessor to ensure fast and reliable data processing. For a simple and inexpensive two-channel ABS controller a single 8-bit microcontroller, such as the Motorola MC68HC11A8, is still the most cost-effective solution. For more sophisticated four-channel ABS systems the requirement for faster and more accurate calculations means that highly integrated 16-bit technology is required. A good example is Motorola's MC68HC16Z1 16-bit microcontroller, which incorporates a timer processing unit (TPU) that can process wheel speed signals independently of the CPU.

To illustrate the sophistication of modern ABS controllers, the Teves Mk 20 ECU is taken as an example (Figure 7.7). It combines ABS and ASR (a form of traction control) in a single package and uses electronic brake force proportioning (EBV) to obtain maximum retardation at the locking limit. The ECU circuit board contains two microprocessors, Processor 1 performs the wheel slip and solenoid control calculations whilst Processor 2 is dedicated to signal input and wheel speed evaluations.

Data from the wheel speed sensors arrives at the ECU as sine-wave signals that have a frequency and amplitude proportional to wheel speed. The wheel speed sensor interface circuit filters and amplifies these signals, and then converts them into digital pulses of fixed amplitude, suitable for use by the 8-bit processor. Since the ABS modulator solenoids have response times of 10–20 ms it is essential that the ABS ECU can predict the onset of locking in a shorter time than this. Typically, the ECU must be able to assess wheel speeds every 5–10 ms (i.e. 100–200 times each second) and it is the 8-bit processor which does this. Secondary functions are detection of brake operation (via the brake light switch) and monitoring of the supply voltage (via the ignition feed).

Using the calculated wheel speeds, the 16-bit processor then carries out the anti-lock strategy according to preprogrammed control parameters for the particular vehicle. It computes values for wheel speed, wheel acceleration (or deceleration), control reference speed, wheel slip ratio and pressure reduction speed. The control reference speed is a speed which decreases with time and represents the theoretically optimum deceleration for the vehicle. It is indicative of a wheel rotating at the predetermined target slip ratio where tyre grip is greatest. A wheel rotating at the pressure reduction speed is operating at the maximum permissible slip ratio, just short of locking.

If a measured wheel speed falls below the calculated control reference speed then the ABS system intervenes to prevent the wheel from locking. The processors send solenoid and motor control signals to the respective drive circuits and relays. The relatively high currents required by hydraulic modulator valves are produced by the solenoid valve drive circuits. These are high power stages, incorporating current regulators so that a constant current is delivered to each valve. Monitoring of each solenoid current enables short or open-circuit faults to be detected.

Finally, a communications circuit may be incorporated to allow the ABS ECU to exchange information with other electronic controllers such as the engine and transmission ECUs.

Safety circuits

A major portion of the ECU is concerned with safety. Each time that the ignition is turned on, the ECU performs a self-check of the complete ABS system. This usually involves both a static check (evaluating the steady currents and voltages at various points in the system), and a dynamic check (briefly operating the solenoid valves and return pump). If any signals are not within the design limits for the system, the ECU will switch off the anti-lock function and put on the dashboard warning light. With most ABS systems the brakes will still function conventionally, but the vehicle must be repaired as soon as possible.

ABS ECUs also incorporate diagnostic software that enables fault data to be stored and retrieved. Faults can be static, dynamic or intermittent and must be read using either an off-board diagnostic tool or by reading a 'flash code' from the ABS warning light.

The ABS control cycle

Figure 7.8 shows the signals associated with an anti-lock control cycle for a single wheel of a typical ABS system. When the brakes are first applied, the ECU uses wheel sensor data to calculate the actual vehicle speed, a hypothetical control reference speed and a pressure reduction speed.

As the brake pedal is depressed, the fluid pressure quickly rises, braking force increases proportionately and wheel slip increases. If the wheel speed falls below the control reference speed then the slip ratio is likely to become excessive and so the ECU sends a pressure HOLD signal to the solenoid valve to prevent the wheel from locking. If the wheel continues to decelerate its speed will drop below the pressure reduction speed and it will lock. The ECU therefore operates the pressure DECREASE valve and switches on the return pump motor, thereby reducing the braking force and allowing the wheel to speed up again.

When the wheel speed rises above the control reference speed, the solenoid valve is moved to the

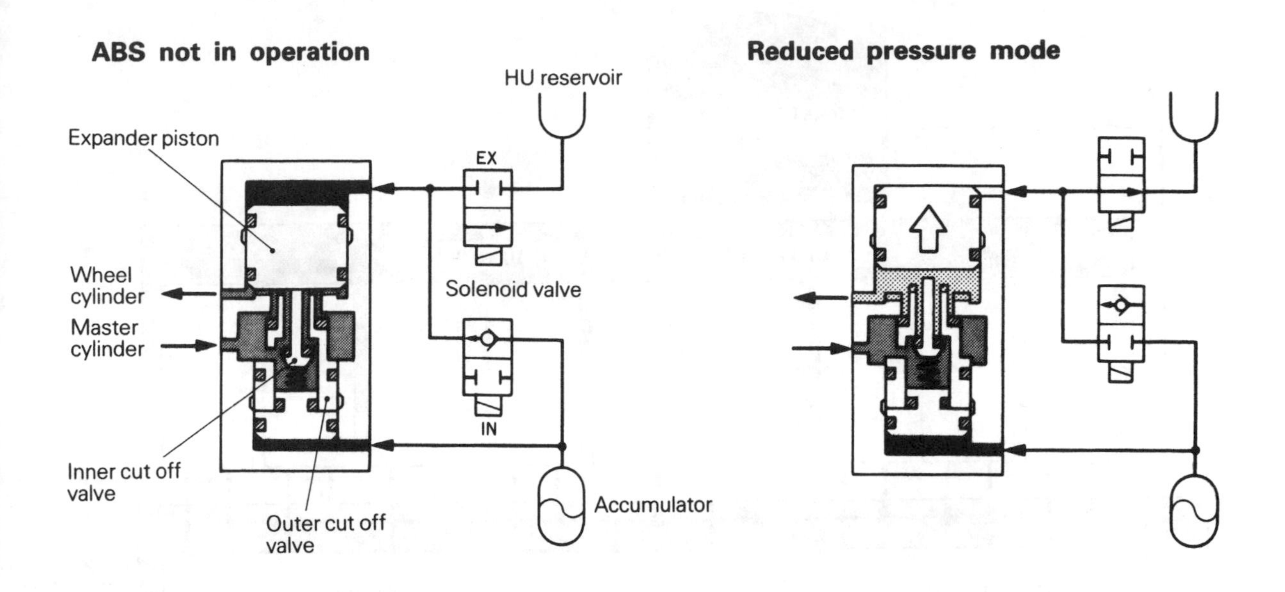

Increased pressure mode

Held pressure mode

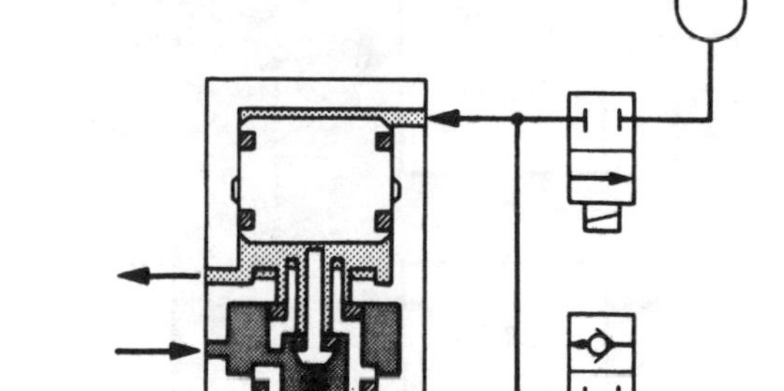

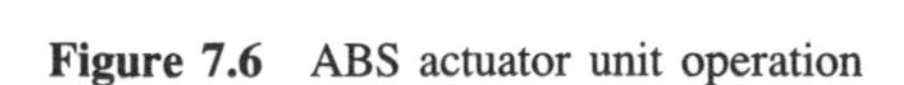

Figure 7.6 ABS actuator unit operation

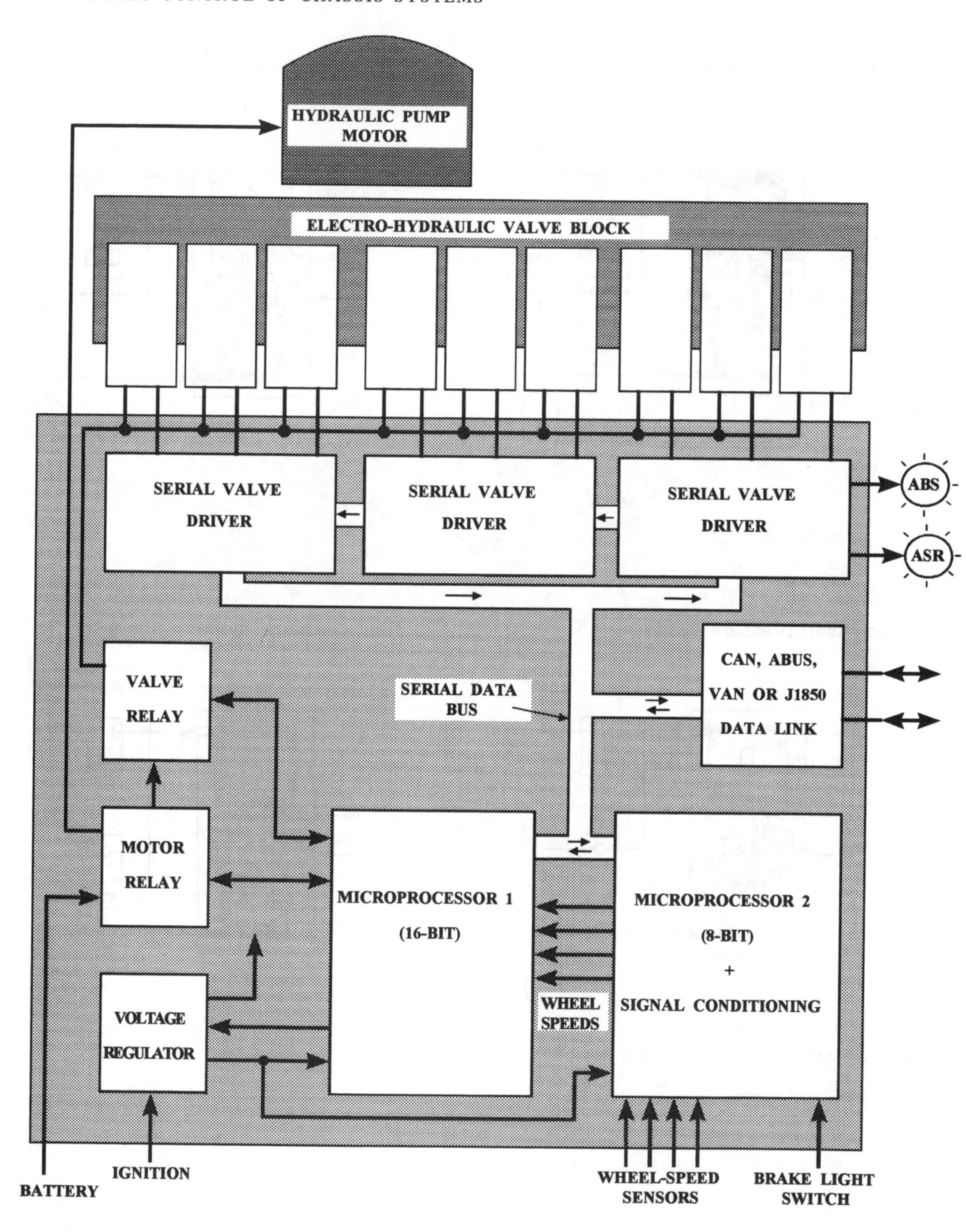

Figure 7.7 Block diagram of ABS ECU employing two microprocessors

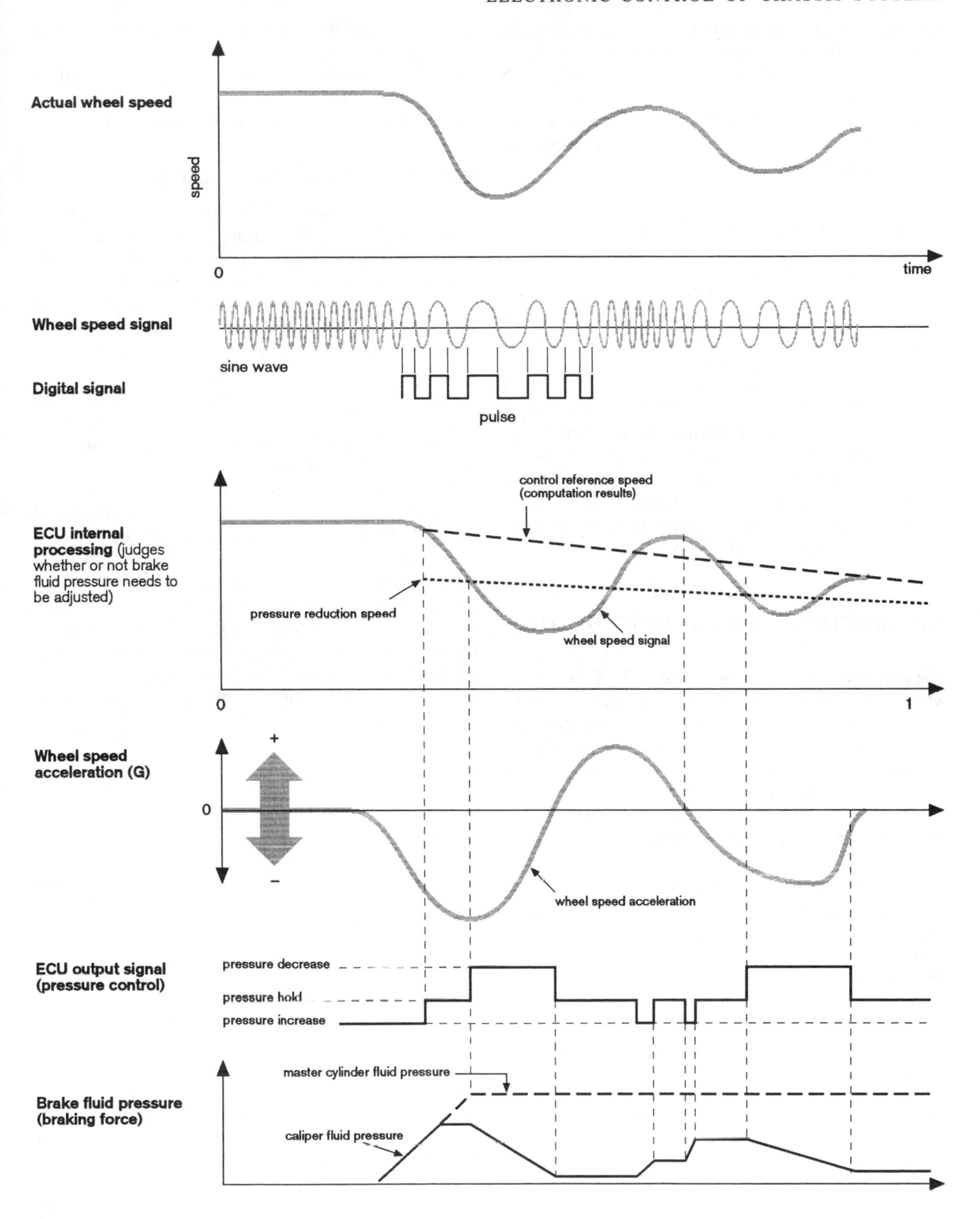

Figure 7.8 Anti-lock control cycle for a single wheel

pressure INCREASE position and the control cycle starts again. In this manner the wheel speed oscillates either side of the control reference speed and so the tyres are operated at the optimum slip ratio. The vehicle is therefore decelerated as quickly as the laws of physics allow, with no wheel locking and full steering control.

Electronically controlled brake servo (EAS)

EAS is an electronically controlled brake servo system developed by Lucas to be used in conjunction with ABS. Research conducted by Lucas has shown that in an emergency situation, about 70% of drivers do not use the full potential of a car's braking system because they apply insufficient pressure to the foot pedal. EAS overcomes this problem by detecting a harder than normal brake application and then applying a disproportionately greater amount of servo assistance. Further developments of EAS include a 'hill-hold' facility and vehicle load sensing so that a given pedal pressure always results in the same retardation, irrespective of vehicle payload.

7.2 TRACTION CONTROL SYSTEMS (TCS)

The aim of a traction control system (TCS) is to limit wheelspin when starting off or accelerating on a slippery road surface. In many ways TCS is the opposite of ABS, working to maximize grip when accelerating, rather than when braking. It intervenes to limit the torque applied to the driving wheels, thereby maximizing the friction coefficient, μ, by maintaining the tractive slip ratio at an optimum value (see Figure 7.1).

Although not a replacement for mechanical four-wheel-drive, TCS is a useful safety feature in that it increases the lateral stability of the vehicle in slippery conditions, making the vehicle much easier to handle. On a front wheel drive car it permits acceleration with full steering control, whilst on a rear wheel drive car it stops the tail breaking away when the driver uses too much power. These advantages are illustrated in Figure 7.9.

Although there are a variety of TCS systems in production they all function by electronically reducing the driving wheel torque when excessive slip is detected. This is achieved by one, or more, of the following interventions:

(i) Partially closing the throttle valve to reduce engine torque.
(ii) Reducing the quantity of injected fuel to reduce engine torque.
(iii) Retarding or cancelling the ignition spark to reduce engine torque.
(iv) Applying the brakes to the driven wheels.

TCS via throttle and brake control

In this configuration TCS is provided by electronic control of the throttle and application of the brakes. Brake application is particularly effective since under split-μ conditions it can provide the level of traction control offered by a limited-slip differential, but with better precision. TCS using the braking system is also cost-efficient in that many components can be shared with the ABS system. Specifically, the wheel speed sensors, hydraulic actuator unit and ECU can be made to perform a dual function. With the addition of an Electronic Throttle System (ETS) a sophisticated TCS system may be provided.

It is unfortunate that although elegant, this strategy is not particularly well suited to vehicles fitted with automatic transmission. Applying a braking torque to the driven wheels of an automatic car could cause excessive engine power to be dissipated in the transmission's torque converter, leading to overheating and damage. Cars fitted with automatic transmission are therefore often only equipped with TCS systems that control engine torque. This leads to poorer performance when accelerating on split-μ surfaces since no 'limited-slip' facility is provided.

TCS brake control

When a front wheel drive car sets off on a slippery surface, the driven wheels will start to spin. The TCS ECU computes a target driven wheel speed by taking a vehicle reference speed from the (undriven) rear wheels and adding an amount for the required slip ratio (say, 20%). If either front wheel exceeds this target speed then it is deemed to be suffering excessive slip and the appropriate brake is applied. This reduces slip on the low-grip side and transfers some torque, via the action of the differential, to the higher grip side to give maximum traction

As a refinement, some TCS systems allow a small amount of wheelspin when accelerating at low speeds. This gives the car a sporting feel and can be useful in that it may allow the spinning tyre to burn through thin ice and grip the road beneath.

During TCS operation, some engine power is dissipated in the braking components which become hot. Since excessive use of TCS could lead to overheating of the brakes, the TCS ECU may limit brake

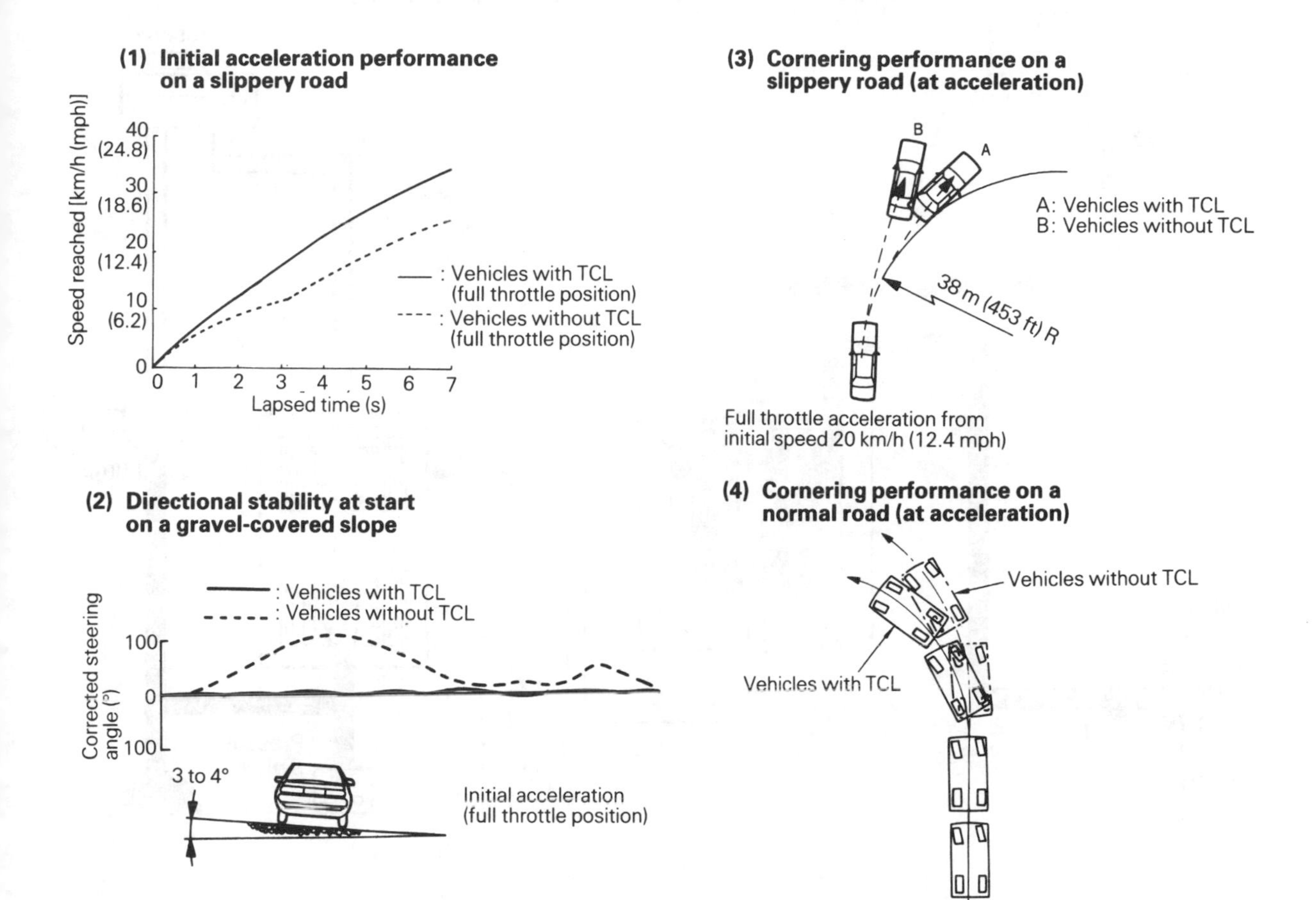

Figure 7.9 Advantages associated with Mitsubishi's 'TCL' electronic traction control (*Mitsubishi*)

application to a certain length of time within a set period. If TCS is required beyond this time then engine intervention alone is used.

Figure 7.10 shows one way in which TCS brake operation may be achieved. Hydraulic pressure is generated by the pump and maintained in the accumulator (shared with ABS). Under normal driving conditions the system is inactive and the TCS ECU keeps the inlet valve closed and the exhaust valve open. Under these conditions the boost piston is held at the bottom of its bore by the return spring. This means that the poppet valve is open, connecting the master cylinder to the wheel cylinder and allowing normal brake operation.

When excessive slip is detected, the TCS ECU changes the solenoid supply currents to open the inlet valve and close the exhaust valve. Accumulator pressure forces the boost piston upwards, closing the poppet valve and raising the wheel cylinder pressure to apply the brakes. This pressure may be held by closing both valves, or reduced by closing the inlet valve and opening the exhaust valve.

TCS throttle control

To control the output torque of the engine an electronic throttle system (ETS) can be used. Although described here as a component of TCS, ETS may be shared with the engine management system and the cruise control system.

An ETS uses a potentiometer-type sensor (Figure

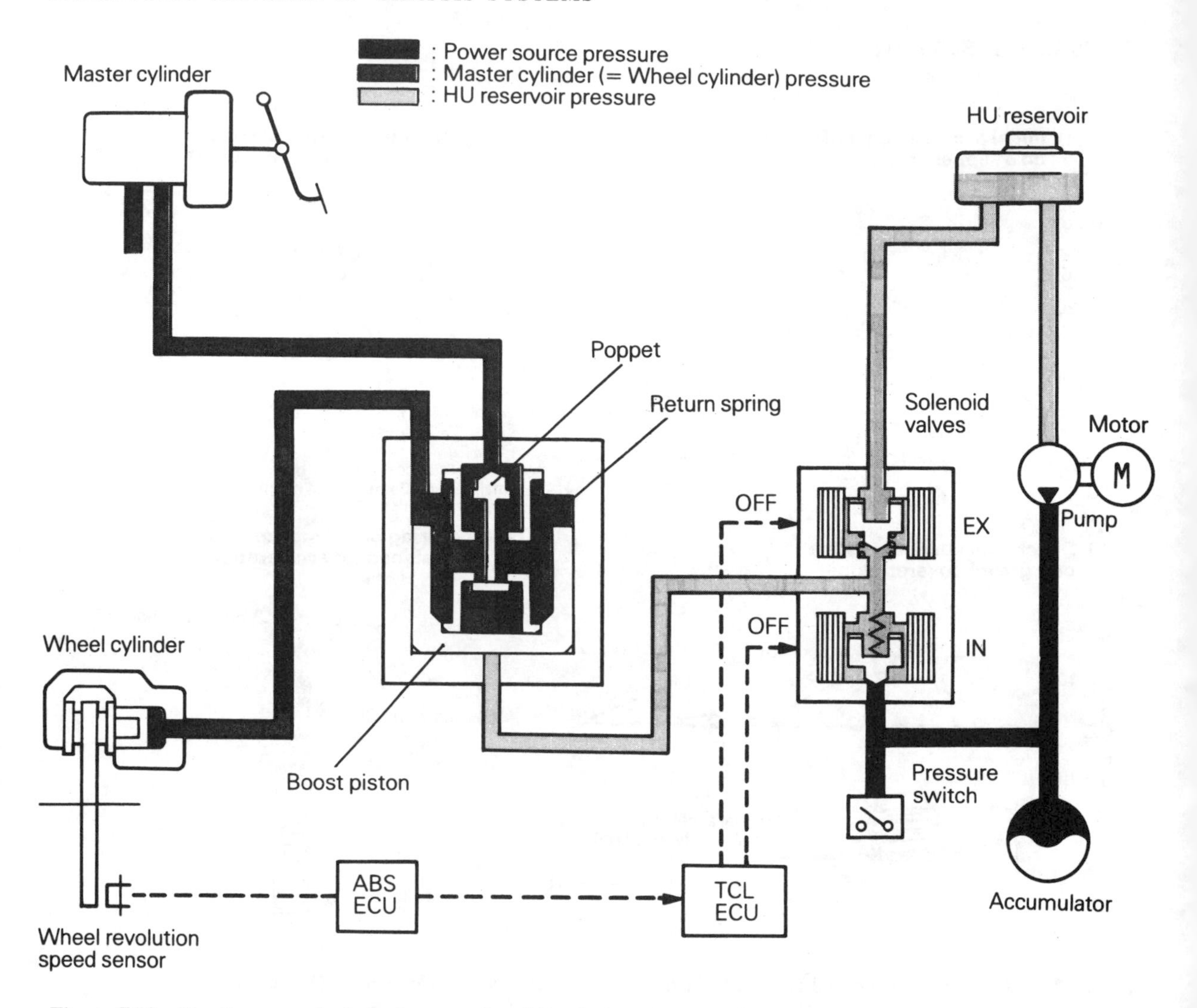

Figure 7.10 Traction control via brake operation (*Mitsubishi*)

7.11) to convert the accelerator pedal position into a control voltage for the ETS ECU. The ECU then produces drive voltages to operate a motor which provides very precise positioning of the throttle valve. Stepper motors and dc servomotors are used, the latter fitted with reduction gears and a position feedback potentiometer. ETS thus replaces the accelerator cable, although a cable is sometimes retained for 'limp-home' purposes.

To monitor for faults, safety switches are incorporated into the ETS pedal position sensor and throttle valve assembly. When the throttle valve is opened beyond a certain angle in response to accelerator depression, the throttle angle safety switch closes. Simultaneously, at the corresponding pedal depression, the pedal safety switch should open. Thus, the ECU can monitor proper operation by checking that on no occasion are both switches closed or both open. If either of these circumstances does arise then the ECU disables ETS and throttle operation reverts to 'limp-home' mode.

If, during TCS operation, the TCS ECU judges that wheel slip is excessive on both driven wheels it sends a request to the ETS ECU to reduce engine output torque. The ETS ECU then acts to reduce the amount of throttle by overriding the driver's demand and partially closing the throttle valve. Thus the resultant fall in engine output limits wheelspin. Once wheel slip drops to an acceptable level, the ETS returns full throttle authority to the driver.

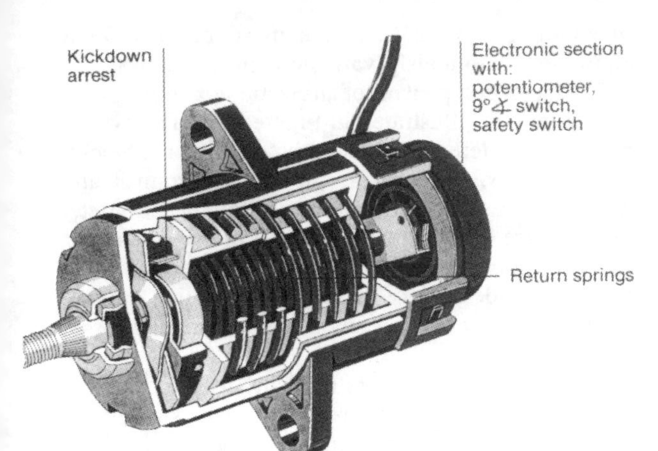

Figure 7.11 Throttle pedal position sensor used with ETS (*BMW*)

Electronic control unit

To reduce cost and complexity the TCS electronic control circuitry is generally incorporated into the ABS electronic control unit, as in Figure 7.7.

ETS is carried out as a separate function and uses a separate ECU. When wheel speed signals suggest that the driven wheels are slipping beyond the target slip ratio, the TCS ECU intervenes to operate the brakes and reduce the engine torque. Since the engine responds relatively slowly to throttle closure, the ECU must initially apply the brakes to stop the slip and give time for the engine torque to fall. Communication between systems may be achieved using a data link such as CAN. Where engine torque reduction is implemented using injection or ignition intervention, the TCS ECU communicates directly with the engine ECU.

7.3 ENGINE DRAG TORQUE CONTROL (EDTC)

Engine Drag Torque Control (EDTC) is a control system which may be integrated with TCS. EDTC prevents excessive slip of the driven wheels due to engine drag torque (engine braking). This may occur when driving on a slippery road surface if the driver changes to lower gear or suddenly lifts his foot off the throttle.

The EDTC system uses ABS/TCS wheel speed sensors to obtain wheel speed data from the driven and undriven wheels. If excessive wheel slip due to engine braking is detected, the EDTC system intervenes to limit the braking effect. This is achieved by:

(i) Commanding the ETS to slightly open the throttle valve to increase engine output torque.
(ii) Commanding the engine management ECU to slightly advance the ignition timing to increase the engine torque.

Despite the use of EDTC, sufficient engine braking for safe driving is still available. Additionally, if the brake pedal is depressed, EDTC is switched off and ABS control takes over.

7.4 ELECTRONIC CONTROL OF SUSPENSION

The function of the vehicle suspension system is to minimize the transmission of road surface irregularities to the vehicle body. At the same time the tyres must be kept in contact with the road and made to behave in a controlled fashion. This improves the ride comfort and minimizes undesirable motions of the body, resulting in increased stability and safety.

All suspension systems comprise springs, dampers (often inappropriately called 'shock absorbers') and locating arms to keep the components correctly aligned. It is the combined characteristics of the springs and dampers that determine the ride and handling qualities of a vehicle. Unfortunately the fundamental laws of mechanics dictate that the requirements for both good ride and good handling cannot be simultaneously achieved with a single value of spring stiffness and damping resistance. Good ride comfort, for example, demands soft springs to allow for generous vertical wheel movement on an uneven road. Good stability, on the other hand, demands firm spring and damper characteristics to limit undesirable body motion, such as rolling when cornering, diving when braking and squatting when accelerating. Suspension design therefore inevitably involves the adoption of a 'ride-handling compromise', where ride comfort and vehicle handling are traded against each other to provide the best compromise for the individual vehicle model.

In general, the ride-handling compromise of modern cars is so good that many automotive engineers would question the need to introduce the refinement of electronic suspension control. Nevertheless, during the 1980s many manufacturers introduced 'semi-active' suspension systems on luxury models. These systems work to automatically alter the suspension characteristics according to the prevailing driving conditions. For example, on a twisting road the suspension becomes firmer to give better handling, while on a smooth and straight highway the suspension softens to give a more comfortable ride.

Spring and damper systems

When a wheel strikes a bump, it is deflected upward, compressing the spring and storing energy in it. Once the bump has passed, the spring will attempt to release this stored energy by extending again. With no damper fitted, this results in the car body bobbing up and down in an oscillatory motion, since there is nothing to dissipate the stored spring energy. The function of the damper is therefore to dissipate energy by offering a resistance to spring motion, quickly eliminating any oscillation.

Most dampers are of the telescopic type consisting of a piston, fitted with tiny valves, which is forced to move in a sealed cylinder filled with oil. The piston divides the cylinder into two compartments and the valves allow the oil to pass from one to the other as it is worked up and down. In this way the damper provides resistance to motion, and energy is absorbed by the oil.

The amount of resistance provided by a damper depends upon the diameters of the valve orifices. When these are very small the damping force is high and the vehicle is said to be firmly damped. Conversely, large valve orifices allow the fluid to move freely between the cylinder compartments, so the damping force is low and the vehicle is said to be softly damped. Firm damping is good for vehicle stability in that it provides increased suspension stiffness during transient conditions (for example, turning into a corner or momentary sharp braking). It does not, however, increase the static suspension stiffness (the deflection of the suspension for a given steady load), since this depends on the spring stiffness.

Soft damping promotes good ride comfort because the springs are allowed to deflect more easily when encountering bumps. However the ride comfort will still very much depend upon the spring stiffness, since a soft spring will permit a greater deflection than will a hard spring and this will reduce the disturbance to the vehicle body.

Electronic damping control

Pioneered by the damper manufacturer Boge, electronic damping control allows the selection of different amounts of damping according to the prevailing road conditions. The system monitors various input parameters (vehicle load, road condition and driving style) and selects the most appropriate damper setting. The result is optimum damping over the full range of road conditions and driving styles. Most manufacturers allow for at least three settings; 'soft', 'medium' and 'firm', and the most recent systems allow for continuously variable damper control.

The major components of an electronic damping control system are illustrated in Figure 7.12. The system may be considered in terms of three functional blocks.

(i) *Sensors*. The sensors provide the control unit with the necessary information to allow the computer to recognize and evaluate a variety of driving styles, road conditions and vehicle loads. Figure 7.13 illustrates the wide variety of sensor inputs that may be used. In practice, only a few of the illustrated sensors need be used.

Acceleration sensors are normally used to measure forces on the car body. Front and rear axle sensors measure vertical accelerations at each end of the vehicle.

Longitudinal acceleration may be measured directly, or can be inferred from brake system pressure and throttle opening angle. This information is then processed to estimate the forces on the car body when the vehicle is braked or accelerated.

A steering angle sensor is used to determine the rate at which the steering wheel is turned, and hence allow the ECU to estimate the lateral forces on the vehicle.

Road speed information is generally obtained indirectly, by taking a digital signal from the instrument pack ECU.

(ii) *Electronic control unit*. This microprocessor-based ECU takes in data from the sensors and processes it to decide the most appropriate damper setting for the instantaneous road conditions. Software is used to calculate the frequency of body oscillations and then actuate the damper control valves to provide an optimum setting. The ECU is programmed to permit two damping strategies, selected by the driver via a dashboard rocker switch. 'COMFORT' allows the ECU to switch between all three damper settings, while 'SPORT' locks out the softest setting and allows switching between 'medium' and 'firm' settings.

(iii) *Dampers*. The electronically controlled dampers are illustrated in Figure 7.14. Each damper is fitted with two ON–OFF fluid control solenoids that are used to select the different damping settings. With no power applied, the solenoid valves are held shut by strong springs and so the dampers operate in the conventional manner, with 'firm' damping. The two additional characteristics ('medium' and 'soft') are achieved by opening the respective bypass solenoid valve to reduce the flow resistance. A fourth characteristic ('super-soft') may also be offered and is obtained by opening both bypass valves.

Versions with only one solenoid valve are also available, these have just two damping characteristics.

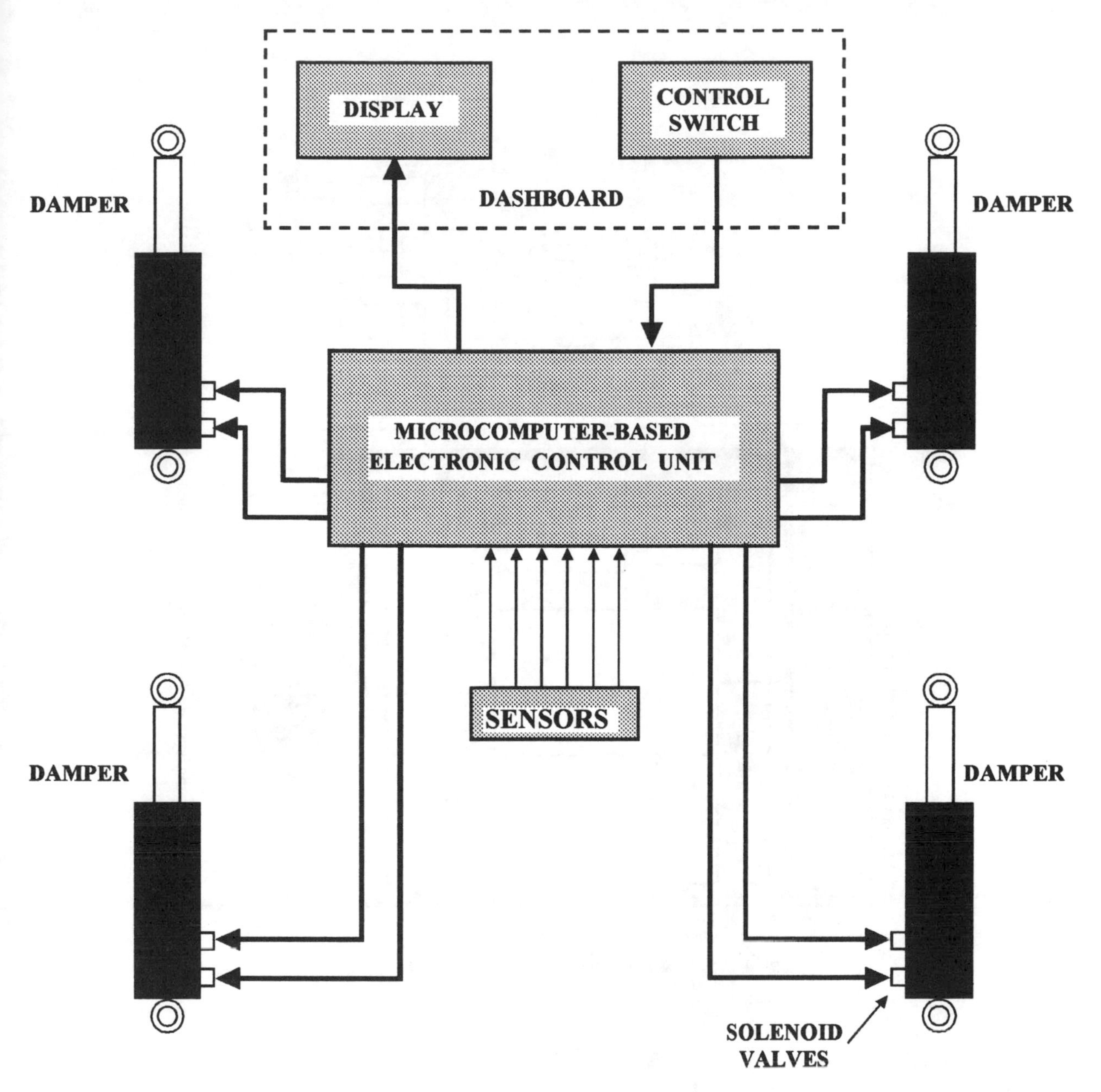

Figure 7.12 Organization of electronically controlled damping system

An alternative style of damper construction features the use of an oil-immersible dc motor which is located in the damper's piston rod. This construction has the advantage that it gives a compact package, compatible with all types of damper, including those with supplementary hydropneumatics. The motor controls the flow of oil to the piston by operating a rotary valve. With three orifices on the piston and two openings on the valve, one of three damping forces can be selected when the motor rotates the valve and either opens or closes the orifices.

Control strategy

Depending on the setting of the program switch (COMFORT or SPORT) the ECU uses sensor signals

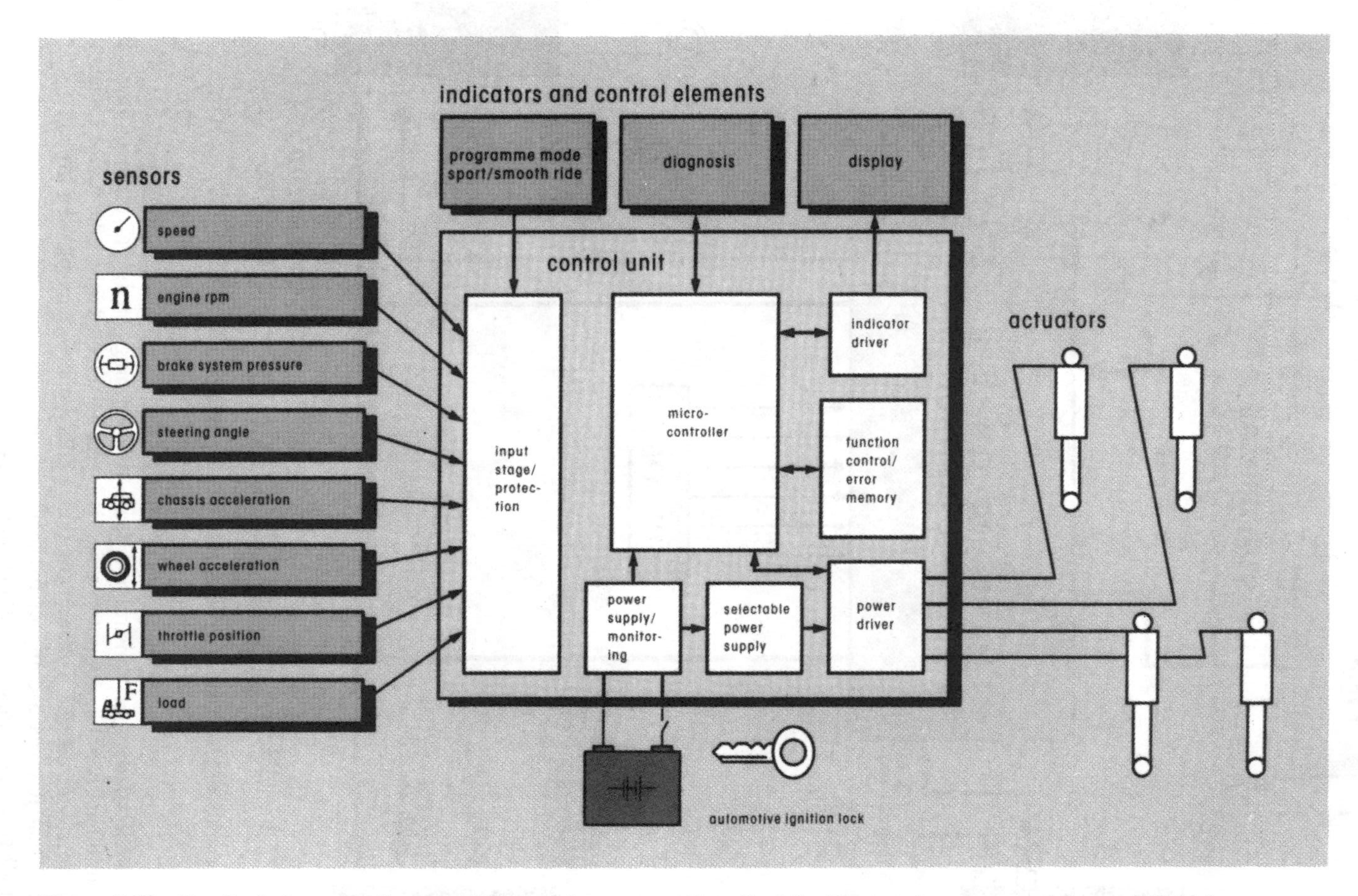

Figure 7.13 Detail of electronically controlled damping system. Not all of the illustrated sensors need be used (*Sachs*)

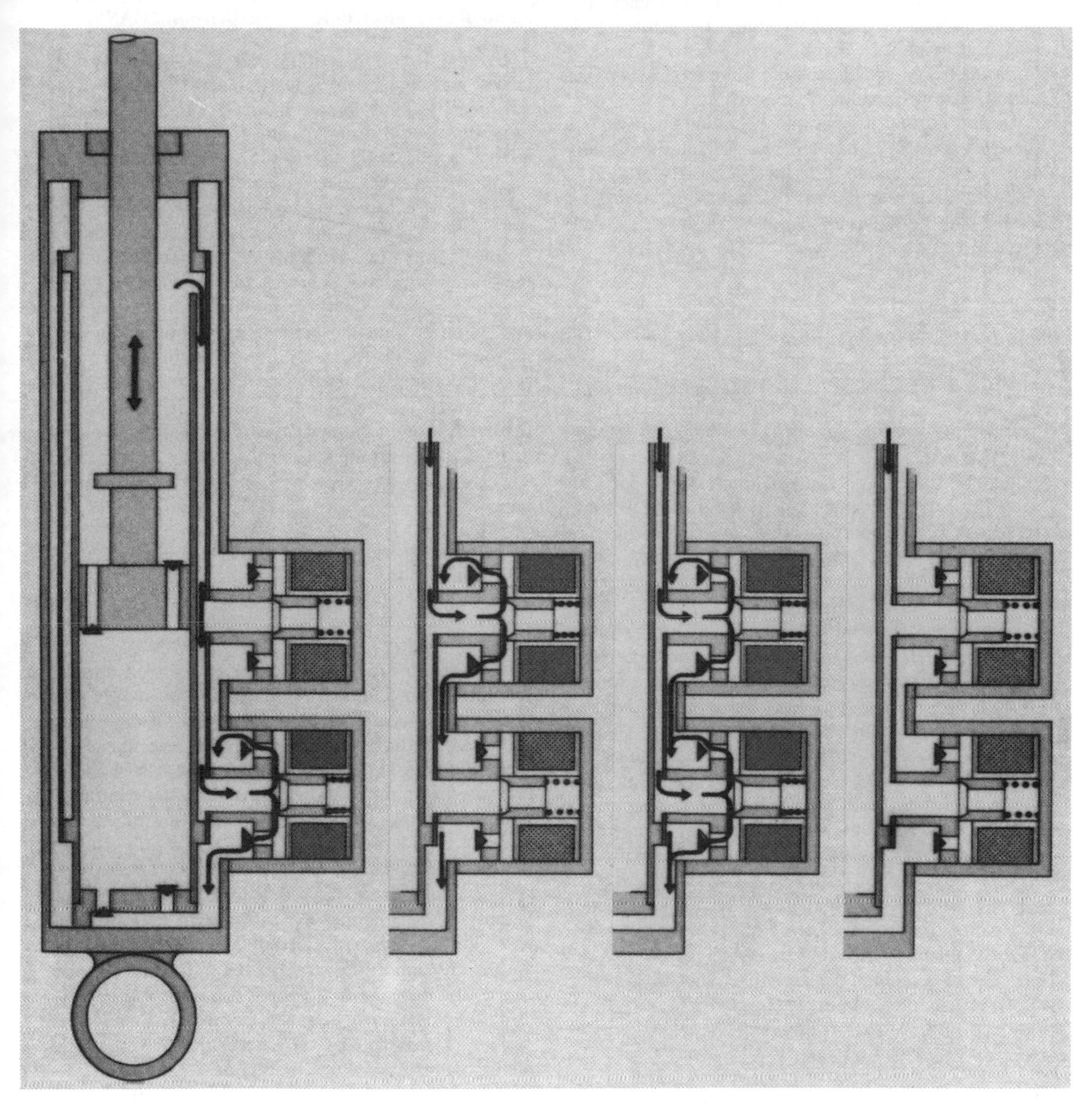

Figure 7.14 Operation of electronically controlled damper at different settings. From left to right – normal, soft, super-soft, firm (*Sachs*)

to judge the vertical body movement frequency. It is then kept within the target range by actuating the solenoids to alter the damping force. Since the switching occurs very quickly (within milliseconds) it passes unnoticed by the vehicle's occupants.

During normal driving the operating characteristic for each axle is computed and actuated separately. However, upon hard braking, hard acceleration or sharp steering manoeuvres, the dampers on both axles are controlled simultaneously. This reduces pitching and rolling of the body, improving comfort and stability.

Alternative electronic damping systems

The type of adaptive damping control system described so far offers three or four damper settings. A logical extension of such a system is continuously variable adaptive damping, in which an electromechanical actuator is used to give any setting from 'super firm' to 'super-soft'. The advantage being that the damping force can be precisely tailored to the prevailing requirements, with no need for compromise. Suitable actuators include dc servomotors, stepper motors and duty-cycle controlled solenoids.

Another recently introduced system is the piezoelectric damper. This uses a stack of about 100 piezoelectric discs that are incorporated into the piston rod to both detect road condition and alter the damping force. During normal driving on a smooth road the dampers are set to 'firm', giving good handling and stability. When the vehicle traverses a step or ridge, the piezoelectric discs are subjected to a compressive force and so generate a small output voltage. The ECU measures this voltage and, together with data from other sensors, it makes a decision to apply a voltage to the piezo elements. Through the action of the reverse piezoelectric effect the discs then become slightly displaced, lengthening the piston rod and opening an additional damper valve. This has the effect of decreasing the damping foce as the vehicle traverses the step, giving better ride comfort.

Electronically controlled air suspension

An air suspension system uses a volume of compressed air instead of, or sometimes in addition to, a conventional steel coil spring. The advantage of the air spring is that it provides a variable spring rate which leads to a constant suspension frequency for all load conditions. Ride quality is improved and the system can be enhanced to give ride height control, or combined with damping control to give improved ride comfort.

Land Rover electronic air suspension (EAS)

The Land Rover Electronic Air Suspension system (EAS) was developed by Dunlop and Land Rover for use on four-wheel drive vehicles. An ECU constantly monitors the ride height using four sensors mounted at each corner of the vehicle. Any variation in sensor output which lasts 12 seconds or longer is detected by the ECU, which responds by raising or lowering the appropriate corner of the vehicle. The vehicle height is thus automatically controlled to take account of the number of passengers and the load. A fixed height improves stability, maintains headlight aim and helps prevent the vehicle from 'bottoming out' on rough terrain. As an additional feature, the driver may select non-standard ride heights for special circumstances such as loading, towing or driving through floods. Although the system is designed to level the vehicle, its basic characteristics do not change and so the system is not of the 'active' or 'semi-active' type.

Components for EAS

Figure 7.15 shows the locations of the major components of the EAS system.

Air springs are used in place of the coil springs found on a conventional vehicle. They are simply rubber air bags, located on aluminium end plates (Figure 7.16). Depending on the ride height selected, compressed air is supplied to the spring at a working pressure of about 6–10 bar via a pneumatic harness. A conventional damper is mounted alongside the air spring.

The microprocessor-based ECU is sited in the passenger compartment and receives inputs from the ride height sensors. It operates ride height control solenoid valves which are located in the valve block. These valves are either ON or OFF to either increase or reduce the pressure in the air springs, raising or lowering the vehicle respectively. Compressed air for the system is provided by an electrically driven single piston compressor which maintains a 9.4 litre air reservoir at a pressure of between 7.5 and 10.0 bar. A pressure switch, fitted to the reservoir, controls the cut-in and cut-out points of the compressor via the ECU.

Three dashboard-mounted control switches are connected to the ECU and allow the driver to select alternative height programs. The vehicle can be lowered for loading ('access' position), elevated for off-road driving, or held at the standard ride height for trailer towing (Table 7.1).

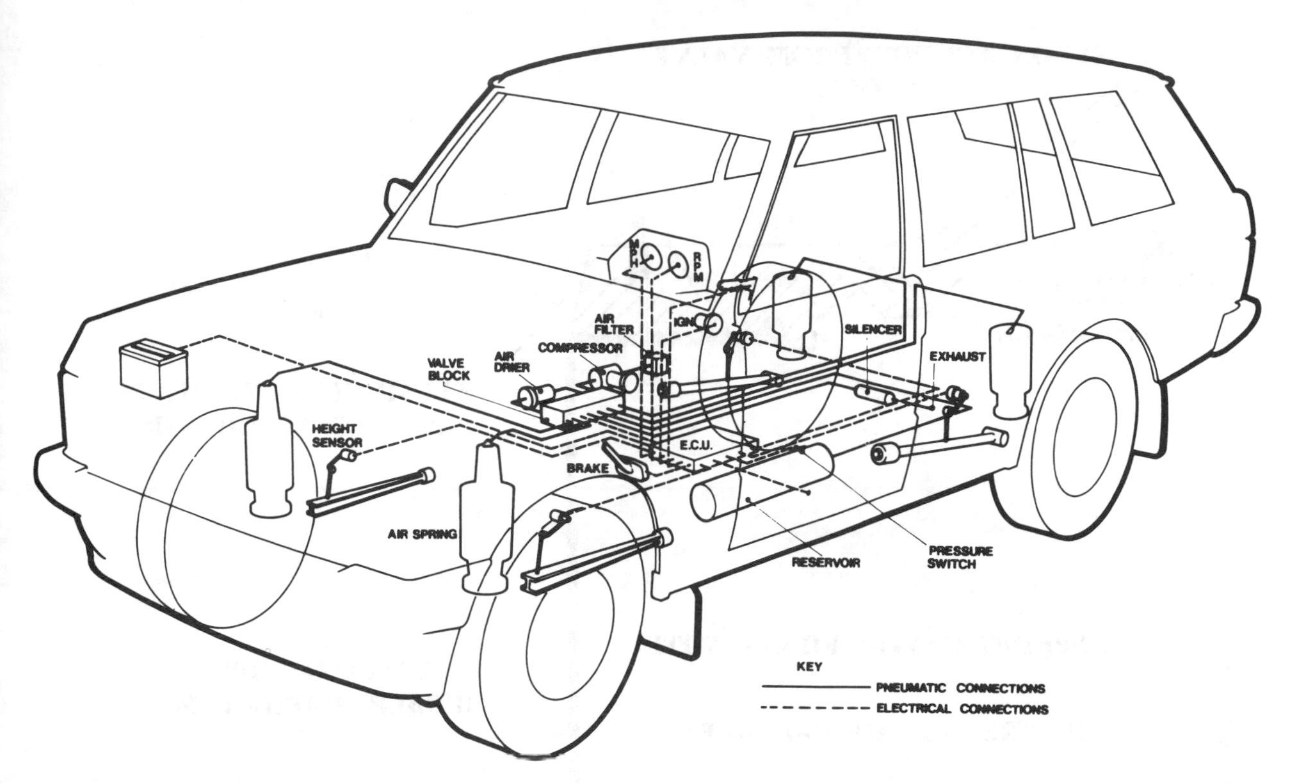

Figure 7.15 EAS component location (*Land Rover*)

ECU inputs and outputs

The EAS ECU monitors ride height by indirectly measuring the chassis-to-suspension displacement. Road surface imperfections mean that the body is constantly moving up and down, and so the number of different heights measured within a fixed time interval (about 12 seconds) are counted. Using this information, the ECU can calculate the percentage of time that the vehicle is at a particular height, and correct accordingly to keep the height within a target range. For example, if the height sensors suggest that a corner of the vehicle is 'excessively high' for 75% of the time, then the ECU opens the outlet valve and lowers that corner. Lowering is stopped when the height is excessive for only about 10% of the time.

The inputs and outputs for the EAS ECU are illustrated in Figure 7.17.

(i) The battery provides a 12 V feed to the ECU, it operates from a 20 second delayed turn off relay so that the ECU can follow a shut-down procedure when the ignition is switched off.

(ii) Engine speed is obtained from a phase tap on the alternator. The ECU normally requires that the engine is running for a height change to take place, this stops the compressor from draining the battery. Additionally, the ECU will inhibit compressor operation if the engine speed is less than 500 rpm.

(iii) Road speed is obtained from the instrument pack via a buffer circuit. When the road speed is above 80 kph (50 mph) the ECU lowers the ride height by 20 mm to reduce drag and improve stability.

(iv) An input from the handbrake switch (manual transmission) or the selector microswitch (automatic transmission) is used by the ECU to check that the vehicle is stationary before entering the Access mode.

(v) The footbrake switch is used to indicate to the ECU that the brakes have been applied. During braking at speeds above 8 kph the ECU inhibits height changes or levelling.

(vi) An input from the interior lamp switch informs the ECU that a door is open and the ECU inhibits height changes.

(vii) Three driver function switches (UP, INHIBIT and DOWN) are used to select height programs from the ECU when in manual mode.

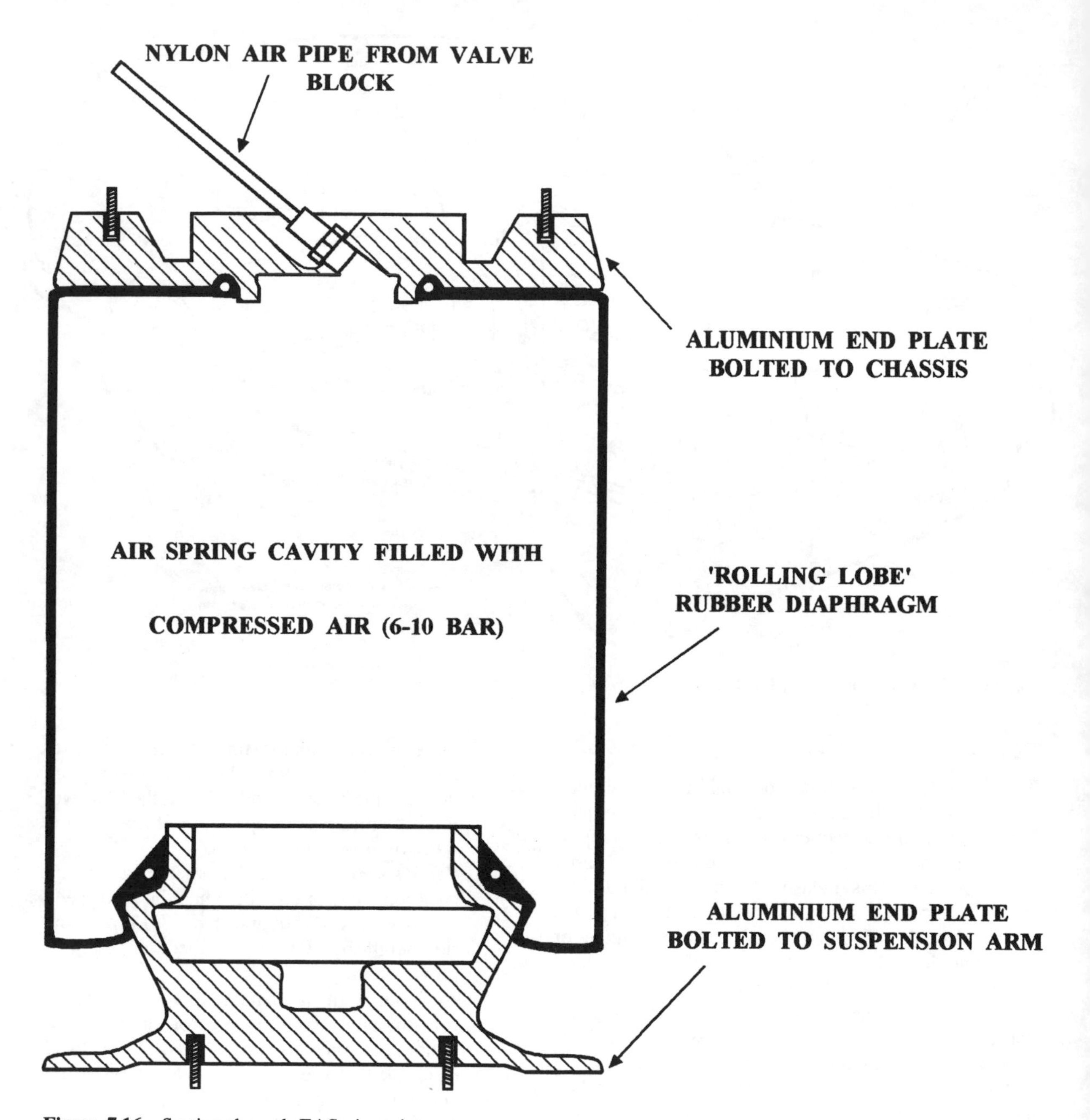

Figure 7.16 Section through EAS air spring

Table 7.1 EAS operating modes

Mode	Height	Condition
Standard	790 mm ± 7 mm	Inhibit on or off, 790 mm on rear, average of the front
Low profile	20 mm Below standard	Above 80 kph (50 mph) for more than 30 secs Inhibit off
High profile	40 mm Above standard	UP switch pressed Below 56 kph (35 mph) Inhibit off
Extended profile	20–30 mm Above high profile	Automatically selects when the chassis has 'bellied out' and the axles are hanging Will de-select after 10 mins
Access Can be selected with the engine running or for 20 seconds after turning the ignition off Auto in P MTX in N	60 mm Below standard	DOWN switch pressed Vehicle stopped Park/handbrake on Footbrake off All doors closed Inhibit off

The INHIBIT switch is used to lock the system at standard height.

(viii) The air pressure switch senses the pressure in the reservoir and provides a signal to the ECU to switch the compressor ON (7.5 bar) and OFF (10 bar).

(ix) The ride height sensors, illustrated in Figure 7.18, measure the chassis-to-suspension displacement at each corner of the vehicle. They are simply rotary potentiometers, supplied with a 5 V feed from the ECU. As the ride height changes, each potentiometer supplies the ECU with a voltage proportional to the displacement at its corner.

(x) The ECU's outputs are the drive signals to the solenoid valves and the compressor relay. The ECU is also able to illuminate a warning lamp in the event of a system fault. An additional connector is used to communicate with a test and calibration unit.

EAS pneumatics

Figure 7.19 shows the arrangement of the EAS pneumatic system. Air for the system is drawn in through the intake filter, compressed, and then dried by a silica gel drier unit prior to storage in the reservoir.

When a height sensor indicates that one corner of the suspension is too low, the ECU responds by opening both the inlet valve and the relevant ride height valve. Air then passes from the reservoir to the appropriate air spring, increasing pressure and raising that corner of the vehicle. Once the ECU detects the correct voltage reading from the height sensor, both solenoid valves are closed. Conversely, if a height sensor indicates that a corner of the vehicle is too high, the ECU opens the appropriate ride height valve, the outlet valve and the exhaust valve. Air is thus expelled from the spring via the drier and silencer. When the ECU detects that the height is correct, the valves are closed.

Mitsubishi electronically controlled suspension (ECS)

Mitsubishi's ECS is an air suspension system that has similarities with the Land Rover EAS system, but with the addition of electronic damping force control. Since the ECU can dynamically alter the suspension characteristics, the system is described as 'semi-active'.

The main components of ECS are shown in Figure 7.20. The arrangement of the suspension components is essentially the same as with a conventionally sprung system, but with the addition of an air spring which is coaxially arranged around each electronically controlled damper. Since the air spring acts in parallel with a conventional coil spring, variations in air pressure can be used to alter the ride height and spring rate.

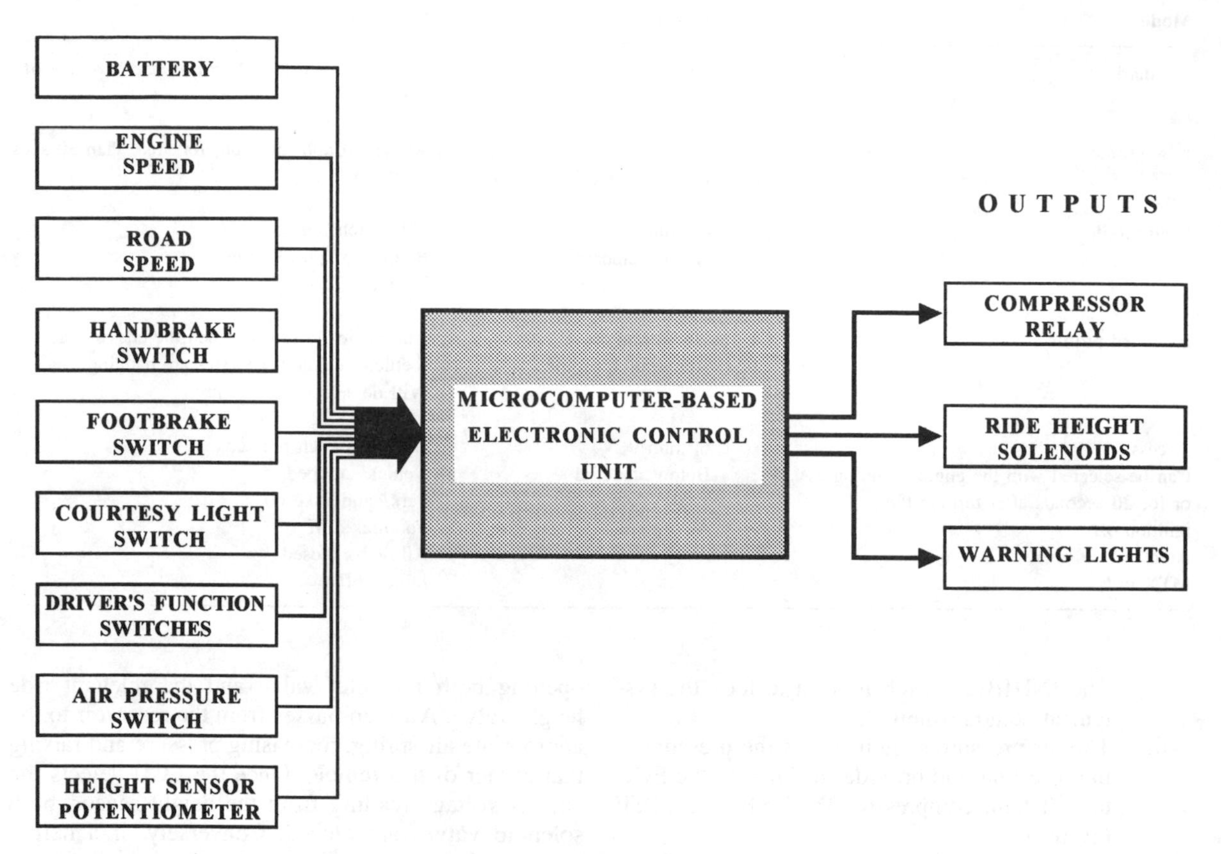

Figure 7.17 Inputs and outputs for EAS ECU

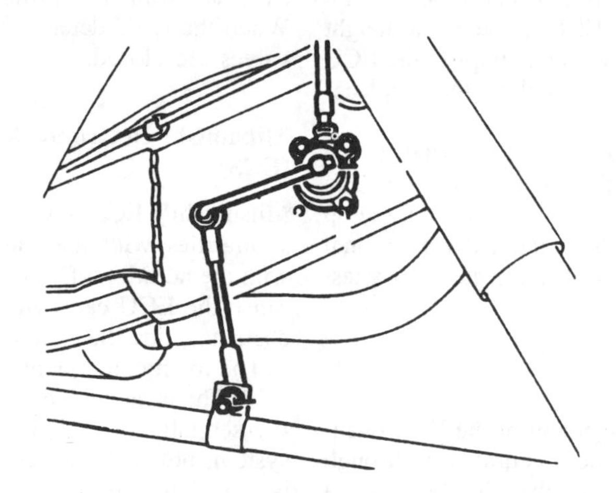

Figure 7.18 EAS ride height sensor (*Land Rover*)

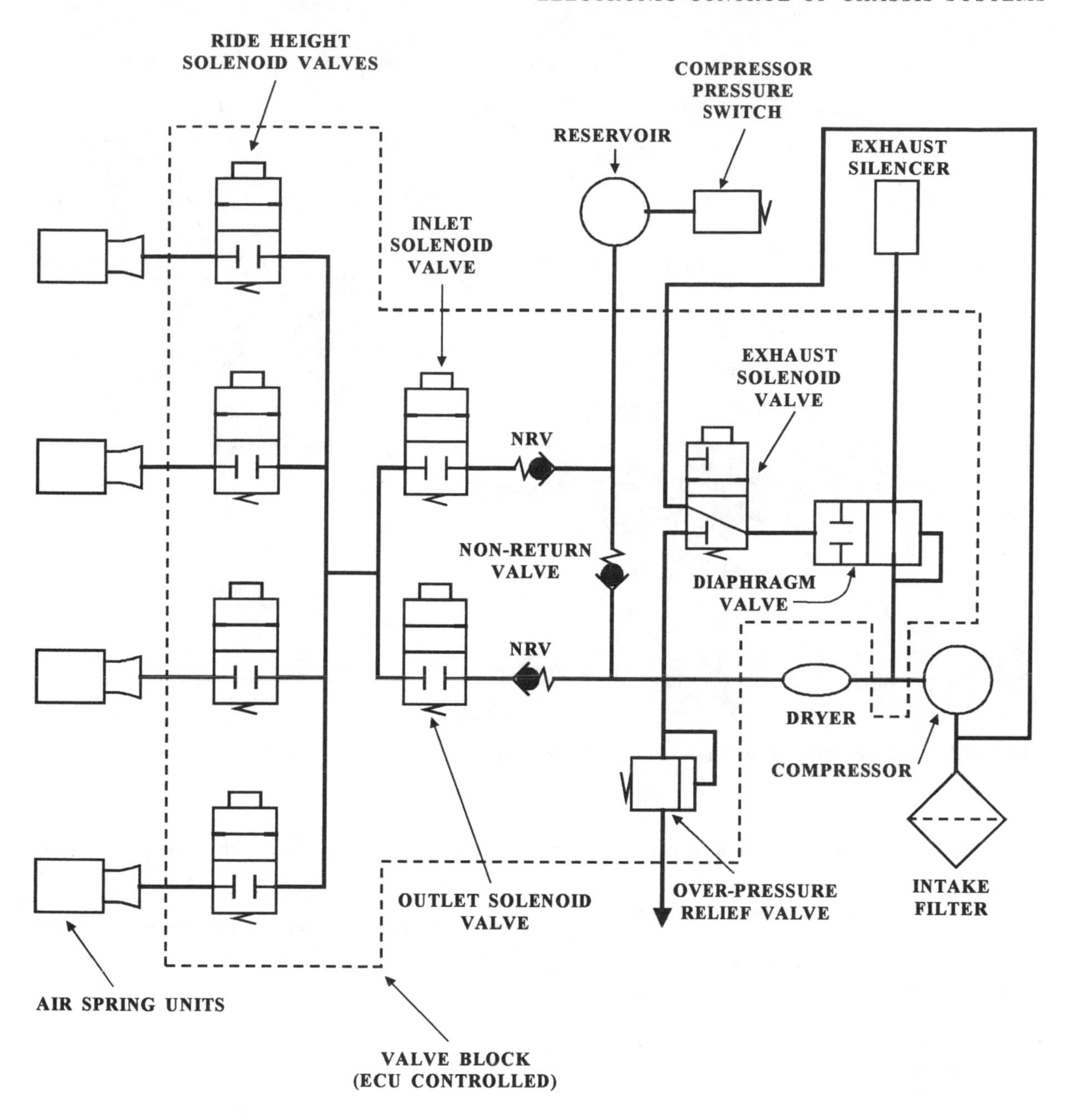

Figure 7.19 EAS pneumatic system

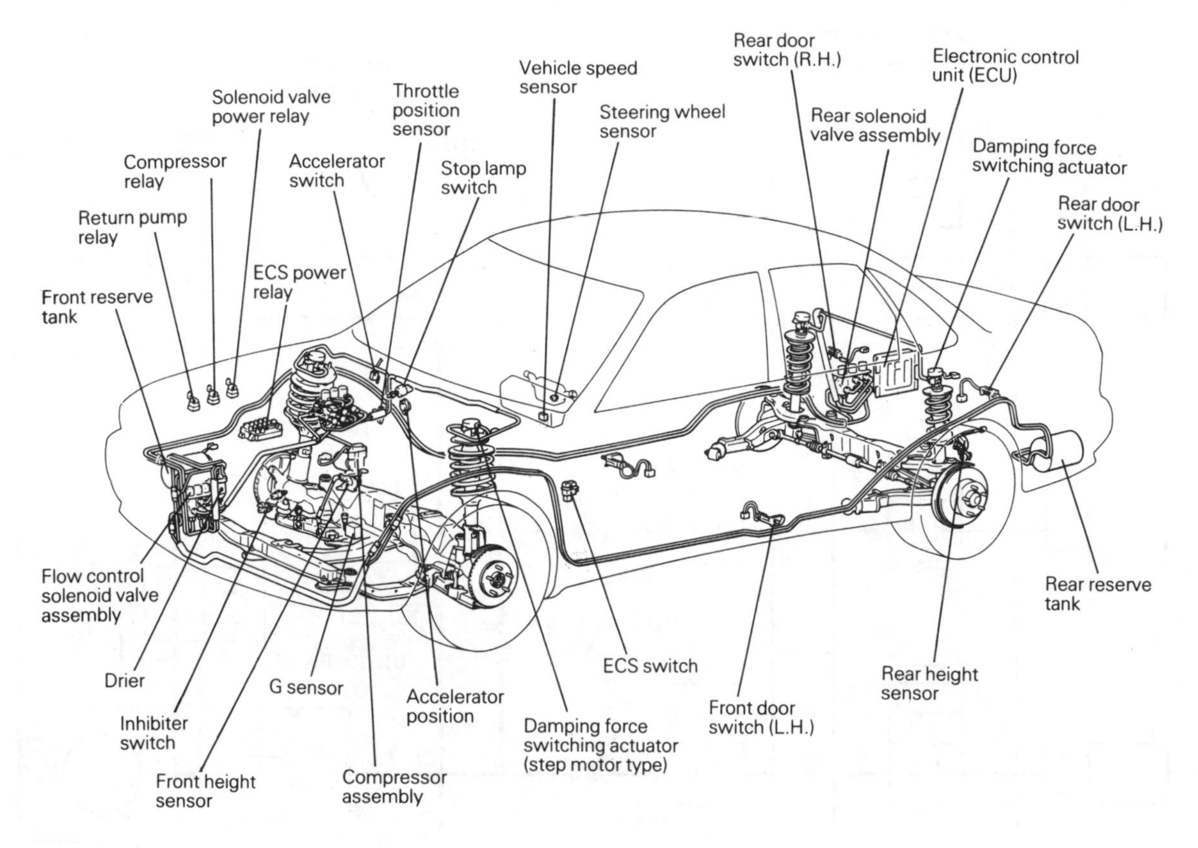

Figure 7.20 Main components of Mitsubishi ECS (*Mitsubishi*)

Driving conditions are monitored by using five types of sensors:

(i) a steering wheel sensor to detect steering manoeuvres;
(ii) a throttle position sensor to detect acceleration;
(iii) a suspension stroke sensor to measure ride height;
(iv) a G-sensor to detect lateral acceleration during cornering;
(v) a pressure sensor to detect air pressure in the air springs;

Using information from these sensors, the ECU actuates the nine solenoid valves to control the pressure in the air springs and so keep the vehicle flat to the road at the correct height, even when turning or braking. Figure 7.21 shows how anti-roll control is carried out, anti-dive and anti-squat control programs are also used.

Damping force is also controlled by the ECU; one of four settings (from 'firm' to 'super-soft') is selected by means of a stepper motor fitted to the top of each damper unit. In common with most such systems, two control modes are possible. 'AUTO' allows the ECU to select any one of the four damper settings to obtain the best ride-handling characteristics. 'SPORT' allows switching only between 'medium' and 'firm' damper settings, which biases the suspension towards good handling at the expense of comfort.

Sensors

Since the height and steering sensors are subject to almost continuous use they must be virtually wear free. Therefore contactless sensors, operating on the photo-interrupter principle, are used. By way of example a steering sensor is illustrated in Figure 7.22. It consists of a fixed detector block and a slit disc that is fitted to the steering column and rotates with the wheel. The disc moves past the photo-transistors in the detector block and alternately exposes them to light from the LEDs (light-emitting diodes). When the steering wheel is turned, a train of logic pulses is generated by each interrupter and these signals can

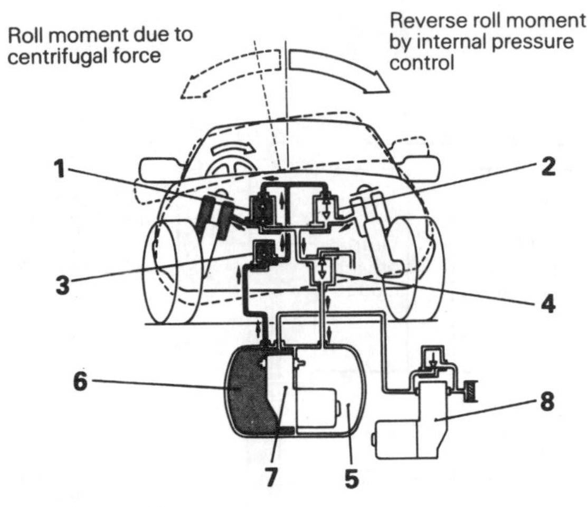

1. Right solenoid valve
2. Left solenoid valve
3. Intake solenoid valve
4. Exhaust solenoid valve
5. Low pressure chamber (0.7 kg/cm^2)
6. High pressure chamber (9.5 kg/cm^2)
7. Attitude control compressor (return pump)
8. Vehicle height control compressor

NOTE
Indicated is the control process when turning left.

Features of Each Mode

<table>
<tr><th>Mode / Function</th><th>AUTO</th><th>SPORT</th></tr>
<tr><td>Anti-roll control</td><td colspan="2">Provides roll-free turning by controlling the air spring internal pressure for the inner and outer wheels that are turning according to steering speed and vehicle lateral acceleration.
Outer wheel: Internal pressure increased
Inner wheel: Internal pressure decreased
The amount of roll stiffness progresses in this order: AUTO > SPORT</td></tr>
<tr><td>Anti-dive control</td><td colspan="2">Keeps the vehicle in a level position by supplying air to the front air springs and exhausting air from the rear air springs according to longitudinal acceleration during braking.</td></tr>
<tr><td>Anti-squat control</td><td colspan="2">Provides control that is reverse to anti-dive control in the mode of supplying air to and exhausting air from front and rear springs.</td></tr>
<tr><td>Pitching and bouncing control</td><td colspan="2">Provides control according to the extension/compression of the shock absorbers (vehicle height change). (Air is supplied to the extended end and is exhausted from the compressed end.)</td></tr>
</table>

Figure 7.21 Anti-roll feature of ECS (*Mitsubishi*)

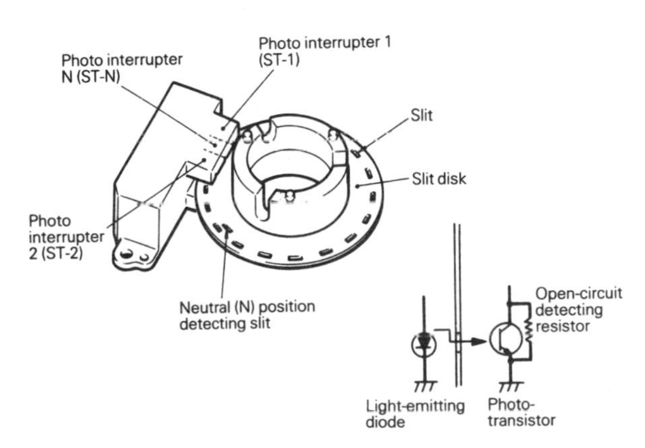

Figure 7.22 Steering sensor based on photo-interrupter principle

then be processed to detect steering angle and rate of turn. Two interrupters (ST-1 and ST-2) must be used so that the direction of rotation can be detected. A third interrupter (ST-N) can be added to give an indication of the neutral (straight ahead) position.

The G-sensor is a small semiconductor accelerometer. It is mounted near to the front of the car to give a quick response to lateral acceleration when the vehicle enters a turn.

The high-pressure and low-pressure switches are used to turn the compressor on and off by detecting the air pressure in the reservoir tank. They each consist of a diaphragm that moves against a microswitch.

The rear pressure sensor uses a sprung diaphragm, moved by variations in the air pressure. These movements are conveyed to a small potentiometer, which

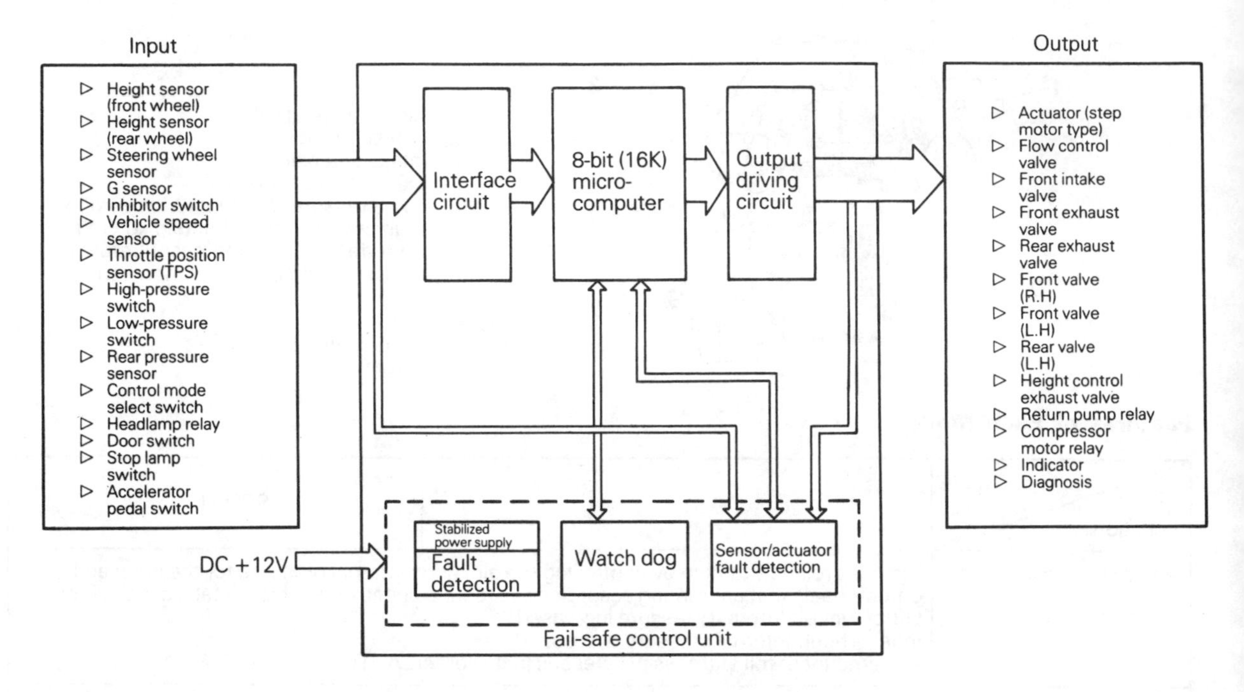

Figure 7.23 ECS control system block diagram (*Mitsubishi*)

converts them to a varying voltage for use by the ECU.

ECS electronic control unit

Figure 7.23 shows the ECS ECU, together with its input and output signals. The ECU comprises an 8-bit microcontroller, together with an input interfacing circuit and an output driving circuit. Fail-safe circuitry is also included, as is an interface for diagnostic equipment.

The ECU makes decisions based on information received from various sensors and selects the appropriate pre-programmed control mode. Control modes for a particular ECS variant are shown in Table 7.2.

Citroën hydractive suspension

Citroën was a pioneer of sophisticated suspension design, launching their unique self-levelling hydropneumatic system on the DS model of 1955. Hydropneumatic suspension works on the principle of having a nitrogen gas spring (very similar in construction to the accumulator used in an ABS actuator unit) connected to the suspension arm by means of a hydraulic ram. The ram can be lengthened or shortened (and hence the ride height altered) by changing the pressure of the fluid supplied to it. Pressurized fluid is provided by an engine-driven pump and supplied to each spring through a network of small-diameter pipes. Pressure, and therefore ride height, is regulated by mechanical control valves that sense the body-to-axle displacement at each end of the vehicle and distribute the fluid accordingly.

Although hydropneumatic suspension was a spectacular success when first launched, its superiority became less evident as conventional suspension systems were progressively improved. To restore the appeal of their hydropneumatic system, Citroën took the decision to add computer control. This system, first seen on the XM model, is called Hydractive suspension. It gives exceptional ride comfort, combined with precise wheel control when the need arises.

Hydractive operation

The major components of the Hydractive system are illustrated in Figure 7.24. It is a semi-active

Table 7.2 Control modes for Mitsubishi ECS

Control function			Function description
Selection of control mode			• AUTO or SPORT can be selected by the control mode select switch. • AUTO, HIGH and EXTRA HIGH mode can be selected by the vehicle height (HIGH) switch.
Damping force and active attitude control	1	Anti-roll control (minimizes roll at time of cornering)	Map control (controlled by computer data) according to steering angular velocity, lateral acceleration and vehicle speed. Supplies air to outside air springs and discharges air from inside air springs.
	2	Anti-dive control (minimizes nose-dive during braking)	Controls when the brake pedal switch is ON and the vehicle longitudinal acceleration is 0.2 G or more. (Damping characteristics HARD) At start: Supplies air to the front air springs and discharges air from the rear air springs. At restoration: Discharges air from the front air springs and supplies air to the rear air springs.
	3	Anti-squat control (minimizes squat at time of starting)	Controls according to throttle opening/closing speed and vehicle speed. At start: Discharges air from the front air springs and supplies air to the rear air springs. At restoration: Supplies air to the front air springs and discharges air from the rear air springs.
	4	Pitching and bouncing control) (minimizes vehicle body pitching and bouncing due to road irregularities)	Controls by detecting suspension stroke and operation frequency with height sensor. Suspension in extending stroke: Supplies air Suspension in compressing stroke: Discharges air
	5	A/T gear shift squat control	Controls with accelerator pedal switch, brake pedal switch and vehicle speed change signal (inhibitor switch) and according to vehicle speed. Damping characteristics are switched to HARD by means of A/T shift lever.
	6	Damping force selection control according vehicle speed	Controls by detecting high speed driving with vehicle speed sensor.
Vehicle height adjustment	1	Normal vehicle height adjustment	Three control models of AUTO, HIGH and EXTRA HIGH
	2	Quick vehicle height adjustment [quick increase of vehicle height (in approx. 2 sec.)]	The flow rate changeover valve is turned on by rough road detect (front height sensor) or HIGH switch.
Compressor and return pump drive			Controls by high- and low-pressure switches.
Diagnostic and fail-safe functions			Upon detection of fault, the alarm lamp is turned on, the fail-safe operation is performed, and diagnostic code is output.
Service data indication and actuator test			Controls according to the multi-use tester (MB991194) command.

suspension system that offers two spring rates ('Sport' and 'Comfort') and two damping forces ('Soft' and 'Firm'). Like the original hydropneumatic suspension system, Hydractive uses hydraulic rams acting on nitrogen gas springs. The big difference is the inclusion of a third gas spring for each axle. During normal driving the microcomputer-based ECU actuates solenoid valves to bring the 'third spring' into the hydraulic circuit of each axle. This increases the volume of compressible gas by 50% and so reduces the suspension stiffness, improving ride comfort. Damping force is also reduced, since each solenoid valve additionally opens an orifice to permit fluid to flow between all three springs on each axle.

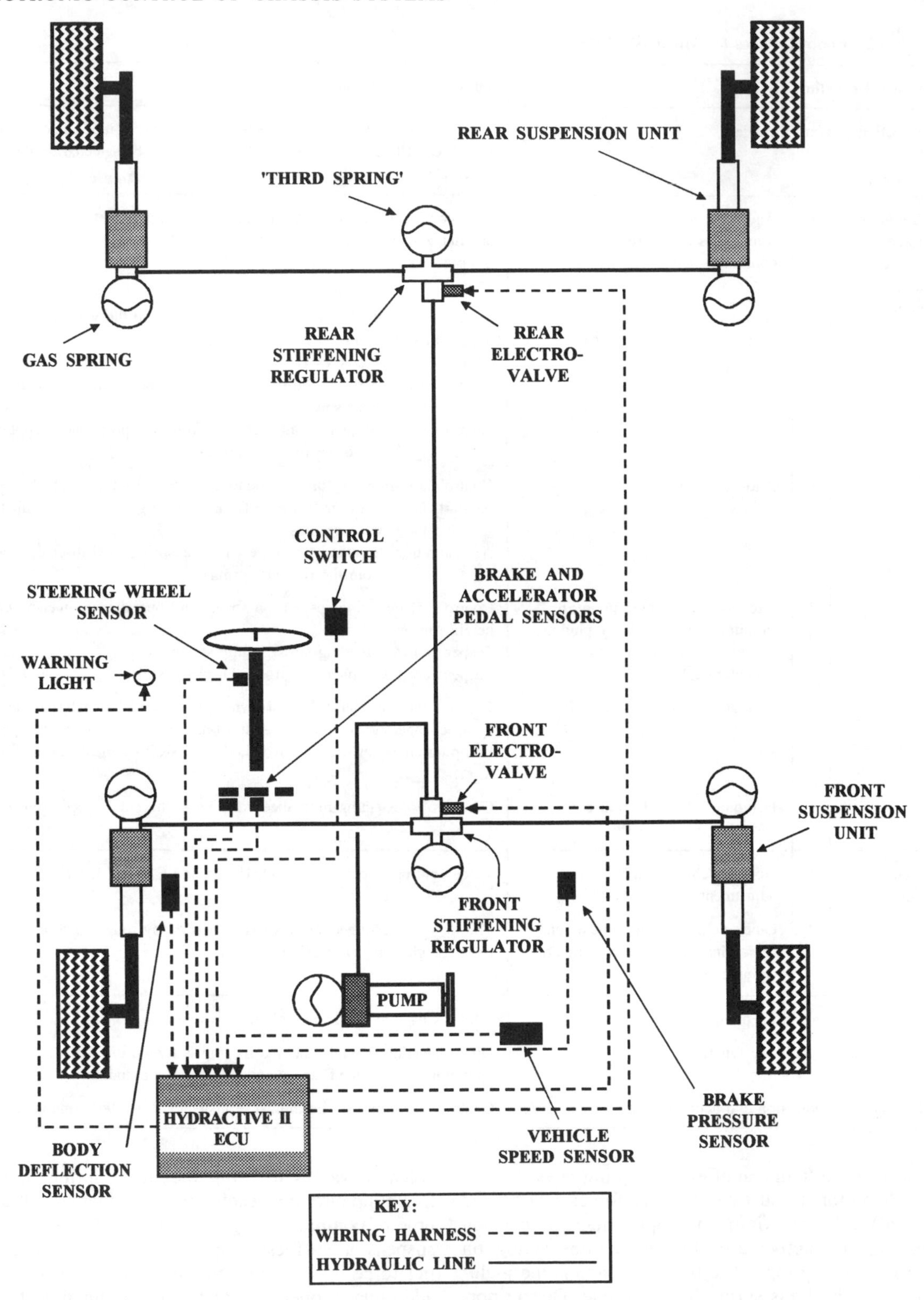

Figure 7.24 Citroën Hydractive II hydropneumatic suspension system

Conversely, when the ECU decides that driving conditions demand a firmer suspension setting, it actuates the solenoid valves to isolate the 'third spring' assemblies from the system and stop the transverse flow of fluid between the main suspension struts. This results in an increased spring stiffness, firmer damping and an anti-roll characteristic.

Sensors

The ECU responds to eight signals; position of the control switch (SPORT or COMFORT), steering wheel position and rate of turning, vehicle speed, accelerator pedal movement, braking force, body displacement and door and boot light switches. The steering wheel and body displacement sensors are of the twin photo-interrupter type and operate in a similar way to those described for the Mitsubishi ECS system.

The vehicle speed sensor, which is shared with the instrument pack, is a simple Hall-effect detector installed in the final drive housing. A take-off shaft, which is driven by the gearbox output shaft, is fitted with a polar ring (a magnetic ring composed of several alternate north and south poles). The Hall-effect detector faces the polar ring and produces a small Hall voltage pulse (about 1 mV) each time a north pole moves past it. The gearing is arranged such that one pulse is obtained for each 20 cm of vehicle travel, giving a highly accurate method of speed measurement.

Accelerator pedal position is derived from the throttle position sensor; a conventional rotary potentiometer is used. Braking force is measured indirectly, using a pressure switch. When the fluid pressure in the braking circuit exceeds 35 bar the switch is open circuit and so the ECU input rises to 5 V. This indicates firm braking. If the brake pressure is below 35 bar, the switch grounds the ECU input, indicating a low braking force.

The SPORT switch is a dashboard mounted switch that allows the driver to select a suspension control program that is more suited to fast or 'sporty' driving.

Actuators

The Hydractive actuator units are solenoid valves, termed 'electro-valves', which are fitted to each 'third spring' (Figure 7.25). With no power applied, the electro-valve is held in the OFF position by a strong return spring. This isolates the third spring and so places the suspension in the FIRM mode (two springs per axle) with FIRM damping. To switch to the SOFT state (three springs per axle, mild damping), the ECU energizes the electro-valve. This is accomplished by supplying the coil with maximum voltage (about 13.5 V) for a half-second to 'pull-in' the plunger, and then 'chopping' the current at 1 kHz to obtain a 'hold-in' current which averages to about 0.5 A. To safeguard against open or short-circuited solenoid windings the ECU uses 'smart' transistors which continuously check for the specified coil resistance (about 5 Ω). The output current is shut off if an incorrect resistance is detected, and the system defaults to the 'FIRM' mode.

Hydractive ECU

The Hydractive ECU is a sealed unit mounted in the engine compartment. It uses two microprocessors to validate, compare and perform calculations on data obtained from the sensors to estimate longitudinal, lateral and vertical body accelerations. To determine the optimum suspension setting for the prevailing driving conditions, all sensor values are continuously compared with pre-programmed thresholds (which are adjusted in proportion to road speed). Normally the ECU maintains the electro-valves in the ON state and so the suspension operates in the SOFT mode. However, when a sensor value exceeds its threshold, the microprocessors make a decision to switch the suspension to the FIRM state and the electro-valves are de-energized. This process occurs very quickly (less than 2 ms).

Once the sensor value drops below its threshold, and after a pre-set time delay, the suspension is returned to the SOFT setting.

When the driver moves the SPORT control switch to the SPORT position, the ECU lowers the signal value thresholds by 33%. Thus the suspension system still switches between SOFT and FIRM modes, but it switches to FIRM far more readily.

Electronic power-assisted steering control

Power-assisted steering (PAS) has been a standard fitment on large cars for many years and has recently become a common fitment on small cars. Its basic function is to multiply the force of the driver's steering inputs and so give a lighter and more direct feel to the steering. Traditionally this has been achieved by using high-pressure hydraulics; more recently PAS using electric motor assistance has been developed.

In a basic PAS system, the amount of assistance remains constant at all vehicle speeds. This presents major shortcomings, since if the level of assistance is set to give light steering at low speeds, the system will suffer over-sensitivity at high speeds. Such a system can make a vehicle difficult and tiring to drive on motorways because the driver receives very little

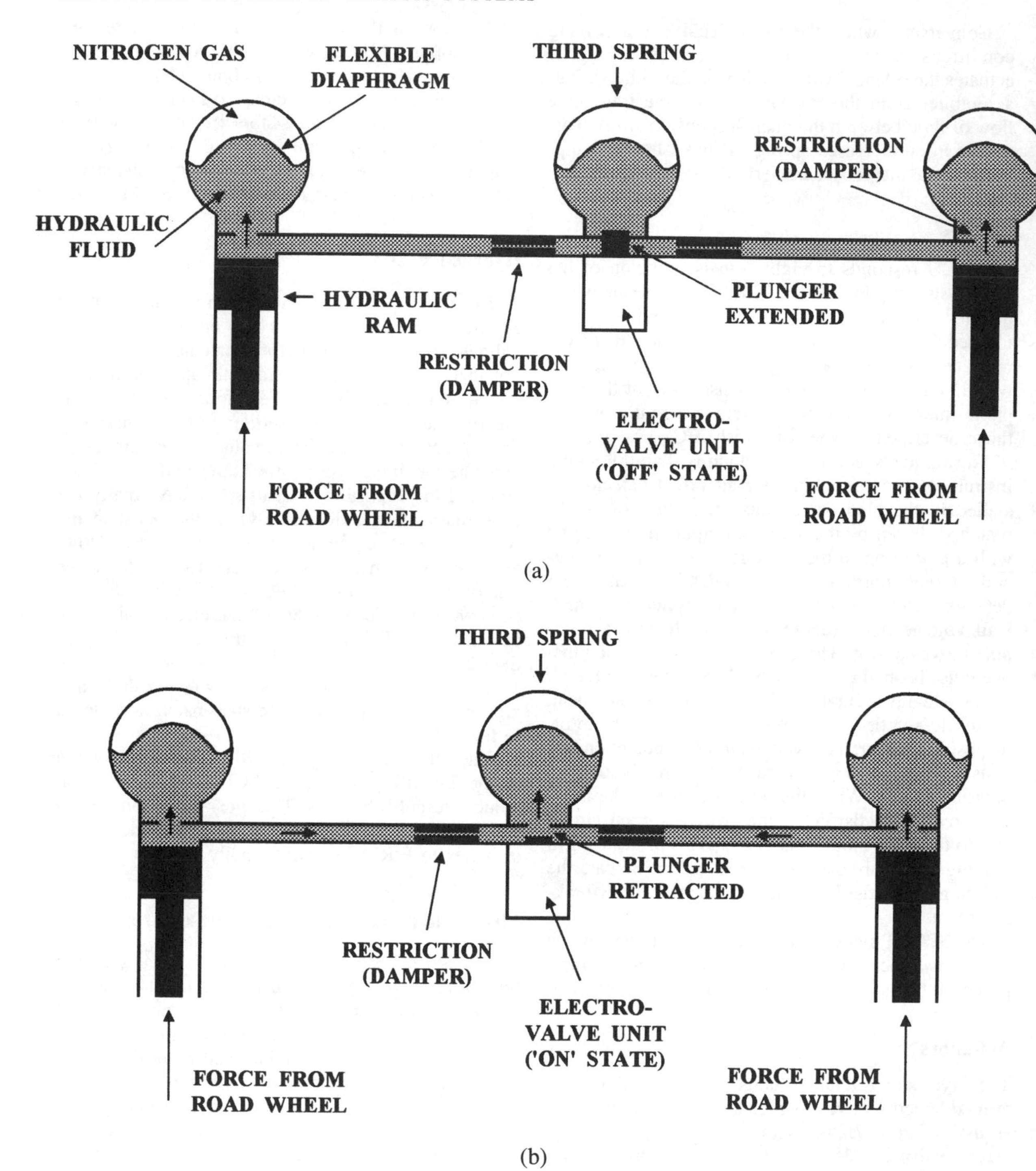

Figure 7.25 Operation of Hydractive II electro-valve to give: (a) firm springing and damping; (b) soft springing and damping

feedback ('feel') from the front wheels and constantly over-corrects.

Ideally, a PAS system presents the driver with a level of assistance that reduces in proportion to road speed. A high level of assistance when moving slowly makes for easy manoeuvring, while a low level of assistance at high speeds gives optimum feel and sensitivity. This requirement has led to the development of PAS systems in which the level of assistance is controlled by an ECU which electronically senses road speed.

Electronically controlled hydraulic PAS

Most speed-sensitive PAS systems are of this type, although there are many minor variations in design. The system comprises a fluid reservoir, an engine-driven hydraulic pump, the power-steering gear, a duty-cycle controlled solenoid valve, a vehicle speed sensor (shared with the speedometer) and an ECU.

When the engine is running, the hydraulic pump produces a flow of hydraulic fluid at high pressure (typically 30–40 bar). This fluid passes along pipes to the power-steering gear, which consists of a conventional rack and pinion together with a hydraulic control valve and a ram. If the steering wheel is turned, the control valve is mechanically opened by an amount corresponding to the steering force applied by the driver. The partially opened valve allows high-pressure fluid to enter the ram, which in turn helps push the rack to steer the wheels.

Since the amount of power assistance is proportional to the pressure of the hydraulic fluid, it is possible to reduce the steering sensitivity by reducing this pressure. This is done by having the ECU open the solenoid valve to allow fluid to bypass the ram and return to the reservoir. To provide a speed-dependent response, the ECU uses the speed sensor signal to determine vehicle speed and hence open the solenoid valve by a proportionate amount. Thus, when the car is moving very slowly, the solenoid valve is shut and so the hydraulic pressure, and therefore level of assistance, is at a maximum. As the car accelerates, the ECU detects the increase in speed and so progressively opens the solenoid valve. At high speeds the solenoid valve will be fully open and so there will be little power assistance. The overall result is a steering effort that is more-or-less independent of speed.

Speed-dependent PAS ECU

The ECU for control of speed-dependent PAS performs only a simple function and has just one input (the speed signal) and one output (solenoid valve duty-cycle signal). The speed signal is mapped against solenoid duty-cycle to provide the desired speed-sensitivity characteristic for the particular car model. On certain vehicles the driver has the option to select an alternative map which gives a lower level of power assistance for 'sports' driving.

A large part of the ECU circuitry is concerned with safety monitoring of the system. In the event of an ECU or speed sensor failure the solenoid valve is allowed to move to the fully open position, reducing the amount of power assistance to the minimum value.

Electric power assisted steering (EPAS)

Electric power assisted steering (EPAS) is a PAS system which uses a conventional rack-and-pinion steering system, but with the addition of an electric motor, rather than a hydraulic ram, to aid the driver in turning the wheels. The use of EPAS offers several benefits over the traditional hydraulic PAS, especially in small car applications. In particular EPAS gives:

(i) a cheaper, more compact and lighter installation;
(ii) lower power consumption and therefore greater fuel economy (a 5% fuel saving is typical);
(iii) a more reliable system with lower maintenance requirements;
(iv) the ability to communicate with other electronically-controlled systems via a CAN databus.

During the early 1990s EPAS systems were initially manufactured only in Japan for fitment to small cars, such as the Suzuki Alto, where there is limited room for additional equipment. Other early installations included some Honda cars fitted with four-wheel steering where large hydraulic flow losses to the rear steering assembly meant that the use of an electric motor actuator was essential for fast response. With the realization that drivers would increasingly demand PAS on small cars, many European and American component suppliers developed their own EPAS systems. Typical of these is the Lucas EPAS system, designed for installation on small cars from the mid-1990s onwards.

Configuration of Lucas EPAS

Figure 7.26 shows the arrangement of the Lucas EPAS system. A conventional steering rack is used, with just four additional components:

(i) a torque sensor to measure the steering force applied by the driver;

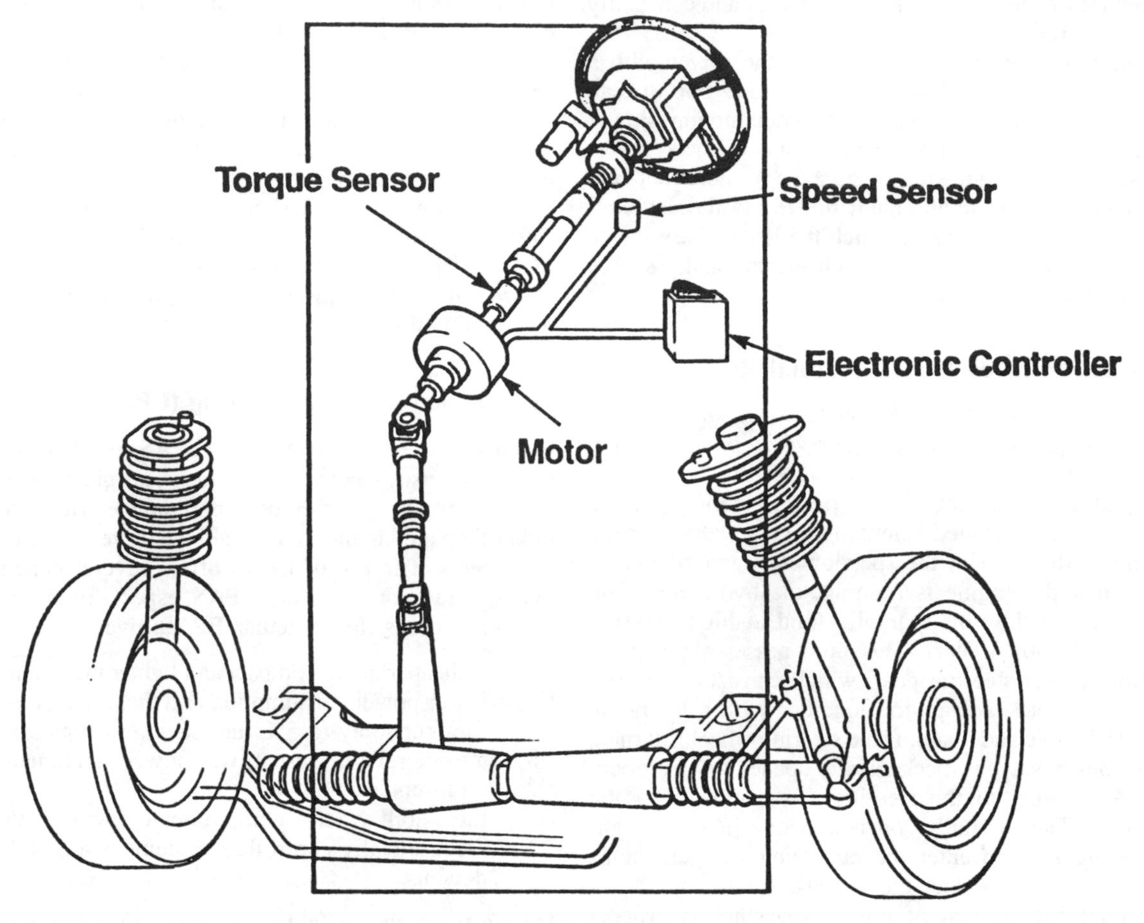

Figure 7.26 Arrangement of electric power assisted steering components (*Lucas*)

(ii) a small electric motor, which drives the steering column through an electronically controlled clutch and gears, to provide power assistance;
(iii) a road speed sensor (shared with other systems) to enable speed-sensitive assistance to be provided;
(iv) an electronic control unit to process torque and speed signals, and drive the motor.

Owing to the small size of these components it is possible for the whole system to be installed in the passenger compartment, around the lower part of the steering column, making it suitable even for very small cars.

The torque sensor is a key part of any EPAS system. For reliable and virtually wear-free operation Lucas use an opto-electronic sensor, consisting of two photo-interrupter discs spaced a short distance apart and connected to each end of a torsion bar which forms part of the steering shaft (Figure 7.27).

Each disc has a pattern of slots forming two tracks of alternating clear and opaque areas (Figure 7.28). The discs are aligned facing each other so that the slots on opposing tracks partially overlap. A light-emitting diode (LED) is arranged to shine on the two tracks and its intensity is detected by a pair of photodiodes. When the driver turns the steering wheel, the torsion bar twists very slightly, moving the photo-interrupter discs slightly out of alignment and so varying the degree of overlap between the slots. This alters the intensity of light falling on each photodiode by an amount proportionate to the steering force. If the torque is applied in one direction, the output from the outer track photodetector increases while that from the inner track detector decreases. When the torque is applied in the opposite direction then the converse happens. An analogue torque signal is thus obtained from each photodetector and reported to the ECU.

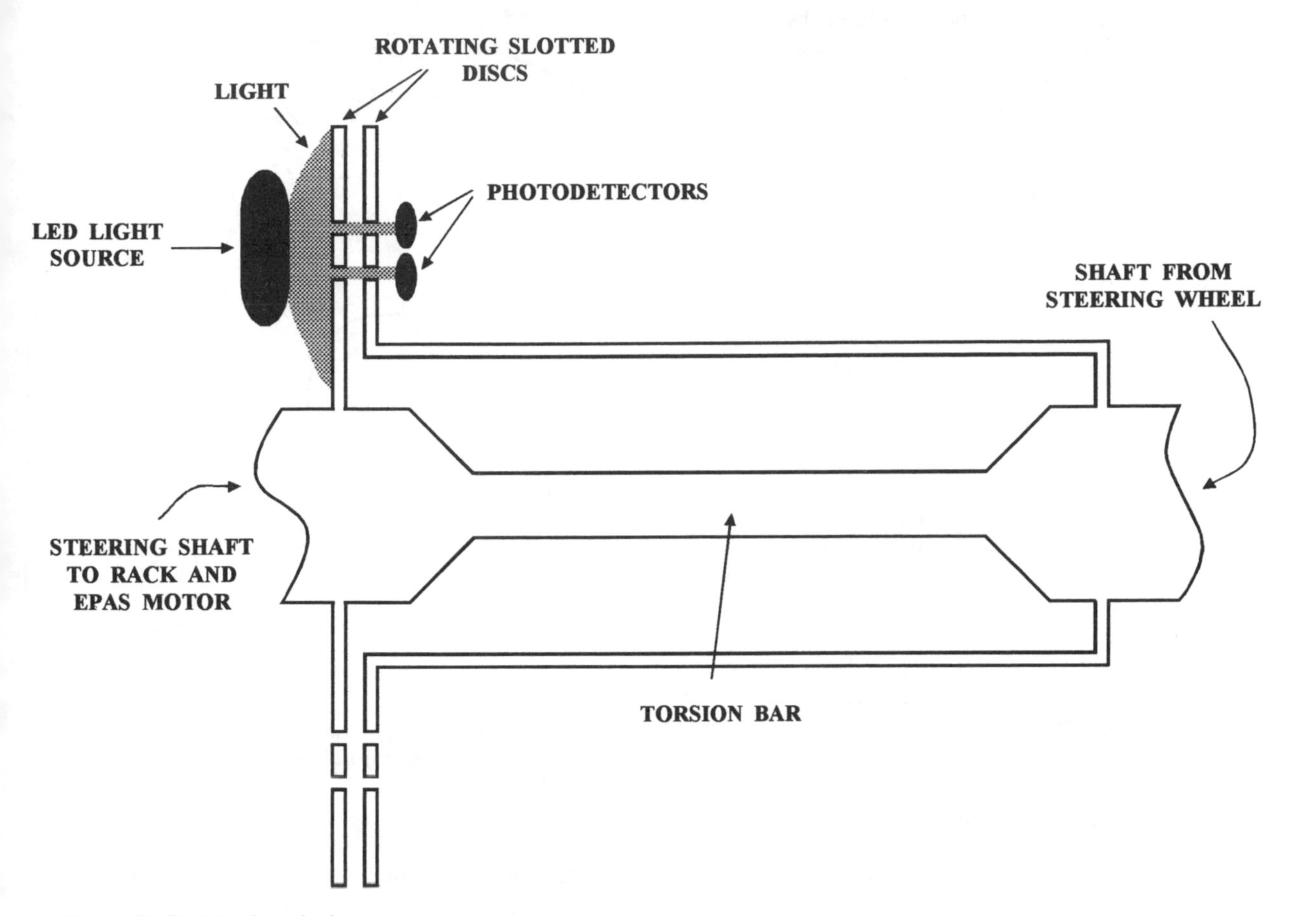

Figure 7.27 EPAS optical torque sensor

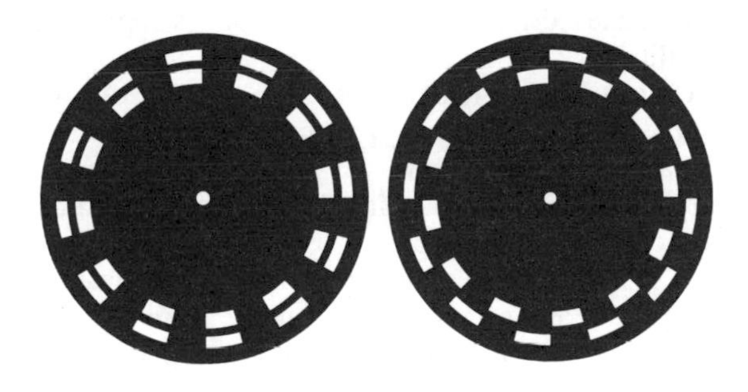

Figure 7.28 Optical sensor discs. Note different pattern on each disc (*Lucas*)

By monitoring steering effort and vehicle speed, the ECU's microprocessor can calculate the direction and level of the drive current to apply to the motor. Different amounts of assistance can be provided simply by changing the software to suit different cars or driving conditions.

The safety and reliability of the steering system are paramount, and it is for this reason that a direct mechanical link is maintained between the steering wheel and the road wheels. So-called 'drive-by-wire' steering, where the mechanical link is absent, have so far only been used in rear-wheel steering applications where the steering angle is small and an electrical failure would be less catastrophic.

7.5 ELECTRONICALLY CONTROLLED FOUR-WHEEL STEERING

The idea of steering a vehicle by turning all four wheels is not new. Indeed, in the early 1900s a variety of cars, buses and trucks were designed with mechanical four-wheel steering systems in an effort to make them more manoeuvrable and easier to drive. Although comparatively successful from a technical viewpoint, the added cost, weight and complexity of such systems meant that they found little favour with the motoring public. So, apart from limited application on some heavy trucks and factory vehicles, the notion of four-wheel steering lay dormant until the 1980s.

During the 1980s, Japanese engineers became aware of the possibility of improving high-speed handling and low-speed manoeuvrability through using four-wheel steering. Honda and Nissan were the first into production; Honda used an entirely mechanical system, with a rear steering assembly linked to the front rack via a long shaft and a planetary gearbox. Nissan took a different approach, and in 1985 launched their HICAS four-wheel steering system which used a hydraulically actuated rear steering rack.

Both companies subsequently improved their systems, and now offer electronically controlled four-wheel steering (E4WS) systems actuated by either high-pressure hydraulics or an electric motor. Other Japanese car makers have also adopted E4WS for their high-performance vehicles and some of these systems now offer very sophisticated control.

European and American manufacturers have been more sceptical of the advantages of four-wheel steering and very few offer it. They prefer to allow the rear-wheels to 'passively steer' by careful design of the rear suspension geometry and the use of soft locating bushes. An exception is BMW, who have introduced a simple E4WS system called 'Active Rear Axle Kinematics' (abbreviated to 'AHK') on certain models.

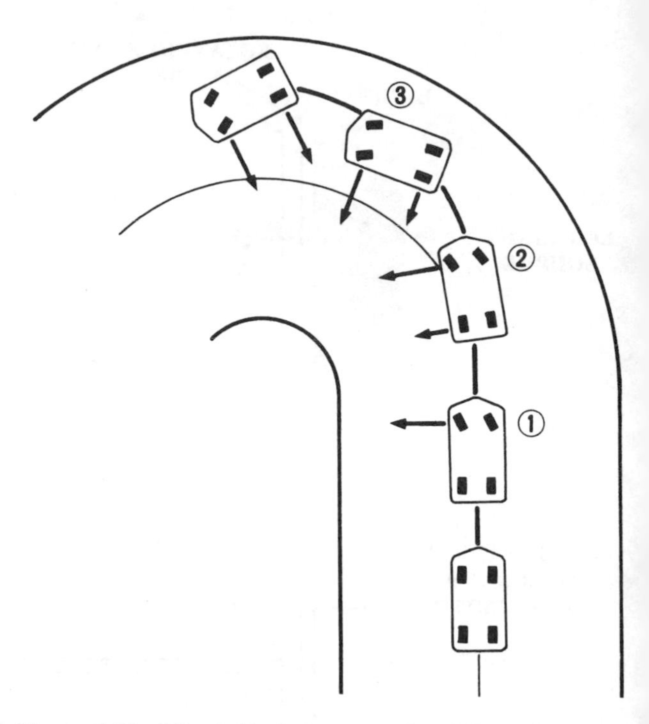

Figure 7.29 Wheel slip during cornering: [1] steering input; [2] car sideslips; [3] full cornering force generated (*Nissan*)

Principles of four-wheel steering

When a car fitted with two-wheel steering is first turned into a corner, the front wheels start to sideslip by a small amount and so generate a lateral force at the front of the vehicle (Figure 7.29). This lateral force causes the vehicle to yaw about its centre of gravity, and so turn into the corner. Since the rear wheels are rigidly coupled to the chassis they only start to generate a side-force once the body has begun turning, leading to a small delay between the creation of cornering forces at the front and rear wheels. This results in a lack of responsiveness and the need for the driver to predict and correct for the delay when making high-speed manoeuvres. Typically, the steering wheel must be turned through a large angle as the vehicle enters a curve (to cause rapid yaw) and then turned back slightly as the rear wheels begin to generate a cornering force. Similarly, after changing lanes on a motorway, the driver has to 'countersteer' slightly because the rear wheels are still sideslipping by a small amount.

Four-wheel steering eliminates the need to predict and correct for the cornering force delay by steering the front and rear wheels in the same direction (Figure 7.30). Using four-wheel steering (4WS) a cornering force is simultaneously generated at both ends of the vehicle, reducing the body's sideslip. The overall steering characteristic is thus modest understeer, with no need to make corrections during cornering. This makes driving easier and improves high-speed stability.

Unfortunately there are also some drawbacks with four-wheel steering. In particular, the understeer characteristic reduces the car's steering response on very twisting roads, meaning that the driver must turn the steering wheel through a greater angle to obtain the required turn-in. This problem can be solved by applying 'contra-phase' (also called 'counter-phase' or 'anti-phase') rear-wheel steering as the car enters a corner, and then switching to in-phase steering to complete the turn (Figure 7.31). The initial contra-phase steering produces a momentary oversteer condition which gives a high yaw rate and a sharp turn-in. On a winding road such a strategy ensures extremely good directional responsiveness, and it is for this reason that most E4WS systems now operate in this fashion.

An additional feature of some E4WS systems is the availability of large-angle contra-phase steering at low speeds. This reduces the radius of the turning circle when performing tight manoeuvres and so is

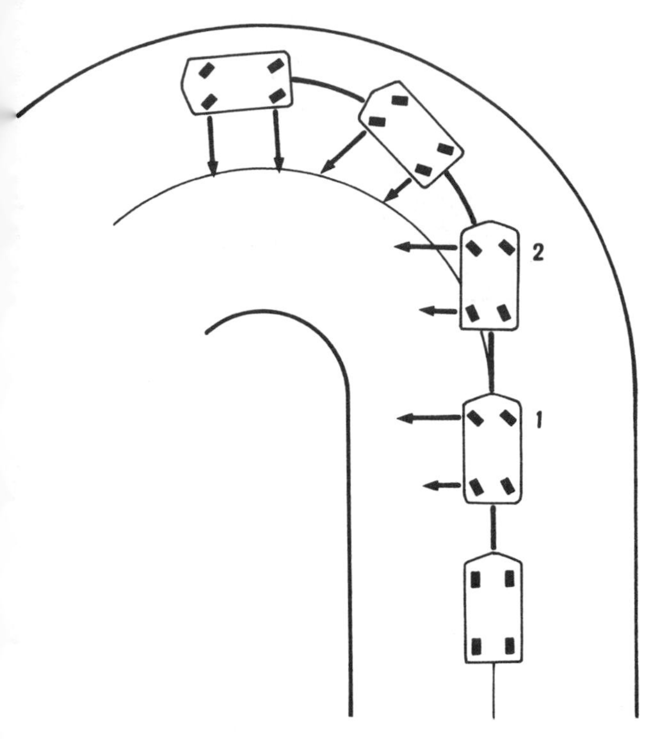

Figure 7.30 Advantage of four-wheel steering: [1] steering input; [2] full cornering force generated (*Nissan*)

useful when driving in town. Unfortunately the sideways body movements that accompany contra-phase steering mean that its use is an acquired skill. Care is required if the car is parked close to an obstruction such as a kerb since the rear of the vehicle will move in the opposite direction to that of the front, possibly resulting in a collision.

Mitsubishi Active-4WS and Nissan Super-HICAS system configurations

All E4WS systems are controlled by a microprocessor-based ECU. The ECU senses the driver's steering inputs and vehicle speed, and so computes the required rear-wheel steering angle in accordance with preprogrammed data (Figure 7.32). The rear wheels are then steered to the desired angle by a hydraulic power cylinder.

Figure 7.33 shows the arrangement of the Mitsubishi Active-4WS hydraulic system (Nissan's Super-HICAS is similar). Hydraulic pressure is generated by a conventional power steering pump which feeds the front and rear steering mechanisms via the flow divider valve. The flow divider valve is a duty-cycle controlled solenoid valve that divides hydraulic pressure between the front and rear steering cylinders in accordance with road speed. When the vehicle is driven slowly, the ECU supplies the valve with a large current (about 1 A) which causes fluid to be preferentially supplied to the front steering cylinder to provide maximum front-wheel power assistance. As the car is driven faster, the valve current is progressively reduced to increase the amount of fluid delivered to the rear-wheel steering cylinder. Maximum delivery to the rear occurs when the valve current falls to about 0.6 A. In the event that the current falls below 0.6 A (a fault condition) the valve enters a fail-safe state and the hydraulic supply to the rear cylinder is cut off. The system is then functional as a conventional 2WS system.

The rear-wheel steering angle is determined by the ECU, which actuates the E4WS control valve assembly. This component consists of two solenoid valves

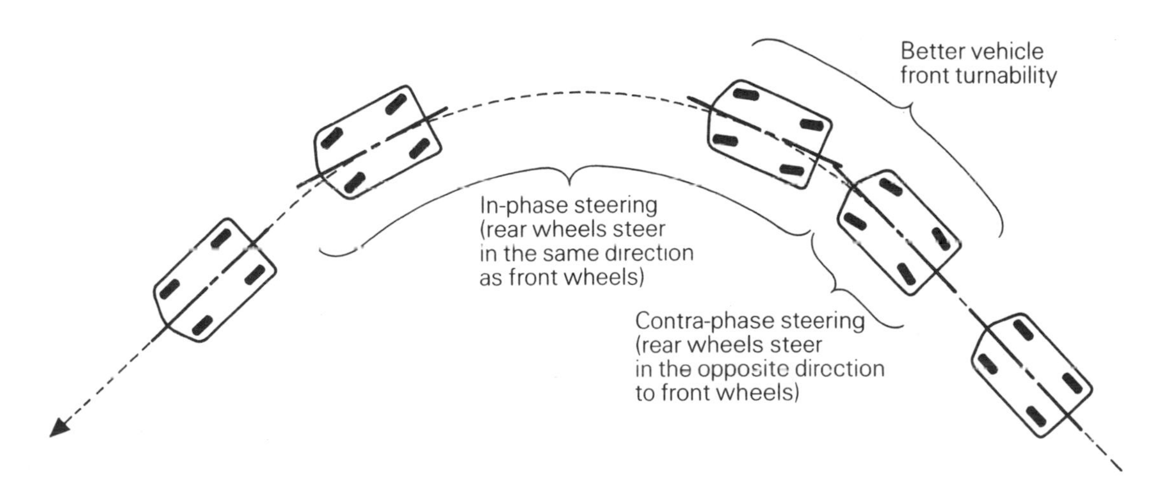

Figure 7.31 Use of contra-phase steering to improve turn-in (*Mitsubishi*)

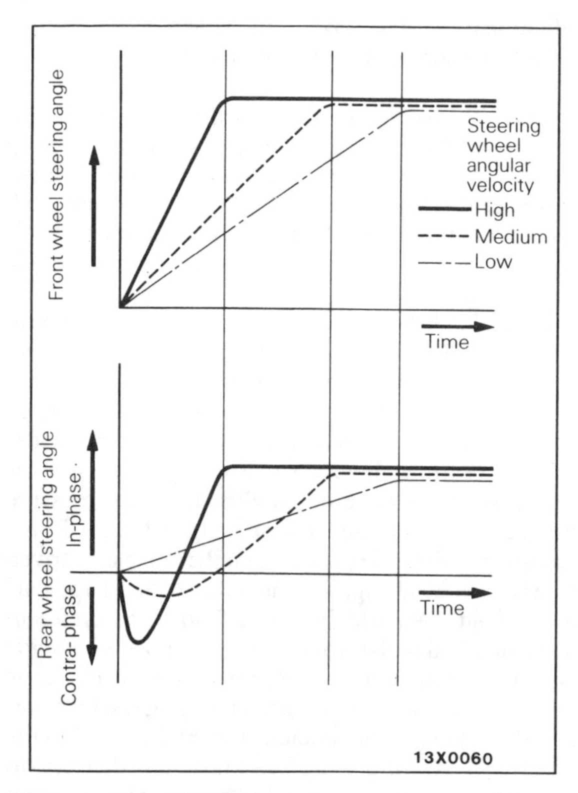

Figure 7.32 Dependence of rear-wheel steering angle on steering speed of front wheels (*Mitsubishi*)

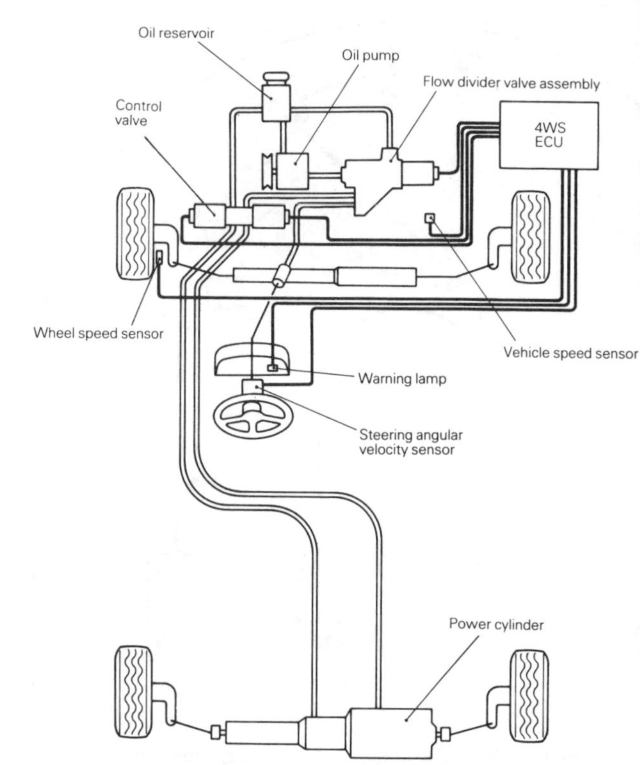

Figure 7.33 Hydraulic system of Active-4WS (*Mitsubishi*)

(for left and right steering angles) which act upon a spool valve. With no current applied to the solenoids, the spool valve is held centrally by return springs and so all of the hydraulic fluid is returned to the reservoir, no fluid reaches the power cylinder and so the rear wheels remain in the straight-ahead position. When current is supplied to either of the control valve solenoids the resulting hydraulic pressure imbalance causes the spool valve to move either to the left or right, and so fluid is directed to either the left or right chamber of the power cylinder. This causes the power cylinder to move against its centring springs and so the rear wheels are steered to the target angle.

Sensors and ECU

The E4WS ECU uses data from a vehicle speed sensor to judge the appropriate drive current to apply to the flow divider valve. Steering actions are then sensed by a photo-interrupter sensor mounted on the steering column, enabling the ECU to calculate the steering wheel angular velocity (rate of turning). A control map is used to obtain the rear-wheel steering angle which is appropriate to the prevailing circumstances. This theoretical angle is then converted into the required control valve solenoid drive current.

Honda E4WS system configuration

Honda's E4WS differs markedly from the Mitsubishi and Nissan systems described above. Rather than using a hydraulic system to steer the rear wheels, it uses an electric motor actuator (Figure 7.34). This has the advantage that hydraulic losses are eliminated and so a more efficient and faster-acting system can be constructed. As with the Mitsubishi and Nissan systems, Honda's E4WS provides in-phase and contra-phase rear-wheel angles according to the driving conditions.

The E4WS ECU monitors road speed and front-wheel steering inputs. Using this data it then calculates the optimum rear-wheel steering angle and supplies an appropriate drive current to a powerful permanent-magnet electric motor, located within the rear steering actuator (Figure 7.35). The motor drives a ball-screw assembly which converts rotary motion into a linear motion of the rack, so steering the rear

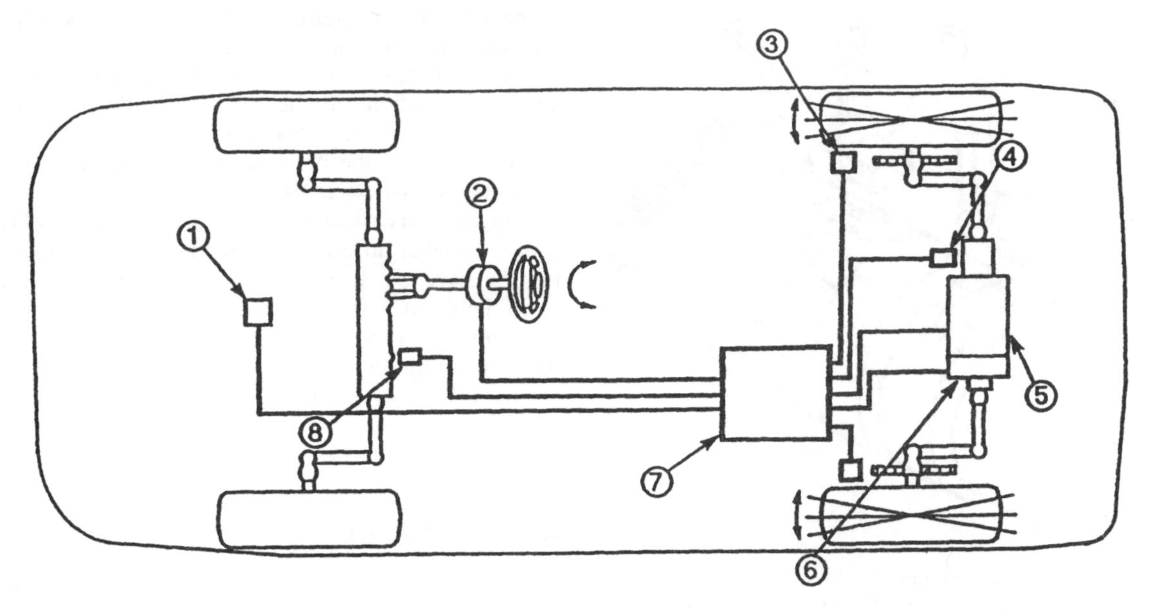

Figure 7.34 Honda E4WS system configuration: [1] speed sensor; [2] main front steering angle sensor; [3] wheel speed sensor; [4] auxiliary rear wheel angle sensor; [5] rear actuator; [6] main rear wheel angle sensor; [7] ECU with power unit; [8] auxiliary front steering angle sensor (*Honda*)

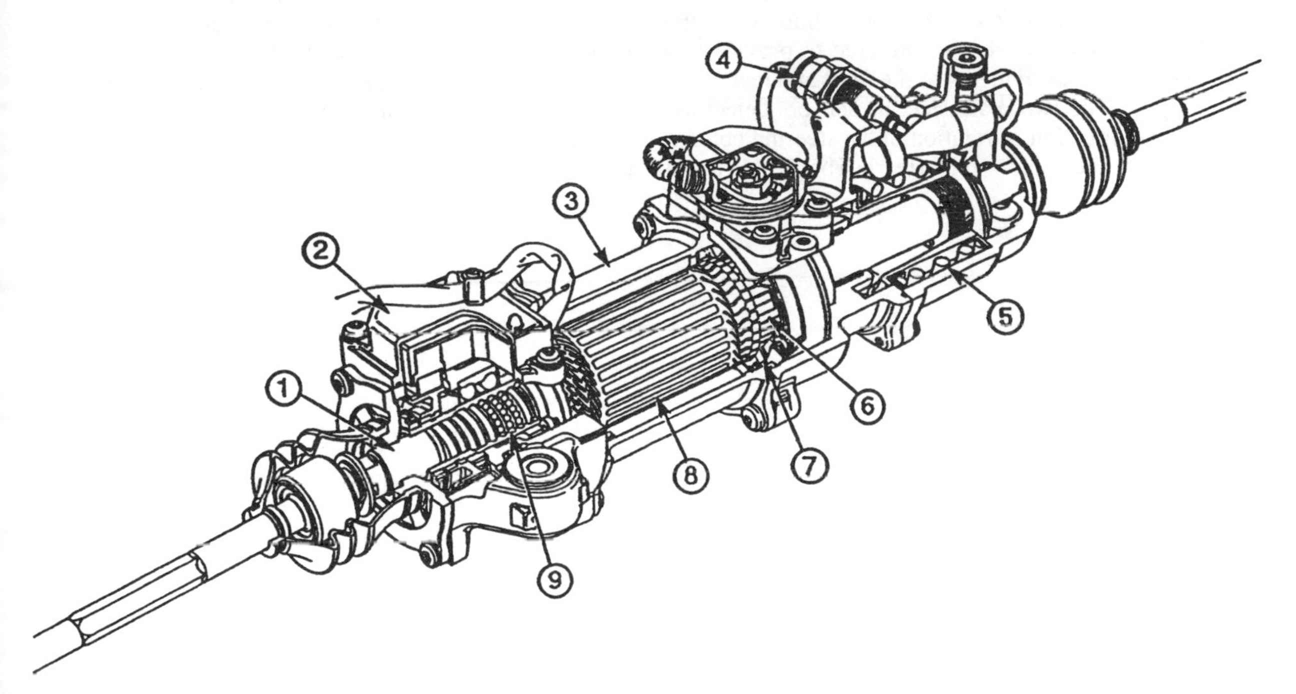

Figure 7.35 Electric motor rear steering actuator used by Honda E4WS: [1] steering shaft; [2] main rear wheel angle sensor; [3] stator; [4] auxiliary rear wheel angle sensor; [5] return spring; [6] commutator; [7] electric motor brushes; [8] motor; [9] ball screw (*Honda*)

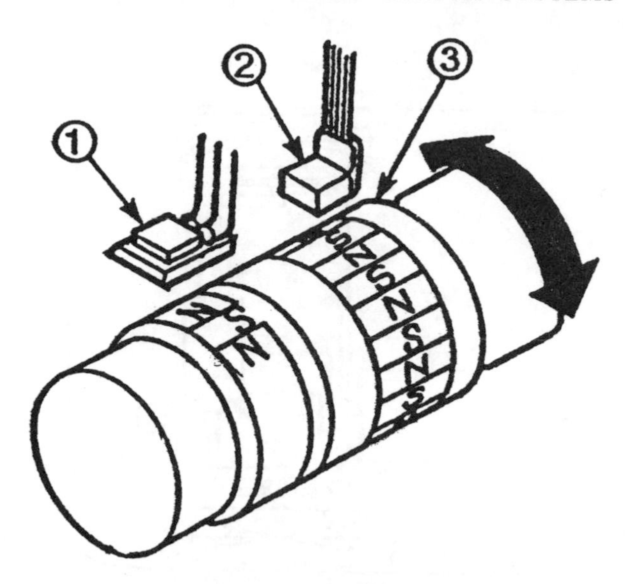

Figure 7.36 Construction of E4WS main steering angle sensor: [1] Hall-IC element; [2] MR element; [3] steering column

wheels. Two sensors (the main and auxiliary rear-wheel angle sensors) provide feedback control of the direction and amount of rear-wheel steering.

In order to ensure the safety and reliability of the system, two separate sensors are used to provide two independent measurements of the front and rear steering angles. The two main steering sensors detect the speed and direction of rotation of the steering column (front) and the motor shaft (rear). They each use a magneto-resistive element (MR element) which is located adjacent to a magnetic polar ring fitted over the relevant shaft (Figure 7.36). As the ring rotates it causes fluctuations in the MR element's resistance. The steering 'neutral' position is detected by a Hall IC which senses the presence of a single south pole on an additional polar ring.

The auxiliary steering-angle sensors are LVDTs (linear variable differential transformers) which function by having a spring loaded rod-shaped core that slides between two coils. The rod rests on a tapered portion of the rack and so progressively moves in or out, depending on which way the rack is moved. This has the effect of varying the degree of inductive coupling between the two coils and so allows the position of the rack to be inferred.

Honda E4WS ECU

The ECU uses two 16-bit microprocessors to independently process signals from the main and auxiliary front-wheel steering angle sensors and the vehicle speed sensor. A control map is then used to obtain a value for the theoretically correct rear-wheel steering angle and the motor is actuated until the rear sensors indicate that this angle has been achieved.

In the event that the two microprocessors give differing results, the system shuts down and a fault indicator light warns the driver of an E4WS malfunction. The motor drive current is cut off and the rear steering rack self-centres under the action of a strong return spring.

8

Body electronic systems

8.1 INTRODUCTION

The body electronic systems are those systems that are fitted within the passenger compartment of the vehicle and enhance the comfort and safety of the occupants.

Until the early 1970s the body electrical system on most cars was limited to the heater motor assembly and some aspects of the instrumentation, such as fuel and temperature gauges. However, in parallel with the enormous growth in the use of engine electronics there has been a similar expansion in the use of body electronic systems and many vehicles are now fitted with numerous luxury and safety features. Commonly encountered systems include electronic instrumentation, central door locking, anti-theft systems, cruise control, air-conditioning and supplementary restraint systems (SRS or 'air-bag'). As the manufacture and sale of cars becomes ever more competitive it is likely that vehicle designers will specify even more sophisticated features such as collision-avoidance radar and satellite-based navigation systems.

8.2 INSTRUMENTATION

All cars are fitted with an instrument pack which usually includes meters for the indication of road speed, engine speed, fuel level, coolant temperature, and a selection of warning lights to inform the driver of critical conditions such as low fluid levels or engine overheating.

Although digital displays have their supporters, research in the field of ergonomics has shown that most drivers prefer simple analogue instruments and find them easier to read at a glance.

Analogue electronic meters and gauges

Speedometer

The analogue speedometer consists of a road speed sensor fitted to the transmission final drive, together with an electronic circuit and meter unit mounted in the dashboard. Several types of speed sensor are in production; most feature the use of a rotating magnet driven directly by the transmission output shaft, with a magnetic field detector that produces electrical pulses at a frequency proportionate to the rotational speed of the magnet. A variety of detector types are used.

(a) *Reed switch.* A reed switch is a miniature switch which closes when subjected to an external magnetic field. The reed switch speed sensor can therefore be used to produce a voltage pulse each time a pole of the magnet passes it.

(b) *Hall-effect sensor*. The operation of the Hall-effect sensor is described in Chapter Two (Section 2.10), it produces a small voltage every time a magnetic pole passes the sensing element.

(c) *Magneto-resistive element (MRE*). The MRE is a small resistive device that shows a step-change in the value of its resistance every time a magnetic pole passes by. When used as one arm of a Wheatstone bridge, the MRE produces voltage pulses at the magnet's rotational frequency.

Speed sensor pulses are routed to the speedometer unit (Figure 8.1) where the electronic circuit filters and shapes the pulses and generates the appropriate currents to operate the cross-coil movement. Another portion of the circuit counts the speed pulses and then supplies accurately timed current pulses to a stepper motor which drives the odometer and trip-meter gears.

The operation of the cross-coil meter movement is illustrated by Figure 8.2. The mechanism consists of a pointer shaft that is fixed to a small permanent magnet housed within two coils wound at 90° to each other. The meter electronic circuit independently supplies each coil with a magnetizing current so that two magnetic fields are produced at 90° to each other. Since the permanent magnet is free to rotate within the cross-coils it aligns itself with the combined effect of these two magnet fields and the pointer is deflected accordingly. It is the task of the electronic circuit to vary the drive currents to each coil in proportion with the speed sensor pulses, thereby changing the direction of the combined field and so controlling the pointer's position.

Tachometer

The tachometer shows the number of revolutions per minute of the engine, either by counting voltage pulses from the ignition coil primary winding or by using the engine management system's crankspeed signal,

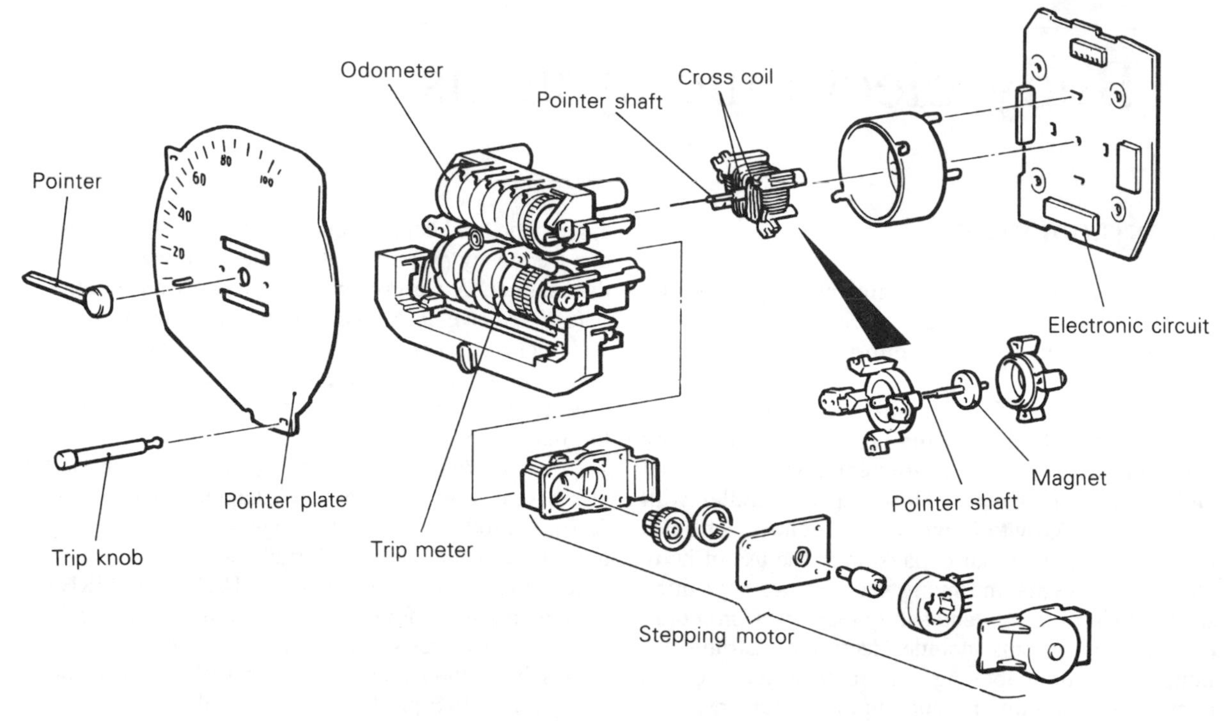

Figure 8.1 Construction of the cross-coil type electronic speedometer

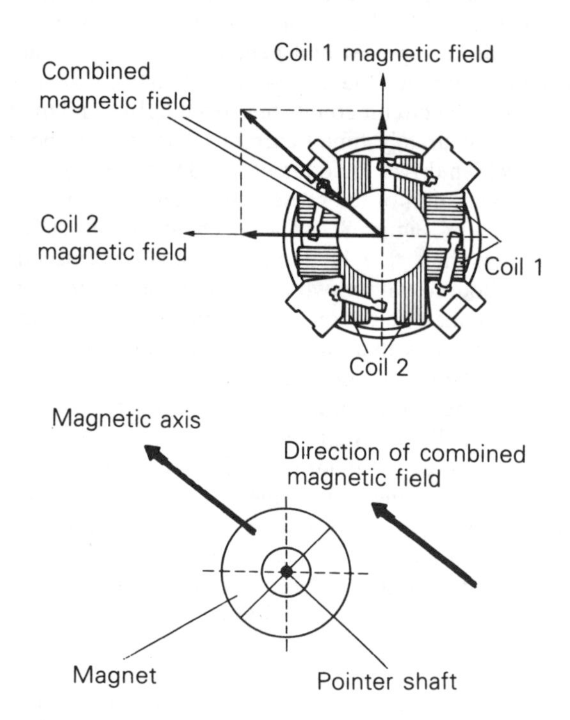

Figure 8.2 Operation of the cross-coil meter movement. In this case the pointer shaft is attached 20° anticlockwise from the magnetic axis

derived from a variable-reluctance type pickup. The rpm signal is then averaged over a period of time to minimize short-term fluctuations and the resulting value is displayed using a cross-coil mechanism.

Coolant temperature gauge

The coolant temperature is sensed using an NTC thermistor which is usually mounted in the thermostat housing (see Chapter Two, Section 2.10). As the coolant temperature increases, the resistance of the thermistor falls and this change is displayed by the gauge unit, which is generally of the cross-coil construction. The sensor signal may also be supplied to a comparator circuit which illuminates a high temperature warning light when the sensor resistance falls below a preset minimum value.

Fuel gauge

The fuel tank contents are sensed using a float which is connected to the wiper of a potentiometer. As the fuel level varies, the float moves up and down, so altering the voltage picked off from the potentiometer track by the wiper.

The gauge unit itself is usually of the cross-coil type, but sometimes features a mechanism filled with a viscous silicone fluid to damp out pointer fluctuations as the fuel sloshes about in the tank. An incidental advantage of the fluid-filled gauge is that it retains its reading even when the ignition is switched off, thereby giving a continuous fuel level indication.

As an alternative to the cross-coil mechanism some manufacturers fit a slow-acting thermal-type gauge which uses a brass-steel bimetallic strip to move the pointer. The strip is heated by a small coil of wire which is wrapped around it and supplied with current from the tank potentiometer. Since brass has a thermal expansion about twice that of steel, the bimetallic strip bends markedly when heated. With a full tank, the potentiometer resistance is large (about 250 Ω) and so the heating current is small, allowing the pointer to remain at the 'full' end of the gauge. As the fuel level falls so does the potentiometer resistance and so the heating current rises, causing the bimetallic strip to bend and drive the needle towards the 'empty' end of the gauge.

To warn the driver of a very low fuel level most vehicles are fitted with a 'low-fuel' warning light, usually activated by the output of a voltage comparator when the sensor resistance falls below a predefined limit. The warning light circuit is often fitted with an 'anti-slosh' timer so that the bulb is only illuminated when a low fuel level is continuously recorded for several seconds.

Oil pressure gauge and warning light

All vehicles are fitted with a warning light to alert the driver to a critically low engine oil pressure. Some vehicles are additionally fitted with an oil pressure gauge to give a continuous indication of the engine oil pressure.

The low oil pressure switch is screwed into an engine-block tapping close to the oil-pump output port and contains a small spring-loaded diaphragm that is connected to a pair of contacts. When oil pressure is low, the diaphragm is held down by the spring and so it connects the switch contacts together, completing the warning light circuit. As oil pressure rises the diaphragm moves out against the spring and away from the switch contacts, allowing them to open and so breaking the warning light circuit.

Where an oil pressure gauge is fitted it may be of either the cross-coil or bimetallic strip construction and is supplied with a signal from a sensor unit which is usually located near to the oil pressure switch. The sensor provides an oil pressure signal by measuring the deflection of a diaphragm which is constructed using one of several methods.

(a) *Silicon or alumina strain-gauge sensor*. This type of sensor is described in Chapter Five (Section 5.4) in the context of inlet manifold pressure measurement. When used to sense engine oil pressure, the oil causes a deflecion of the diaphragm, resulting in a change in the value of piezo-resistors that are integral with the diaphragm's surface.

(b) *Thermal sensor*. The thermal sensor uses two contacts; an earth contact fixed to the surface of the diaphragm and a fixed supply contact made from a bimetallic strip and connected to the battery via the gauge unit. When the diaphragm is deflected by oil pressure the two contacts close and the sensing circuit is completed. Current passes through the bimetallic strip, causing it to heat up and eventually bend away from the earth contact, breaking the circuit. After a short time the strip cools down and straightens out, re-closing the circuit for a short time before heating up and reopening again. In this way the contacts rapidly pulse on and off, supplying a current with an average value which is in proportion to diaphragm deflection and therefore oil pressure. The gauge pointer assumes a position in proportion to this current.

(c) *Potentiometer sensor*. This sensor employs a potentiometer whose wiper is fixed to the diaphragm. As the diaphragm is deflected, the wiper moves along the potentiometer track and so provides an output voltage which varies in proportion to oil pressure.

Digital meters and gauges

Whereas most automobiles use conventional analogue meters to display parameters such as vehicle speed, engine temperature and fuel level, some manufacturers have marketed vehicles which display this information in a digital format. Interest in digital displays reached a peak during the mid-1980s and has decreased since then; this is largely because it is easier for a driver to assimilate information by noting the general *position* of a pointer rather than by making an assessment of *numerical data*. For example, if the pointer of an analogue coolant temperature gauge is not in the red zone then the driver can be sure that the engine is not overheating, but is a digital temperature indication of 95°C normal, or too high?

In order to overcome the shortcomings of digital displays many of the latest designs use display technology which presents data in a 'quasi-analogue' fashion, for example by using bar graphs or pictorial symbols.

Digital display technology

The operation of a digital instrument pack is very similar to that of its analogue counterpart; sensors monitor vehicle parameters and supply data to an instrument ECU in the form of analogue voltage or current signals. The ECU, which usually incorporates a microprocessor, converts the data to a digital form and displays it on the digital display device. Common display devices include light emitting diodes (LEDs), liquid crystal displays (LCDs), vacuum fluorescent displays (VFDs) and cathode ray tubes (CRTs).

(a) *Light emitting diode (LED).* The LED is the earliest solid-state digital display device. It first appeared during the late 1960s in digital wristwatches, soon becoming familiar through its characteristic bright red colour (although other colours are now available). LEDs are only suitable for the display of alphanumeric symbols and bar graphs but have the advantage that they have high brightness, a long life and are very durable.

(b) *Liquid crystal display (LCD).* Liquid crystal displays are based on the phenomenon that certain types of liquids are composed of a regular array of rod-like molecules whose orientation can be changed by the application of an electrical field. This has the effect of changing the liquid crystal's optical properties.

Typically, an LCD segment consists of two thin glass polarizing filter plates, mounted with their polarizing axes at 90° to each other, and separated by a thin layer of liquid crystal material (Figure 8.3). The glass plates are each coated with a microscopically-thin (and therefore transparent) metal film which is connected to the drive electronics. With no voltage applied, the liquid crystal molecules form a vertical array which rotates the plane of light polarization through 90°, so that light that passes through the first plate has its plane of polarization rotated through 90° to pass through the second plate (Figure 8.3(a)). The LCD therefore transmits light when in this mode.

If a drive voltage of a few volts is applied to the LCD then the liquid crystal molecules align as an array of rods positioned lengthwise between the two plates. When this happens the light's plane of polarization is not rotated by the crystal and so light passing through the first polarizing plate cannot pass through the second (Figure 8.3(b)). In this mode the LCD segment acts to block light and so it appears as a black region.

It is important to note that LCDs do not emit any light, they simply either transmit or block incident light. Consequently they can be difficult to read when the ambient light level is low and so normally require 'backlighting' with a filament bulb when used in an automotive application.

A great advantage of the LCD is that is consumes almost no power and can therefore be operated continuously, with no fear of discharging the battery (e.g. on a clock display).

(c) *Vacuum fluorescent display (VFD).* The vacuum fluorescent display consists of a number of electrodes which are enclosed inside an evacuated glass envelope. It functions by having fast-moving electrons bombard a phosphor material which then emits light.

A thin filament-wire is heated to a high temperature by passing a heating current through it, causing it to emit electrons. These electrons are attracted towards a fine metal grid which is connected to a positive voltage supply. Most electrons pass through the holes in the grid and strike the display segments which are coated with phosphor and so emit light.

Through the selective application of a positive voltage to different segments it is possible to display a large range of alphanumeric characters and symbols. The brightness of the display may be controlled by varying the positive voltage on the grid. VFDs are excellent displays and have become widely accepted for automotive use, being robust and reliable, easy to read and available in a wide range of colours.

(d) *Cathode ray tube (CRT).* The cathode ray tube is familiar as the display device used in television receivers and computer monitors. It is extremely versatile, being able to present information in an almost limitless range of styles and colours.

CRTs operate by having a heated filament emit electrons which are accelerated under the influence of a positive grid to strike a phosphor-coated screen, giving a glowing 'spot'. The beam path is controlled by electromagnets which are wound around the tube and enable the spot to be rapidly scanned across the screen, building up an image. If the screen is coated with phosphors that fluoresce in the three primary colours of red, blue and green, then the complete spectrum of colours may be produced by additive mixing.

Although CRTs are good all-purpose displays they suffer from several weaknesses, in particular they are comparatively bulky and heavy, require a high operating voltage (15–25 kV) and need complex drive circuitry. It is for these reasons that they have not been widely adopted in automotive applications.

On-board computer (OBC)

The function of an on-board computer (sometimes called a trip computer) is to provide the driver with detailed journey information. It consists of a keypad and digital display, together with an integral microprocessor control unit (Figure 8.4).

The OBC incorporates an accurate clock/calender circuit and an interface circuit which monitors fuel usage by measuring the accumulated opening time of the fuel injectors. With the addition of speed and

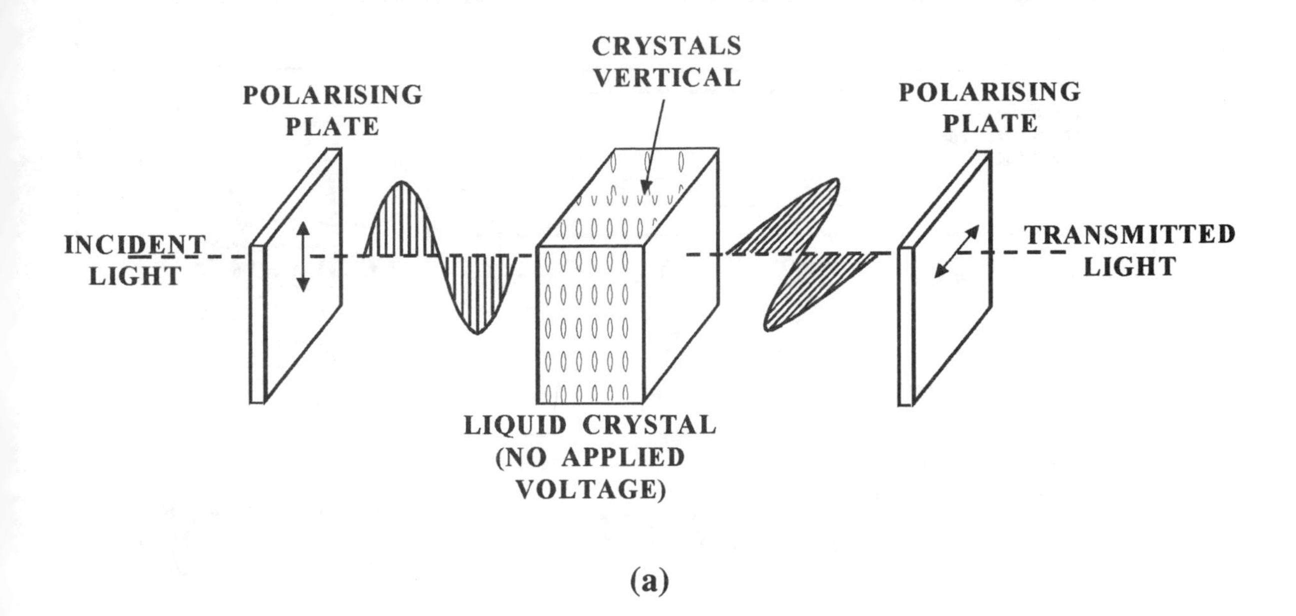

(a)

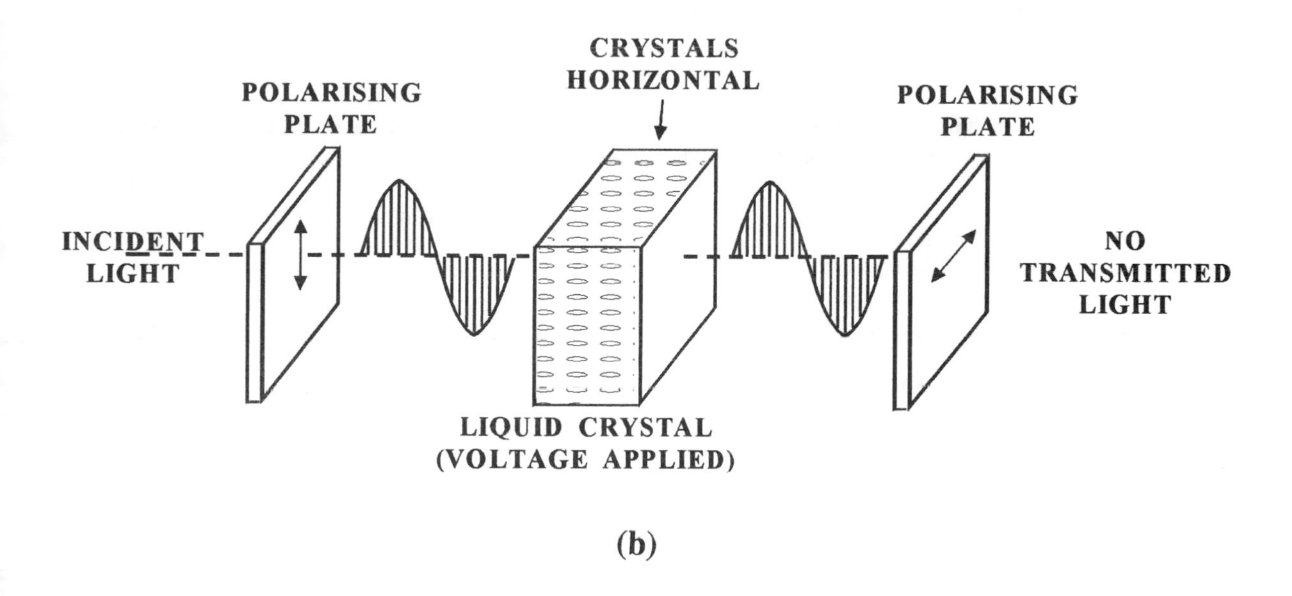

(b)

Figure 8.3 Operation of the liquid crystal display (LCD)

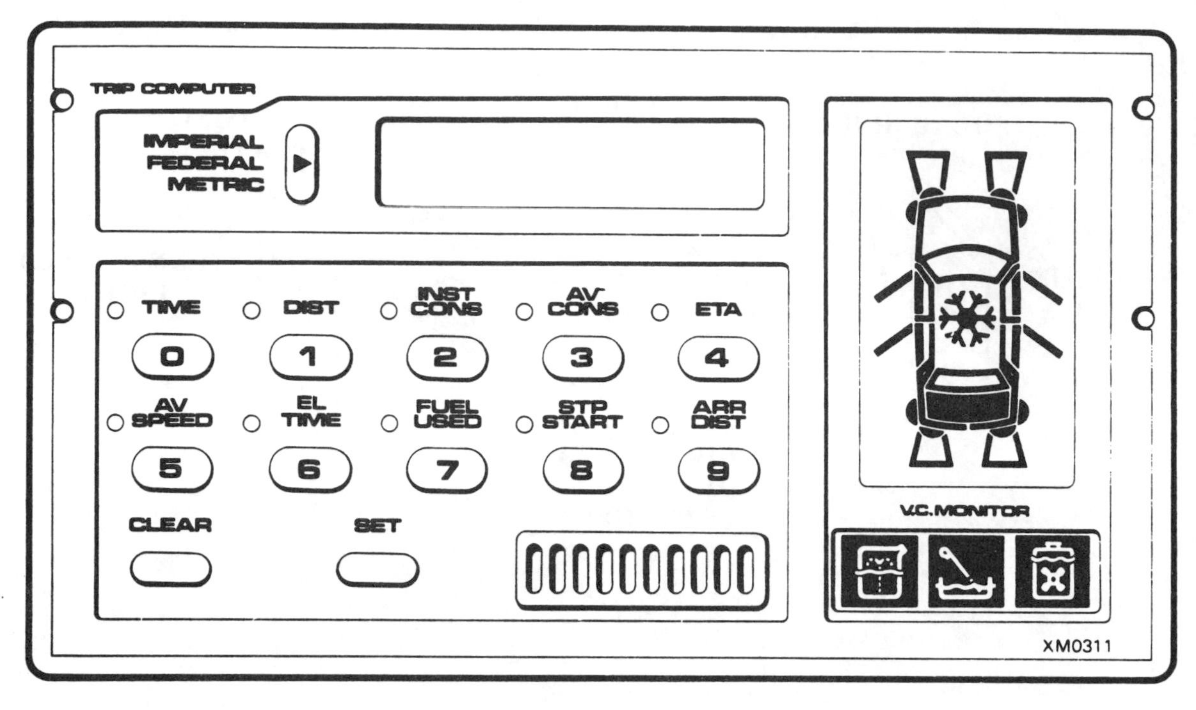

Figure 8.4 Typical on-board computer (OBC) incorporating vehicle condition monitor (VCM) (*Rover*)

distance information obtained from the vehicle speed transducer, the OBC can calculate and display the following data:

(i) Accurate time, day and date.
(ii) Average speed since the journey started.
(iii) The elapsed time since the journey started.
(iv) Average fuel consumption since the journey started.
(v) Instantaneous fuel consumption.
(vi) Fuel used since the journey started.
(vii) Predicted driving radius on the remaining fuel.

If, at the start of the journey, the driver uses the keypad to enter the distance to the destination, then the OBC can additionally calculate:

(viii) The estimated time of arrival at the destination.
(ix) The remaining distance to the destination.

Vehicle condition monitor (VCM)

The vehicle condition monitor may be integrated into the instrument pack, separately-mounted in the centre console or incorporated as a part of the on-board computer system. It is usually implemented as a vehicle 'map' which uses LCD or VFD display segments that are selectively illuminated to indicate the operating condition of the vehicle (Figure 8.4). If a fault is detected, then an audible warning is sounded to draw the driver's attention to the display panel.

The range of functions supervised by the VCM varies according to the vehicle specification, but will include at least some of the following:

(i) Brake light failure.
(ii) Side and headlight failure.
(iii) Opening panel ajar (doors, boot lid, etc.).
(iv) Low outside temperature ('ice warning').
(v) Low engine coolant level.
(vi) Low engine oil level.
(vii) Low screenwash fluid level.
(viii) Excessive brake pad wear.

Lighting circuit monitor

Monitoring of the lighting circuits is usually achieved by detecting the flow of current in the wires leading to the relevant light units. This current is typically sensed in one of two ways:

(i) Placing a very low value resistor (a fraction of an ohm) in series with the lighting feed wire. When the light is illuminated a small voltage drop is measurable across the ends of the resistor. This voltage differential is used to cause a comparator to switch to give a 'high' output state, indicating satisfactory operation. Conversely, if a bulb is blown, then there will be no current through the resistor and therefore no voltage difference to trigger the comparator, which will now give a 'low' output, triggering a warning indication.

(ii) The feed wire to the light unit may be coiled around a reed switch. When current passes through the feed wire, the magnetic field produced within the coil causes the reed switch contacts to close, indicating to the VCM that the light unit is functioning. If a bulb is blown, then no current passes through the feed wire and so the reed switch remains open.

Door ajar warning
To provide a warning of an open door or boot lid, the VCM monitors the state of the courtesy light and boot light switches.

Low outside temperature ('ice warning')
Outside air temperature is measured using an ntc thermistor which is located in an enclosed area away from heat sources (e.g. immediately behind the front bumper). As the air temperature falls, the resistance of the thermistor rises and is monitored by the VCM. When the resistance reaches a predetermined value, corresponding to an air temperature in the region of +4°C, a comparator is triggered and a 'snowflake' symbol is illuminated on the display to warn the driver of the possibility of icy roads.

Fluid level monitoring
Monitoring of the level of critical fluids such as oil, engine coolant and washer fluid, can be accomplished using a magnetic float and reed switch arrangement (Figure 8.5). Normally the reed switch is fixed inside a support pillar and the permanent magnet is enclosed within a sealed plastic float. When the fluid level is correct, the float rises to the upper limit of its travel and so the reed switch closes under the influence of the magnet's field. As the fluid level falls, the float drops and the reed switch opens, causing the VCM to issue a warning.

In the case of engine oil level, the VCM only performs a check for a few seconds before the engine is started. This is because the sump oil level in a running engine is quite low and subject to considerable short-term fluctuation (due to cornering and braking forces) which would otherwise lead to many false alarms.

To safeguard the VCM system against wiring faults, such as open or short circuits, the reed switch sensor incorporates two fixed resistors. When the reed switch is open, the sensor wiring has continuity through the high-value resistor that is connected in parallel with the switch contacts. Conversely, when the switch is closed, the low-value resistor is brought into the circuit. The sensor circuit therefore presents either a high-resistance or a low-resistance to the VCM. An open or short circuit can thus be interpreted as a wiring fault and so a different type of fault indication can be given.

Brake pad wear indication
In order to warn the driver that the brake pads have worn out and require replacement, many vehicle manufacturers fit a brake wear warning light. The simplest systems use a single wire which is buried at a predetermined depth (about 2 mm) from the pad backplate. When most of the friction material has worn away, the wire is exposed and touches the brake disc, providing an earth contact that can be used to illuminate a warning light. The weakness of such a system is that the contact may be intermittent, depending upon whether the brakes are being pressed, and any bad connection in the circuit will render the system inoperative. An alternative solution is to embed a small loop of wire in the pad and pass a small current through it. When the pad is worn down, the brake disc grinds through the loop and so the current in the circuit changes enabling the VCM to illuminate a warning light.

Drowsiness warning system

Statistics indicate that drowsiness at the wheel may be resposible for approximately 3% of all road accidents, with a death occurring in nearly 50% of cases. A system has been commercialized which allows drowsiness to be detected, and the driver alerted before an accident happens.

The system operates on the basis that whereas an alert driver will make continuous, small, steering corrections, a drowsy driver will steer in a distinctly different fashion. A typical pattern is several seconds of inactivity followed by sudden large steering corrections to the left and right. The warning system is

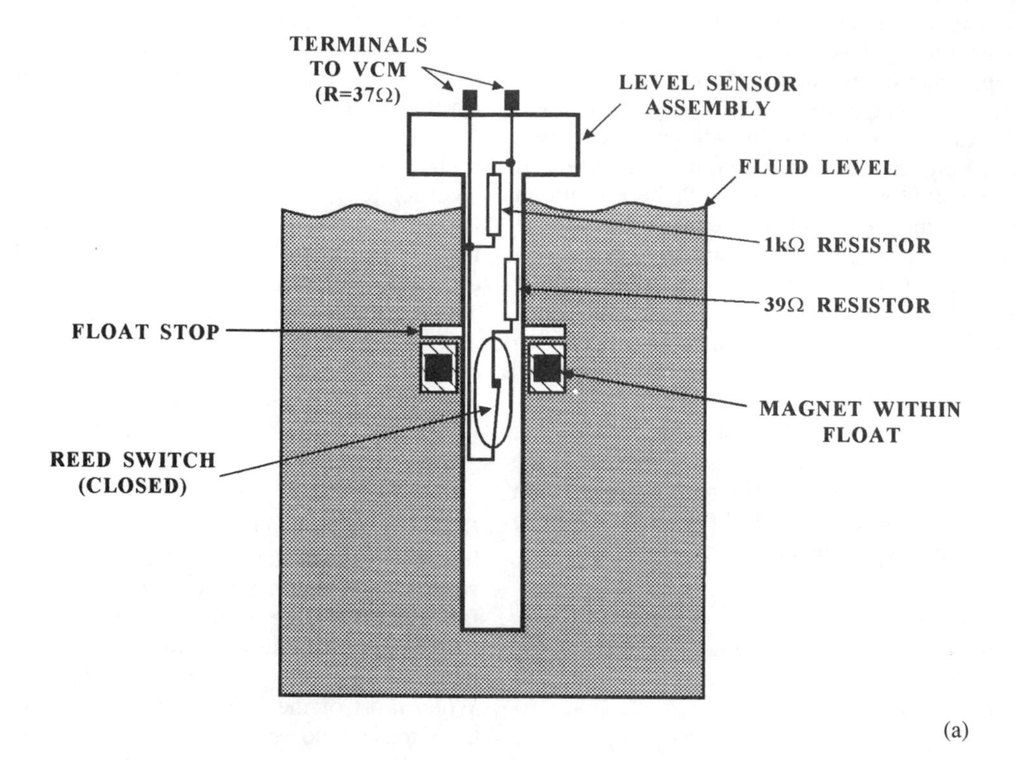

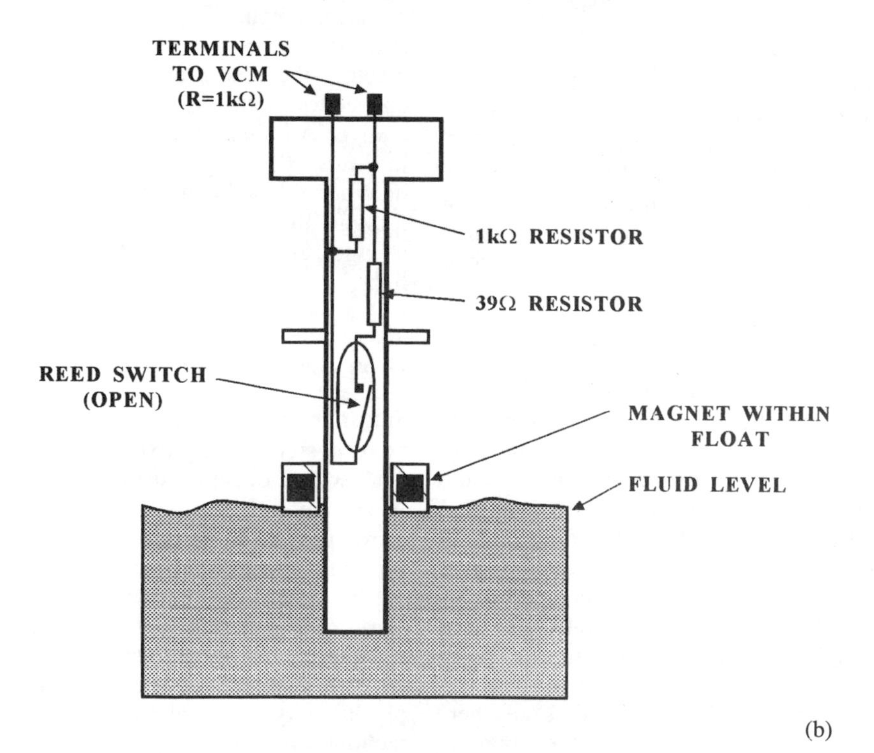

Figure 8.5 Fluid level sensing using magnetic float and reed switch

based around a microcomputer that constantly monitors the driver's steering movements and compares them with steering patterns held in memory which are typical of a drowsy driver. If a match is detected the system issues a warning to prevent the driver falling asleep at the wheel.

8.3 SUPPLEMENTARY RESTRAINT SYSTEM (SRS OR 'AIR-BAG')

The Supplementary Restraint System (SRS) is an air-bag system that works in conjunction with conventional 3-point seat belts to prevent the driver's chest and face from striking the steering wheel in the event of a frontal impact. SRS may additionally be fitted to the passenger's side of the vehicle to prevent impact with the dashboard. 'Side-impact' air-bags have also been developed to protect the upper body and head during a sideways impact.

WARNING: The SRS incorporates a gas-generator module which contains an explosive charge. On no account should the SRS be disassembled or otherwise tampered with. Do not use a multimeter or oscilloscope to probe the SRS connections – only the manufacturer's dedicated diagnostic tester should ever be connected to the SRS. Before scrapping an SRS-equipped vehicle always be sure that the air-bag has been deployed using the manufacturer's deployment tool.

SRS system configuration

Figure 8.6 shows a typical SRS configuration, the function of each component is listed in Table 8.1.

The system consists of a number of deceleration sensors ('D sensors') mounted at the front of the vehicle to detect the onset of a frontal impact occurring within about 30° of the vehicle centre line. An additional deceleration sensor (called a safing sensor or 'S-sensor') is mounted in the passenger compartment and wired in series with the D-sensors. The S-sensor triggers at a much lower deceleration level than the D sensor (2.5 g as against about 15 g) and is used to provide confirmation that a collision has actually occurred and prevent firing of the air-bag due to localized impacts on the wings of the car.

The sensor signals feed into the SRS ECU, often termed a 'Diagnostic Module', which provides constant checking and monitoring of the entire SRS. When a frontal collision is confirmed by both a D-sensor and the S-sensor, the diagnostic module sends a current pulse to fire a gas generator unit contained within the air-bag module which is located in the steering wheel centre pad. The air-bag is then inflated. If the diagnostic module detects a fault with the SRS, an instrument-panel warning lamp is illuminated and the SRS is deactivated to prevent accidental deployment. The driver should then take the vehicle for immediate repair.

SRS crash sensors

The exact type of crash sensor fitted to the SRS varies with the system manufacturer, however all designs work on the basis that they are open-circuit during normal driving but closed-circuit when subjected to a deceleration in the range 15–20 g, corresponding to a 15–20 mph impact with a solid object. Common designs include the 'magnetic bias' sensor and the Rolamite sensor;

(a) *Magnetic bias sensor* (Figure 8.7). This sensor consists of a sensing mass (a metal ball) which is firmly held at the rear of a small cylinder by a powerful bias magnet. During normal driving the electrical contacts are therefore open-circuit. When a collision occurs, the inertial force on the ball overcomes the magnetic bias force and so it rolls forward along the cylinder until it strikes the electrical contacts and completes the detection circuit.

(b) *Rolamite sensor* (Figure 8.8). The Rolamite sensor comprises a roll-spring which is wrapped around a small roller and fitted with an electrical contact pad. The roll-spring is pre-tensioned so that during normal driving the roller is held firmly against a backstop and the detector contacts are open-circuit. In the event of a collision, the inertial force on the roller overcomes the roll-spring's pretension, allowing it to travel forwards until the electrical contacts meet, completing the detection circuit.

To ensure the mechanical and electrical integrity of the crash sensors they are normally housed in airtight metal enclosures that are filled with inert gas to prevent corrosion. Each sensor is wired as a normally-open switch, but with a resistor connected in parallel with the contacts. The resistor allows the diagnostic module ECU to continually monitor the circuit's continuity for connector and wiring defects.

Rotary coupler, air-bag and inflator

The air-bag and inflator are housed in the steering-wheel centre pad and connected to the diagnostic module by a 'clock-spring' electrical rotary coupler which is fitted between the steering wheel and steering column housing.

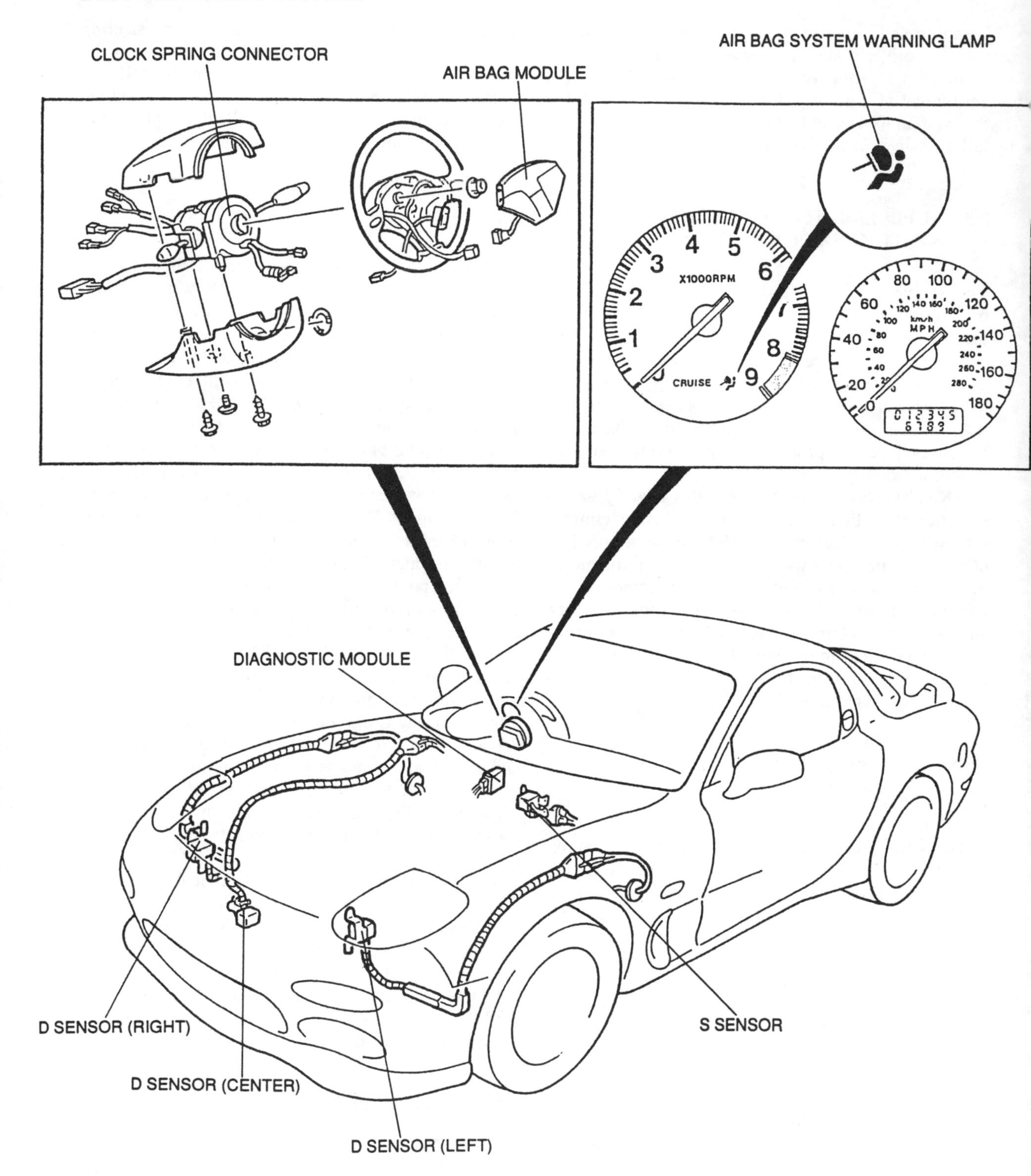

Figure 8.6 Major components of a supplementary restraint system (SRS) (*Mazda*)

Table 8.1 SRS component function

Component		Function	Remark
AIR BAG system warning lamp		Lamp illuminates or flashes if malfunction occurs in air bag system	Located in instrument cluster
Air bag module		Deploys air bag when current flows to integrated igniter	Located in steering wheel hub
Clock spring connecter		Ensures uninterrupted electrical connection to air bag module while allowing turning of steering wheel	Part of combination switch
Crash sensor	D-sensor	Activated (closed) when crash impact detected With S-sensor, completes circuit to inflator	Located in front part of vehicle (left, right, and center)
	S-sensor	Activated (closed) when crash impact detected With D-sensor, completes circuit to inflator	Located on side of heater unit in passenger compartment
Diagnostic module		Monitors components and wiring harnesses in air bag system Indicates system malfunction by flashing or illuminating AIR BAG waring lamp If warning lamp is burnt, sounds warning buzzer Detects short circuit between air bag module and ground or crash sensor malfunction and melts system down fuse to prevent unintended air bag deployment	Contains backup battery

The air-bag itself is a woven nylon bag lined with rubber and folded-up behind a polyurethane cover (Figure 8.9). When the bag is inflated, the cover tears along the crease and hinges open, allowing the bag to deploy. A fully inflated bag has a gas capacity of between 30 and 70 litres, depending upon the vehicle size and market.

The detailed construction of the inflator unit is illustrated in Figure 8.10. Inflation takes place when the diagnostic module supplies a current pulse to the 'squib' igniter. Both electrical connections to the igniter are floating (i.e. isolated from the vehicle ground) so that it cannot be fired by an accidental short-circuit.

Once the igniter rises to a temperature of over 190°C, self-ignition of the ignition intensifier mixture occurs. The heat which is generated causes the sodium azide propellant pellets to undergo rapid combustion, generating nitrogen gas which is filtered and cooled before passing into the air-bag. The bag's internal pressure when fully inflated is comparatively low, about 0.1–0.3 bar (1.4–4.4 psi), but is sufficient to protect the driver as he is thrown against it. The load of the driver on the air-bag causes gas to escape through two large outlet holes in the rear of the bag and further cushions the impact.

Diagnostic module ECU

Reliability is a prime consideration in the design of the air-bag system. If required, it must fire within a few milliseconds, possibly after many years of inaction. The function of the diagnostic module is therefore to ensure that the SRS is always in a state of readiness which it does by:

(i) monitoring the firing and sensor circuits and indicating faults via the SRS warning light. Fault data is stored in non-volatile memory for future recall through the serial diagnostic link.
(ii) processing the crash sensor signals and firing the igniter when a frontal impact is detected.

Since electrical power could be interrupted at an early stage in the collision (for example, by severing of a battery lead), SRS diagnostic modules incorporate some form of power storage. Typically an internal backup battery or a large-value capacitor is kept

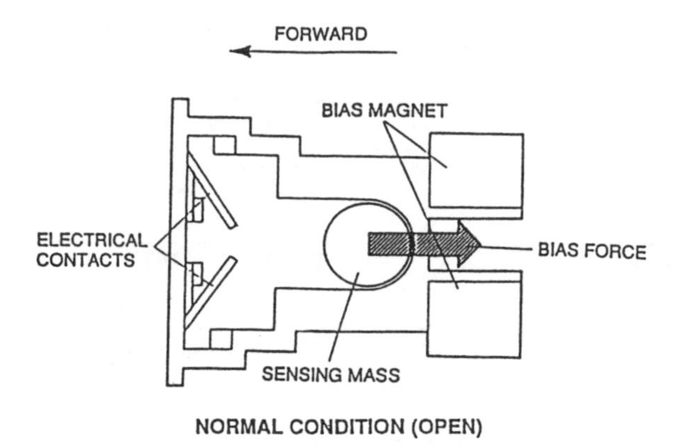

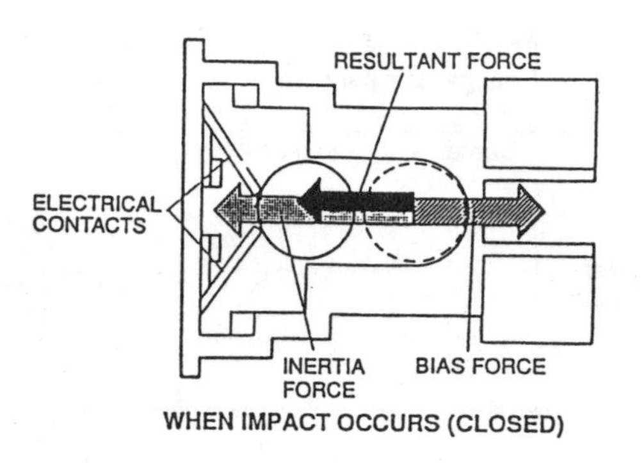

Figure 8.7 The magnetic bias crash sensor (*Mazda*)

charged-up to give about 200 ms of reserve power, sufficient to fire the inflator.

Pyrotechnic seatbelt tightening system

The pyrotechnic seatbelt tightening system complements an SRS installation and uses similar technology and operating principles (Figure 8.11). It is designed to tighten the seatbelt around the driver and front passenger in the initial few milliseconds after an impact, holding them safely and firmly into their seats.

Belt tightening is achieved via a cable which is wrapped around the inertia-reel seatbelt drum. The free end of the cable is attached to a piston that sits at the bottom of a tube mounted in the car's B-post. Beneath the piston is an explosive capsule containing a detonator and propellant. The detonator is fired by a crash sensor that is fitted under the front seat and set to trigger on a frontal deceleration of 5 g or more. The system is designed to provide detonation within about 15 ms of impact, causing the piston to be propelled up the tube. This pulls the cable, which rotates the drum and removes about 10 cm of slack from the seatbelt.

SRS and pyrotechnic seatbelt deployment sequence

The timing of the SRS and pyrotechnic seat deployment sequence is illustrated in Figure 8.12 and is typically as follows:

(i) Time = 0 ms. The vehicle collides with a solid object at an angle within about 30° of the centre line and at a speed greater than about 20 mph.

(ii) Time = 10 ms. The front crash sensors and the safing sensor have moved to the closed-circuit position, causing the SRS diagnostic module to transmit a firing pulse to the igniter. The pyrotechnic seatbelt unit has detonated.

(iii) Time = 13 ms. 3 ms after the arrival of the igniter firing signal, gas generation has started with a loud bang. The driver is still upright in his seat. The seatbelt has been partially tightened.

(iv) Time = 15 ms. The air-bag has partially inflated, breaking the cover crease. The seatbelt is almost fully tightened.

(v) Time = 20 ms. The vehicle is starting to crumple and the driver has moved very slightly forward towards the steering wheel but is being restrained by the seatbelt which is now fully tightened.

(vi) Time = 30 ms. The air-bag is fully inflated and the driver's chest and face are about to come into contact with it. The seatbelt is helping to restrain the driver.

(vii) Time = 80 ms. The load of the driver on the air-bag causes the nitrogen gas to escape through the outlet holes at the rear of the bag and so it starts to deflate. Gas pressure beneath the pyrotechnic seatbelt piston is falling and so it starts to move back down the tube, feeding some slack back into the belt.

(viii) Time = 120 ms. The driver has moved back into his seat and the air-bag is deflating, providing unrestricted visibility and allowing an easy exit from the wrecked vehicle.

At the conclusion of the deployment, the interior of the vehicle will be covered with a small amount of white powder residue from the air-bag inflator, this is mostly sodium carbonate (Na_2CO_3) and a tiny amount of sodium hydroxide (NaOH), neither of which is considered hazardous.

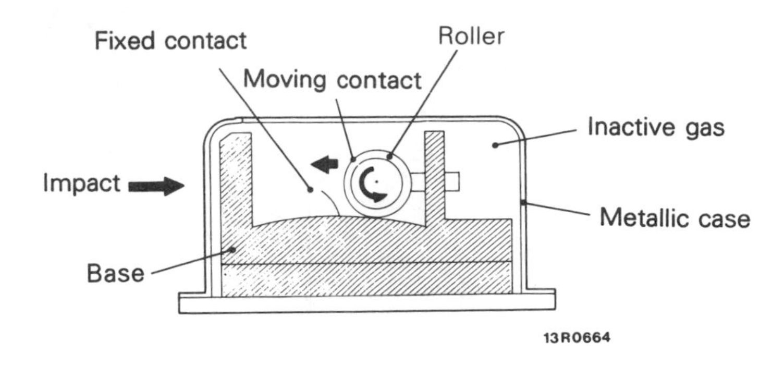

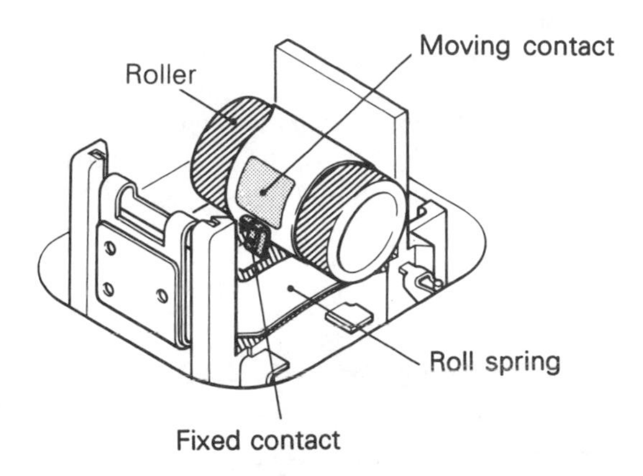

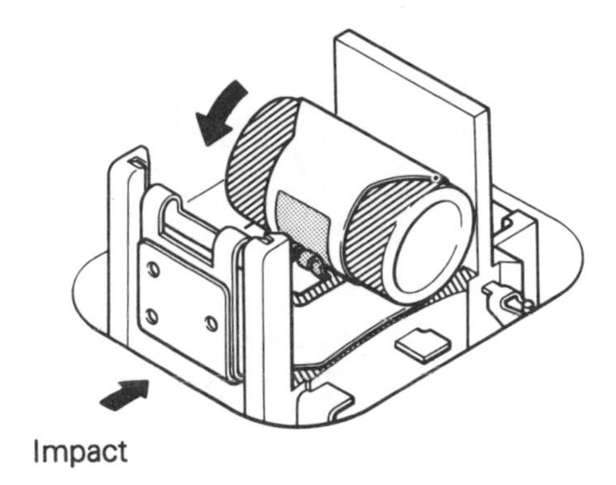

Figure 8.8 Rolamite-type crash sensor (*Mitsubishi*)

Single-module SRS

Some vehicle manufacturers have introduced air-bag systems that integrate all of the SRS elements, including the crash sensors, into a single module mounted within the steering-wheel centre cover. These systems are likely to become increasingly common since there are obvious advantages with regard to minimizing the number of connectors and the length of wiring, improving reliability and reducing cost.

Side-impact air-bags

During the mid-1990s side-impact air-bags were introduced to protect vehicle occupants in the event of a sideways collision. Side-impact air-bag installation varies from one vehicle to another, but generally comprises a 17 litre air-bag module mounted in each door panel plus additional inflatable tubes mounted along the roof rails. Whereas a conventional SRS air-bag system protects the driver by cushioning the secondary impact with the steering wheel during a frontal impact, the side-impact air-bag is designed to gently accelerate a seat occupant into the centre of the vehicle and prevent contact with the door panel. Side-impact air-bags are therefore designed to operate at a higher inflation pressure, and deflate more slowly, than frontal-impact air-bags.

Each side-impact air bag inflator is connected to a diagnostic unit which operates in the same way as a conventional SRS ECU. Triggering of the side-impact air-bags is controlled by lateral acceleration sensors which detect any sudden change in the sideways movement of the vehicle. The construction of these sensors is similar to that of the crash sensors used in an SRS system, although the mounting orientation is obviously different and the trigger level is set at lower ‘g’ level.

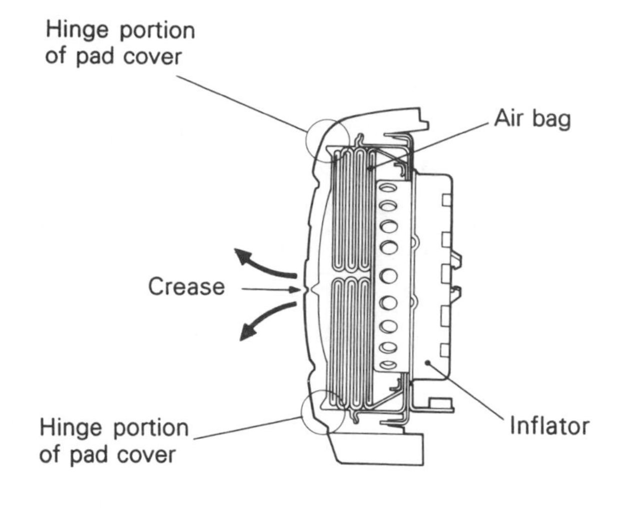

Figure 8.9 Section through air-bag module incorporating inflator, air-bag and polyurethane cover

8.4 CRUISE CONTROL SYSTEMS

Cruise control systems were first introduced during the mid-1960s as a means of reducing driver fatigue on long motorway journeys. They allow the driver to maintain a fixed speed without touching the accelerator pedal.

The system is composed of three main assemblies:

(i) *A driver's switch pack.* This is mounted either on the steering wheel or on a column stalk and incorporates an 'ON/OFF' switch, a speed 'SET' switch and a speed 'RESUME' switch.

(ii) *A throttle actuator unit.* This actuator unit is connected to the throttle valve and takes control of the throttle butterfly position under the authority of the cruise-control ECU. Actuator mechanisms normally employ either a permanent magnet dc motor assembly (Figure 8.13) or a vacuum diaphragm powered by a motor-driven pneumatic pump and controlled by solenoid valves (Figure 8.14).

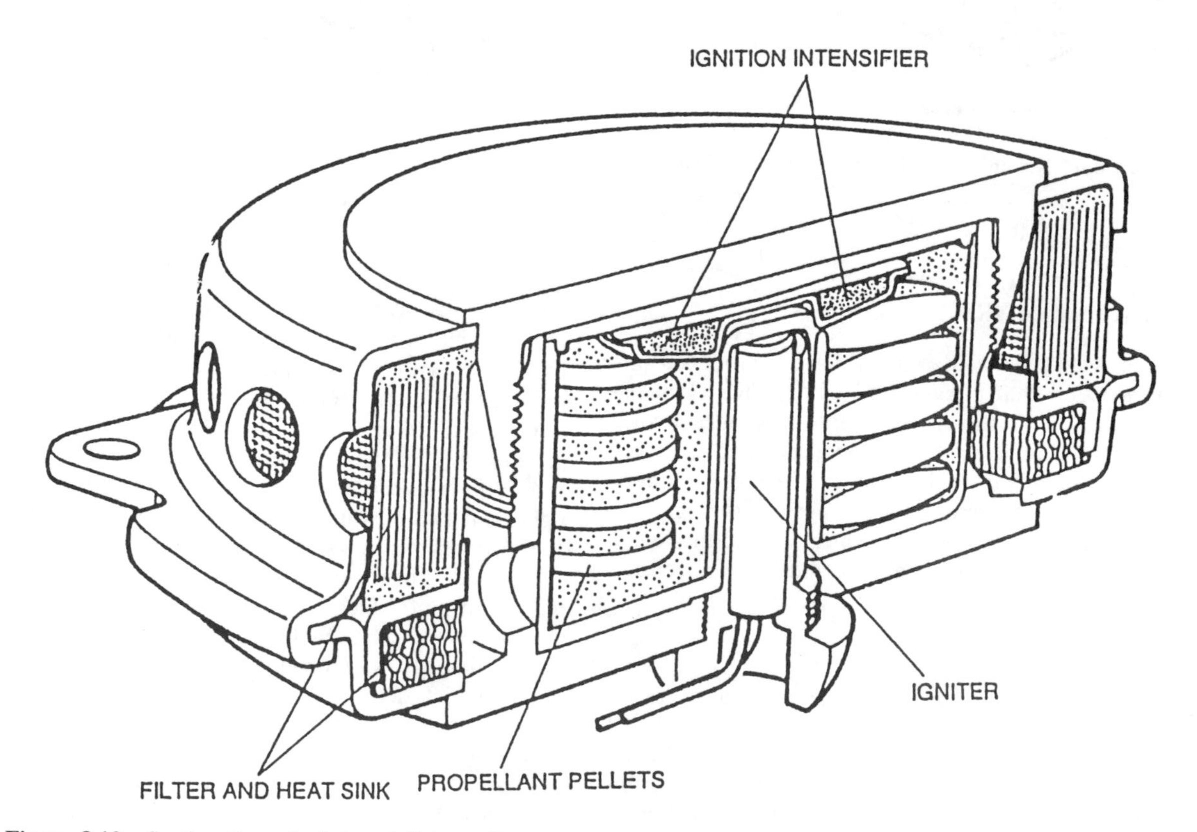

Figure 8.10 Section through air-bag inflator unit

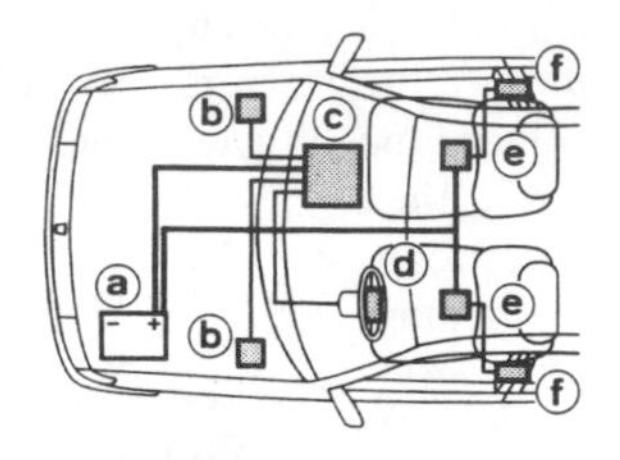

Figure 8.11 Combined air-bag and pyrotechnic seatbelt crash protection system; [a] battery; [b] air-bag sensors; [c] air-bag ECU; [d] air-bag; [e] seatbelt sensors; [f] pyrotechnic seatbelts (*Peugeot*)

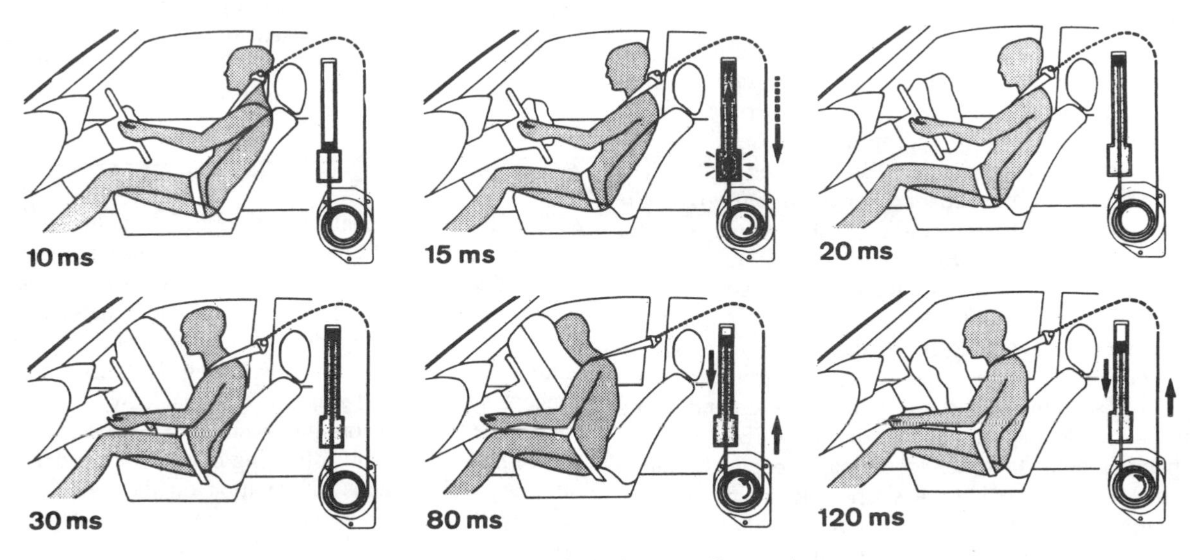

Figure 8.12 Timing of air-bag and pyrotechnic seatbelt deployment (*Peugeot*)

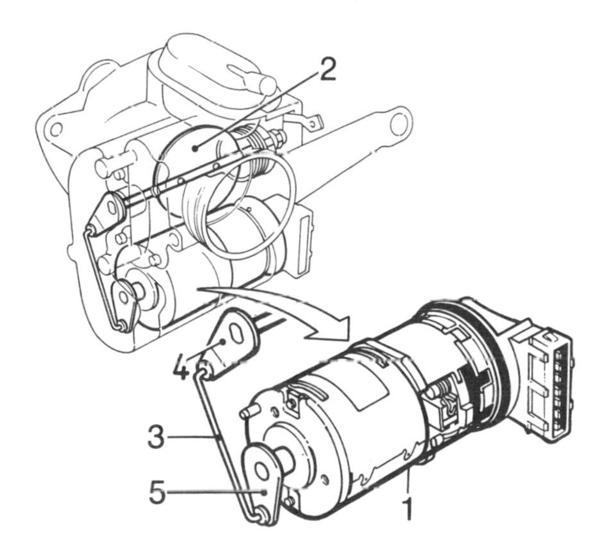

Figure 8.13 Using a motor to control throttle position; [1] motor; [2] throttle valve; [3] link rod; [4] throttle spindle; [5] motor shaft (*Saab*)

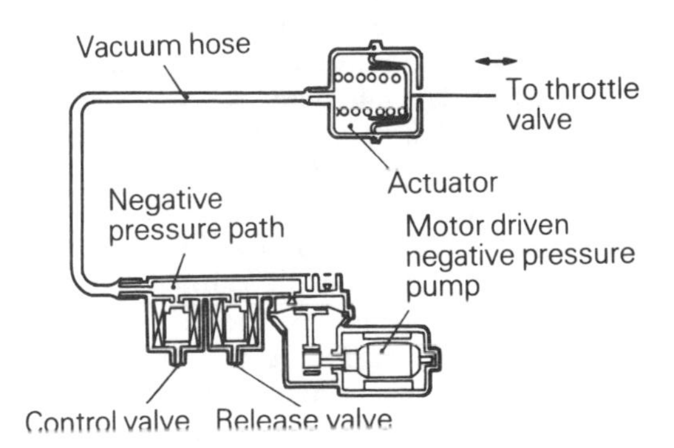

Figure 8.14 Vacuum-type throttle actuator

(iii) *Cruise control ECU*. The ECU incorporates a microprocessor which takes inputs from the switch pack and various sensors to determine the vehicle operating circumstances. It then sends out signals to the throttle actuator to regulate the throttle valve position so as to maintain the memorized cruising speed.

Cruise control system configuration

Figure 8.15 shows the electrical circuit configuration for a cruise control system which uses a vacuum-diaphragm throttle actuator.

When the vehicle is above a minimum cruising speed (typically 25 mph) and the cruise control 'ON' switch is pressed, the system becomes functional.

With the vehicle travelling at the desired cruising speed, the driver presses the 'SET' button. The ECU's speed control microprocessor then energizes the vacuum pump to move the throttle valve actuator diaphragm until the 'SET' speed is maintained without the use of the accelerator pedal and so the driver may remove his foot from the pedal. By continually monitoring the vehicle speed signal, the microprocessor constantly makes changes to the throttle position to take account of variations in road gradients, wind resistance and so on, so that the memorized cruising speed is held. In order to increase speed, the pump motor is energized for a short time to increase the vacuum; to reduce speed the control valve is pulsed open to slightly reduce the vacuum.

If the driver should wish to increase the cruising speed of the car, the 'SET' button can be held down and the ECU will cause the car to smoothly accelerate until the switch is released or maximum cruising speed is reached. The microprocessor then stores the new cruising speed at the instant the 'SET' switch is released.

To overtake another car, the accelerator pedal is pressed in the normal way to increase speed. On release of the pedal the cruise control will automatically return to the memorized speed.

If the brake pedal is pressed, the microprocessor detects the closure of the stop lamp switch and immediately opens the pressure release valve ('dump valve') to rapidly eliminate the vacuum and disengage the system. As a safeguard, the brake pedal is fitted with an additional pair of switch contacts which open to disconnect the positive supply from the solenoid valves, allowing them to open. This action is also achieved on manual transmission vehicles by using a switch on the clutch pedal. On automatic vehicles a relay contact is used so that the supply circuit is only completed when the selector lever is in 'DRIVE'. To re-engage the system the 'RESUME' ('RES') switch is pressed and the ECU operates the vacuum pump to restore the previously memorized 'SET' speed.

Cruise control ECU

Whereas early cruise control ECUs used analogue circuitry, most modern systems are entirely digital, using at least an 8-bit microprocessor. The microprocessor constantly samples the road speed signal and compares it with the 'SET' speed held in memory. When a difference is detected, a proportional-integral (PI) control algorithm is used to generate pump and valve drive pulses such that the 'SET' speed is swiftly restored without overshoot or oscillation.

Autonomous intelligent cruise control (AICC)

Although conventional cruise control systems work well in steady traffic conditions, where constant speeds can be maintained for many miles, they are less successful in congested traffic situations where the vehicle is continually slowing down and then speeding up. Autonomous intelligent cruise control (AICC) is a concept which eliminates this drawback by allowing the vehicle speed to continually adapt to the traffic flow and so maintain a safe following distance. In order to function, AICC needs information on the range and velocity of vehicles immediately in front of it. This can be achieved by fitting a 'headway radar' system which uses a beam of microwaves transmitted from the front of the car and then reflected back to the transmitter by obstructions in its path. Velocity and range information can then be obtained by measuring the radar signal's Doppler frequency shift and reflection delay, respectively.

There are two basic types of headway radar, 'pulse' and 'frequency modulated continuous wave' (FMCW):

(a) *Pulse radar*. These systems rely on sending out a pulse of microwaves at a fixed frequency. The time taken for the pulse to reach an obstacle and bounce back to the transmitter is proportional to the range of the obstacle. If the echo returns after a time, t, then the range, R, of the obstacle is,

$$R = ct/2$$

where c is the speed of the radar waves (speed of light). If there is a difference in speed between the car-mounted transmitter and the obstacle, then the returning radar signal will have a slightly altered frequency (i.e. a Doppler frequency shift). Sophisticated signal processing techniques can convert this frequency change into speed information.

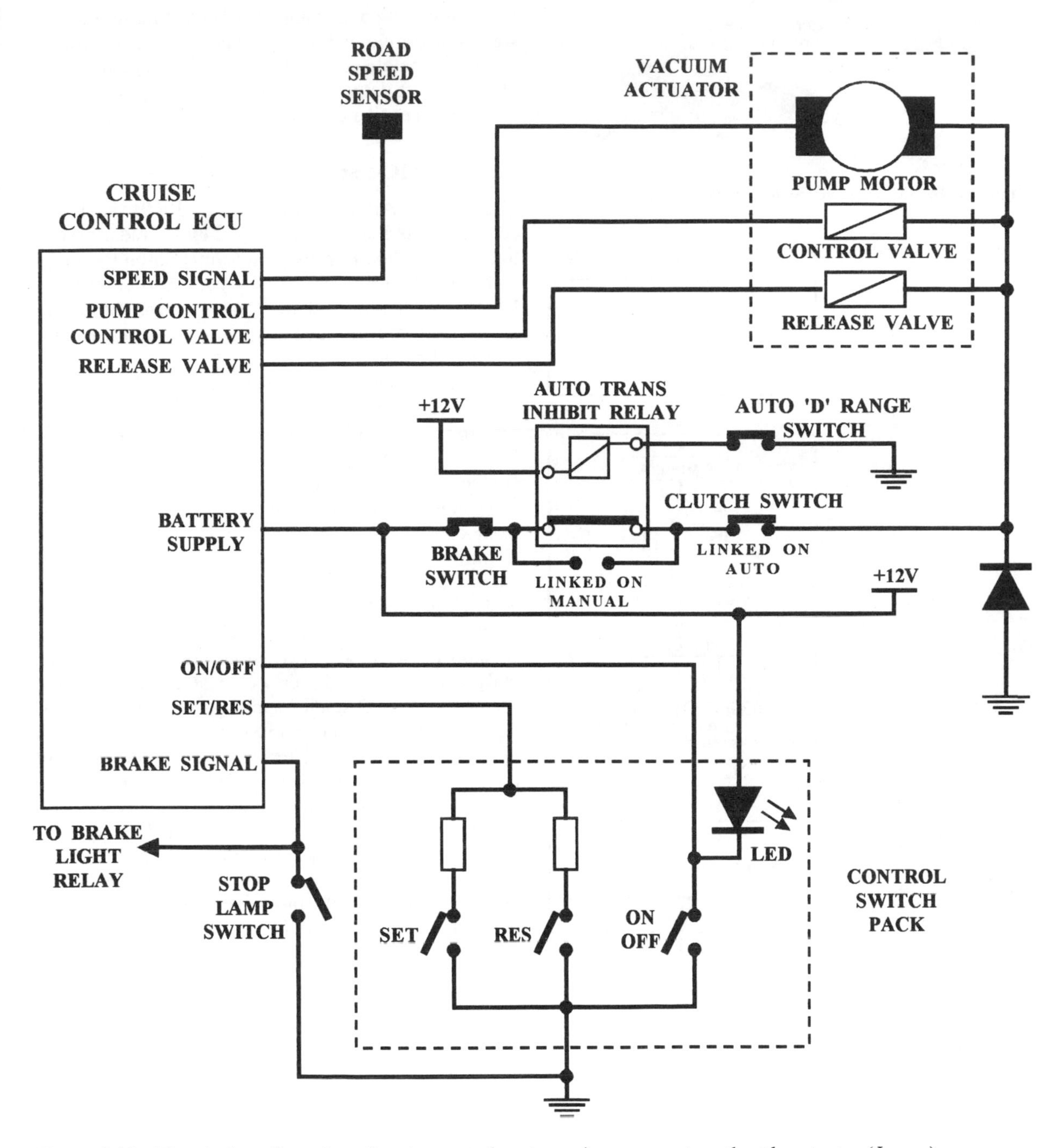

Figure 8.15 Electrical configuration of cruise control system using vacuum-type throttle actuator (*Jaguar*)

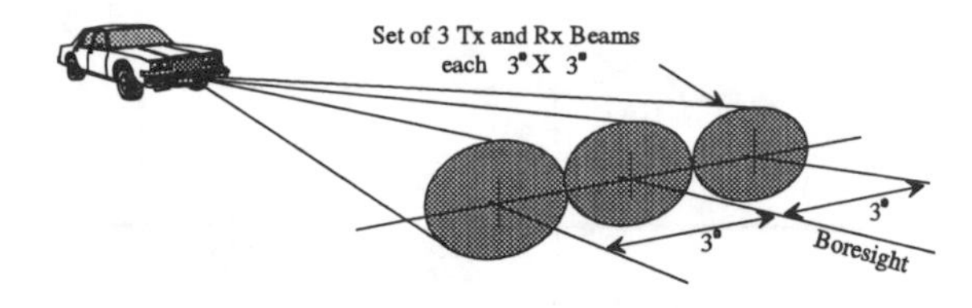

Figure 8.16 Microwave beam geometry (*Lucas*)

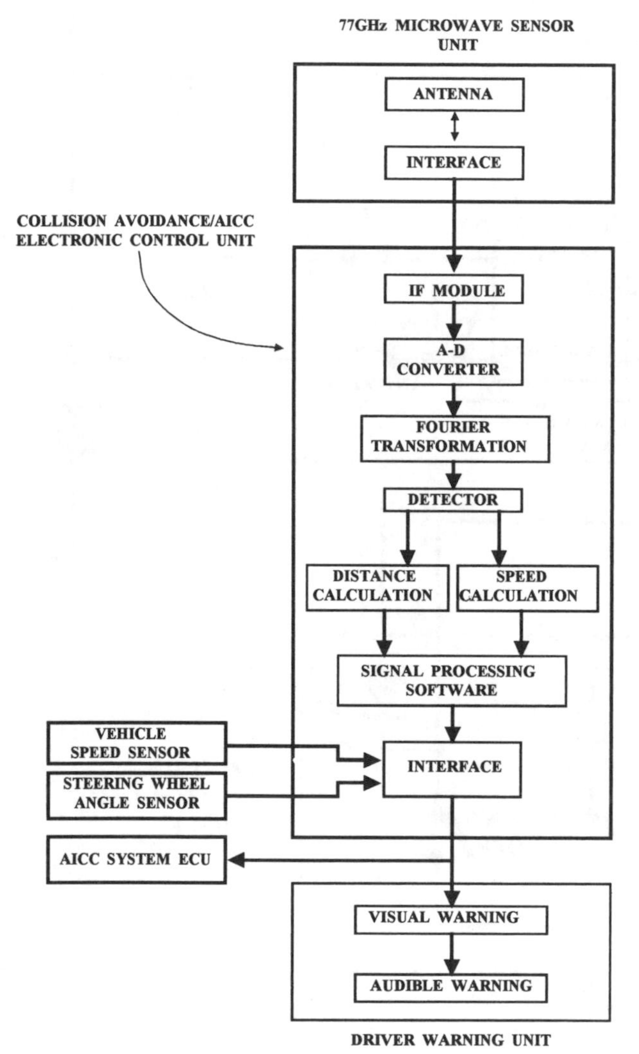

Figure 8.17 Obstacle detection/collision avoidance system used with AICC

(b) *FMCW radar*. With this system a continuous radar signal is sent out, but with a frequency that is repeatedly swept either side of a centre frequency. In the time taken for the transmitted waves to have reached an object and bounced back, the transmitter frequency will have changed slightly. Mixing the two signals will give a 'beat frequency' which is proportional to the distance of the target. Together with Doppler shift detection this strategy can provide both range and speed information for the AICC ECU.

Practical AICC systems

The idea of AICC is not new. In 1969, the Lucas company fitted an experimental 24 GHz radar system to a Ford Zodiac to provide information on the range and speed of traffic in the car's path. Particular problems were encountered with 'false alarms' on bends, where fixed objects such as trees and road signs gave unwanted radar returns.

Modern designs overcome these problems by using sophisticated digital signal processing technology, together with a higher frequency (77 GHz band in Europe) which gives better resolution of targets. A contemporary system typically uses three radar antennae, mounted behind the plastic number plate at the front of the car, to give a $3° \times 9°$ microwave beam (Figure 8.16). These radio signals are reflected by other cars and fixed obstacles and then fed to the ECU (Figure 8.17) which analyses the road situation approximately twenty times each second, taking into account the vehicle's own speed and steering angle. Only events in the driver's own lane are taken into account so that obstacles such as trees and signposts do not give rise to error messages.

Typically, such a system has a detection range of up to 150 m and can resolve an object's position and speed to within 1 m and 1 kmh^{-1}, respectively. This speed and distance information is supplied to the AICC ECU which operates a conventional throttle actuator to maintain a preset safe following distance behind the vehicle in front. In the event of a likely collision the ECU is able to issue audible and visual warnings, enabling the driver to take evasive action.

8.5 AUXILIARY BODY ELECTRONICS

Central control unit (CCU)

Vehicles designed since the mid-1980s incorporate many luxury and convenience features such as central door locking, power windows, heated rear window timer and so on. Since most of these features require a basic timing function, many manufacturers now fit a single central control unit (CCU) which replaces the numerous smaller ECUs that were previously used. The CCU offers numerous advantages; just one microprocessor can simultaneously perform many timing functions, cable harness bulk is reduced, reliability is improved and cost is reduced. Examples of features controlled by a CCU include:

(i) Intermittent windscreen wipe.
(ii) Programmable wash/wipe.
(iii) Rear window wash/wipe.
(iv) Courtesy light delay.
(v) Lights-on alarm.
(vi) Heated rear window timer.
(vii) Headlamp dip/main switchover.
(viii) Headlamp-on delay after ignition switch-off.
(ix) Window lift/sunroof operation with ignition off.
(x) Power seat operation with ignition off.
(xi) Central door lock motor control.
(xii) Security/anti-theft alarm system.

The CCU is connected to all of the switches, relays and actuators associated with each function. The status of each switch can thus be continuously assessed by the CCU's microprocessor, which then uses preprogrammed data to determine the most appropriate output to each relay or actuator.

Central door locking

Central door locking systems are a convenience feature which allow all of a vehicle's door locks to be simultaneously actuated from either the driver's or front passenger's door lock. Most systems also incorporate a boot/tailgate lock and some provide additional functions such as lock disablement if the key is left in the ignition, or automatic central locking when the car first exceeds a preset speed after driving away. For safety reasons all central locking systems allow the locks to be operated from within the car by the conventional mechanical locking button.

The central locking system may either be operated by a dedicated controller or by a CCU. Where a CCU is used it often links the door locking system to a remote-controlled anti-theft facility and automatic sunroof/window closure ('lazy-locking' system).

A simple central locking circuit is illustrated in Figure 8.18. It consists of a digital lock control unit, two actuator/trigger-switch units (one in each front door), an actuator in each rear door and one each in the boot lid and fuel filler flap.

When the front doors are locked or unlocked the switch wiper is moved to provide a ground on either the control unit's LOCK or UNLOCK inputs, respectively. The control unit detects this ground signal and provides a timed LOCK or UNLOCK pulse output, as appropriate, to all of the actuators. In the event of an accident an inertia switch provides a ground signal to the UNLOCK terminals of the controller via the ignition switch and the diodes BD1 and BD2.

Door lock actuators

To lock or unlock the doors, each locking lever must be moved forwards or backwards by a few centimetres; several types of actuator are available to suit this requirement,

(a) *Solenoid actuator*. A special solenoid, with two opposing 'lock' and 'unlock' windings, is supplied with a pulse of current to move the locking rod. Swapping the polarity of supply changes the direction of plunger motion. These actuators were common on early central locking systems but have become unpopular because they are mechanically imprecise in operation and can easily be overcome by a thief.

(b) *Dc motor*. A simple permanent-magnet dc motor is used to move the locking rod. The amount of motion is controlled by feeding the motor with a precisely timed current pulse; the direction of motion is reversed simply by reversing the polarity of supply.

(c) *Stepper motor*. This is the most sophisticated solution. The lock controller feeds the stepper motor with a defined sequence of current pulses such that it steps the locking rod through the correct distance in the required direction.

Electric windows and sunroof systems

Electric window systems are fitted to many vehicles as a luxury feature. Each door has a window-winding mechanism which is very similar to that fitted with manually operated windows, but with the addition of a permanent-magnet dc motor to drive it. The motor is of sufficiently small dimensions to be installed within the door case and is fitted with a reduction-gear mechanism to increase its output torque and reduce the output speed. Motor direction (window 'UP' or 'DOWN') is controlled simply by reversing the connection polarity.

The electric window control circuit on older vehicles is generally very simple; each door is fitted with a three-position rocker switch which activates two relays that are fed from the ignition circuit. The rocker switch is biased to the 'OFF' position and so both relays are normally inactive and the motor is off. When the rocker is pushed to the 'DOWN' position the 'DOWN' relay is energized, connecting the motor to the ignition supply with a polarity such that the motor revolves to lower the window. Conversely, when the rocker switch is pushed to the 'UP' position, the 'UP' relay is energized and the motor is connected with a polarity so as to close the window. Since there is a possibility that the window could stall (perhaps due to icing, or the switch being held down after the limit of window travel has been reached) the motor circuit

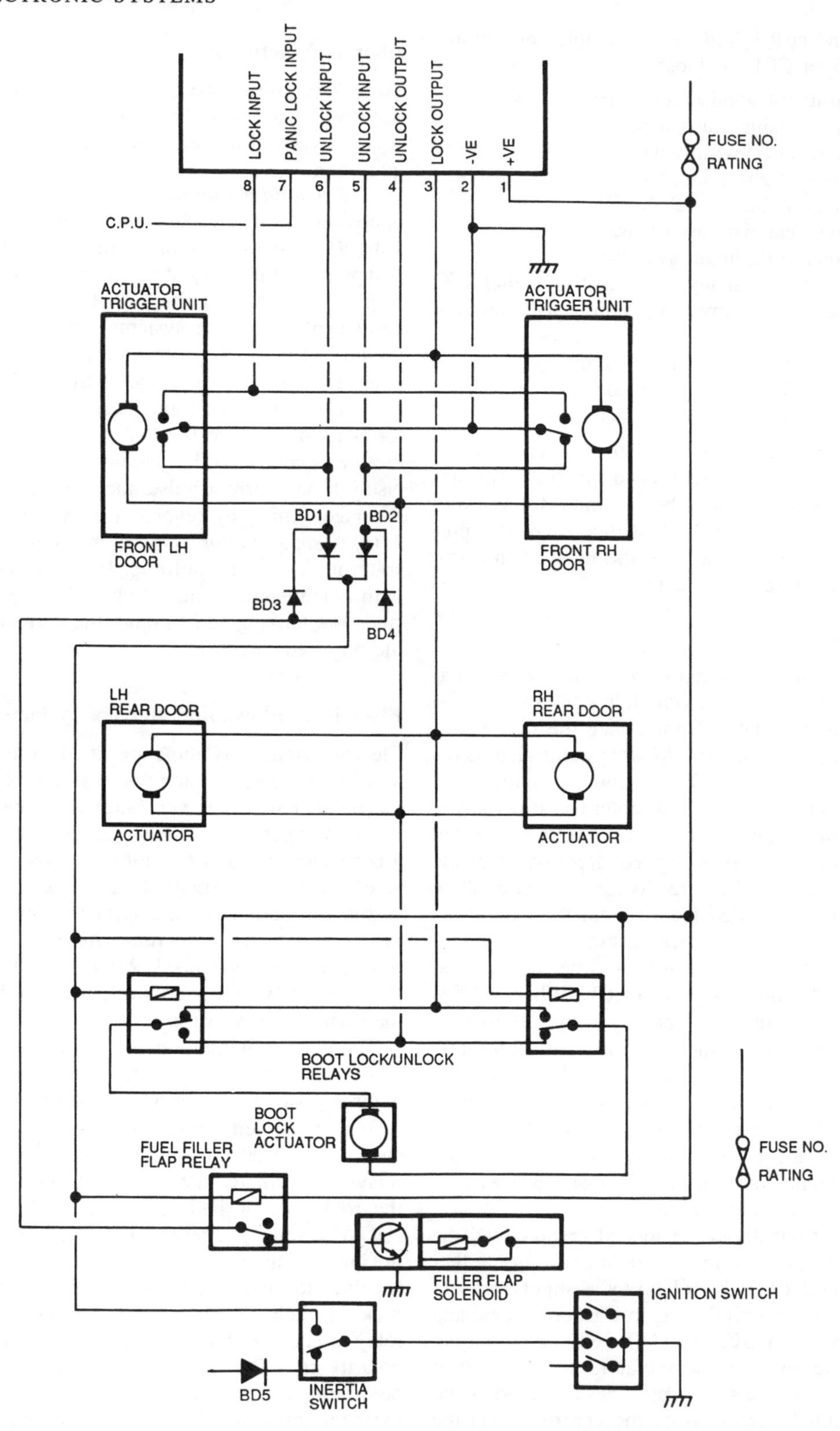

Figure 8.18 Electrical circuit of central locking system (*Jaguar*)

is normally fitted with a cutout to prevent overheating and damage to the motor and wiring. Resettable bi-metallic strip cutouts are most commonly used.

Electric window 'back-off' and 'one-touch' control

Modern vehicles may be fitted with electric window systems that are controlled by either a CCU or a dedicated ECU and offer a number of additional convenience features;

(a) *Back-off control.* Back-off works by having the ECU measure the speed of the motor while the window is closing, enabling it to also calculate the rate of window acceleration or deceleration and to keep track of the window's position within the door frame. If the window speed falls below a preprogrammed limit, and it has not reached the limit of its travel, then it is likely that it has encountered an obstruction. The ECU therefore reverses the winding direction so that the window moves down and the obstruction can be removed.

The ECU measures the motor speed in one of two ways:

(i) Motor brushes. The motor is fitted with an extra commutator contact and two extra brushes which are connected to the ECU. As the motor rotates each brush picks up a pulse train (0–12 V) from the extra commutator, enabling the ECU to calculate the motor's speed and direction.
(ii) Hall-effect sensors. The motor shaft is fitted with a magnetic stud which passes two Hall-effect sensors, mounted 90° apart. As the motor revolves, voltage pulses are generated by each sensor indicating the motor speed and (by comparison of the timing of the two outputs) direction.

(b) *One-touch control.* If the window rocker switch is pressed hard, it moves beyond a detent and the window ECU completes the full 'UP' or 'DOWN' winding operation even if the switch is released.

To implement one-touch control the window ECU operates the motor and counts the motor revolution pulses to determine the window position until the preprogrammed limit of travel is reached.

Some simplified versions of one-touch control use a circuit which operates the motor for a predetermined period of time, judged sufficient to fully open or close the window. Other versions use a low-value sensing resistor, wired in series with the motor feed, which generates a voltage in proportion to the supply current. The motor is operated until the sense voltage rises above a preset limit, at which point the motor is judged to have stalled – the window having reached the limit of its travel.

Electric sunroof operation

Electric sunroof systems are fitted with a permanent magnet dc motor which drives a complex mechanical mechanism to provide a 'tilt-and-slide' movement. The motor is energized through a pair of relays which provide polarity reversal in a similar manner to that described for electric windows. Micro-switches are used as limit-switches to control the movement of the sunroof in each direction.

Electric door mirror systems

Since it is difficult for the driver to reach and adjust the door mirrors (especially on the passenger side) many vehicles are fitted with electrically adjusted door mirrors.

Each mirror assembly consists of an enclosure, fixed to the door frame, and a mirror-glass which pivots on a nylon ball-joint. Movement of the mirror is achieved by having two small dc permanent magnet motors tilt the glass back and forth on the pivot; one motor controls up/down tilt and the other controls left/right tilt. The motors are powered via a small joystick which contains four sets of switch contacts to connect a supply of the appropriate polarity to achieve movement in the desired direction. For example, if the joystick is pushed to the left, the left/right motor is connected with a polarity such that the mirror is angled to the left. Since the motors are small, consuming limited power, it is not normally necessary to provide relays in the circuit.

Heated door mirrors are also included on some models to remove frost and condensation from the glass surface. A length of ptc resistance wire is bonded to the rear of the mirror glass and heats up slightly when connected to the supply. Heating can be arranged to occur only when the heated rear window is switched on, or it can be permanently activated via the ignition circuit.

Electrically adjusted seats

Many luxury cars are fitted with power operated seats which use electric motors to enable precise and effortless adjustment of the seat position from a multiway switchpack. The range of possible adjustments available depends upon the number of motors fitted; horizontal (front/back), recline angle and seat height adjustments are usually the minimum provision.

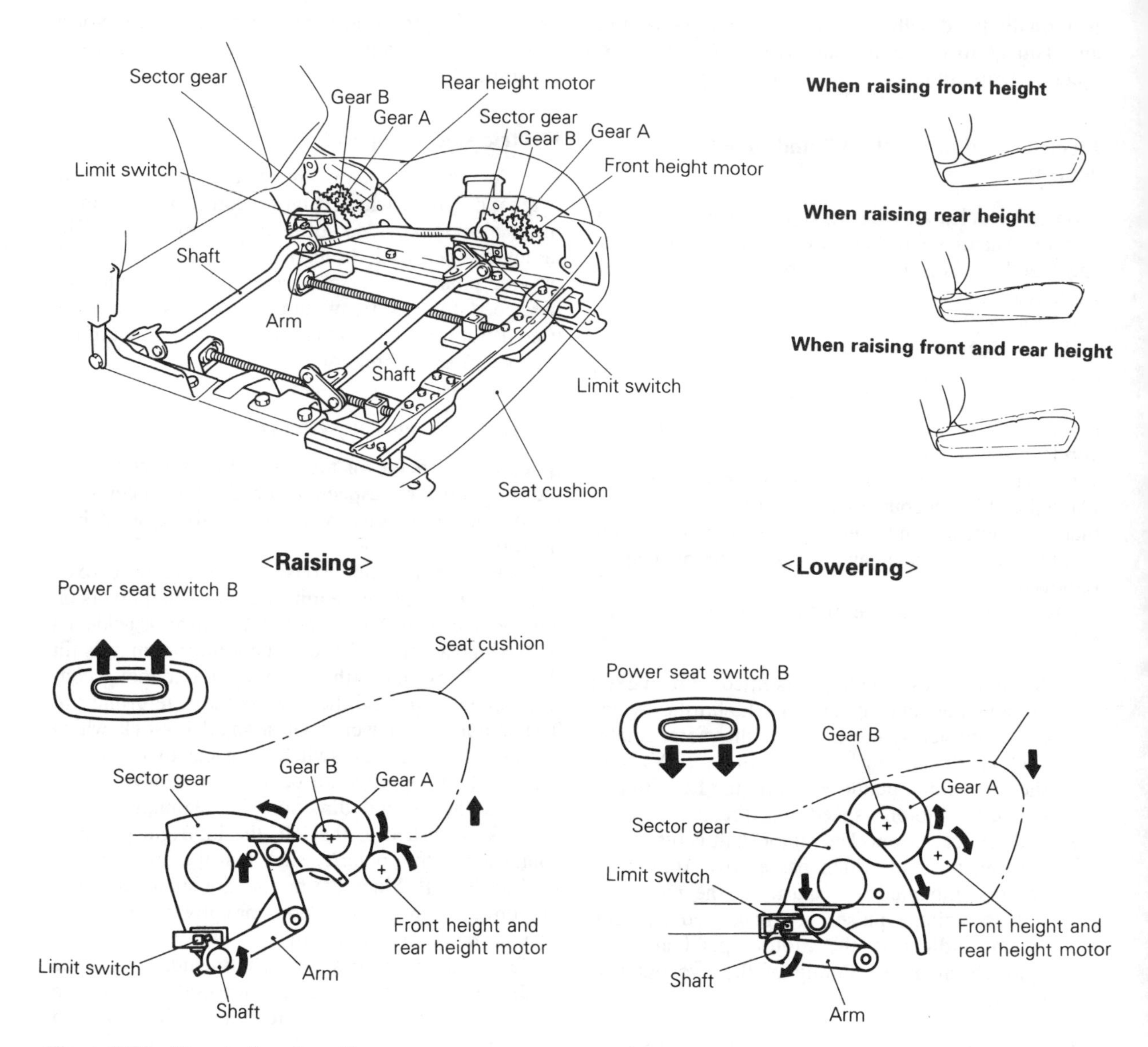

Figure 8.19 Electrically adjustable seat

Each adjustment is achieved using a dc permanent-magnet motor which drives an arrangement of gears to move the position of the seat frame in relation to its mounting rails (Figure 8.19). Every movement axis is fitted with a limit switch, so that when the limit of adjustment is reached the switch is opened and the motor is turned off. Reversal of the adjustment direction is simply achieved by using relays to reverse the motor's connection polarity.

A seat position 'memory' facility can be added to the basic system by fitting each motor with a feedback potentiometer and using an ECU to sample and store each potentiometer output voltage reading for a range of frequently used seat positions. Pressing the appropriate 'recall' button will cause the ECU to interrogate its memory, retrieve the relevant potentiometer voltage data, and energize the motors to drive the seat to the stored position.

Electronically dipped rear view mirror

The electronically dipped rear view mirror automatically darkens when the mirror glass is subjected to

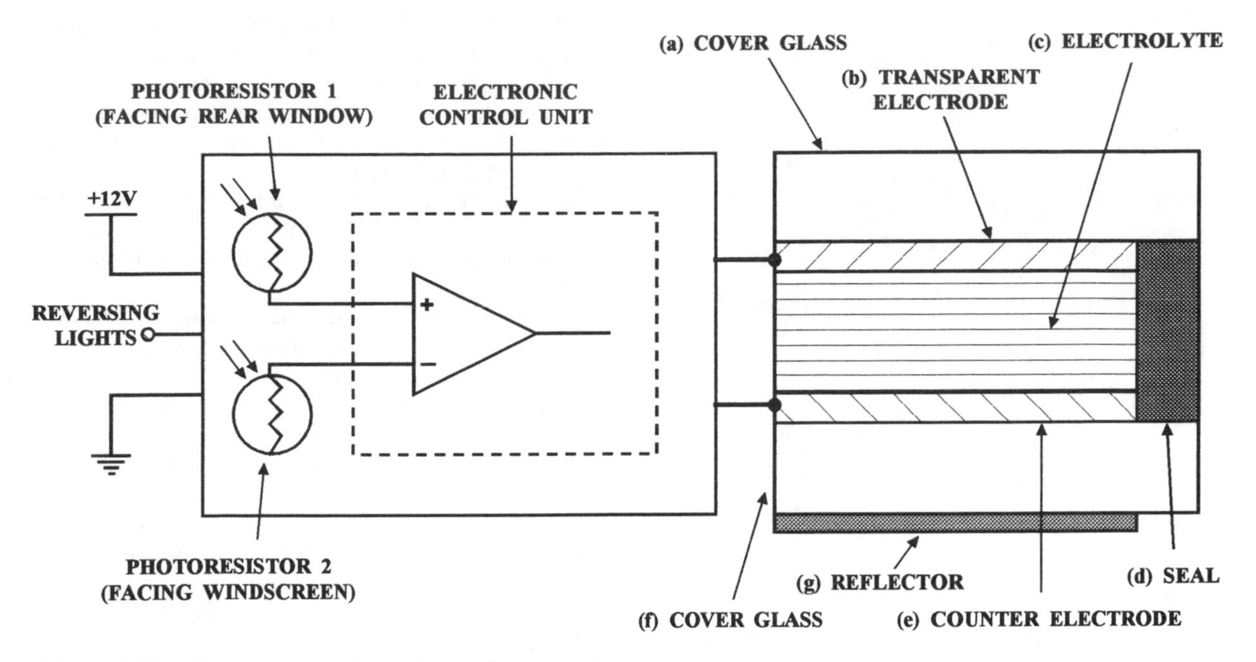

Figure 8.20 Electrochrome mirror glass and control electronics

strong light. It is designed to protect the driver from the glare of following headlights when travelling at night.

The system uses a special electrochrome (EC) glass, whose reflectivity can be electronically controlled. There are no moving parts involved in the unit and therefore it operates noiselessly, without altering the angle of the mirror glass. Figure 8.20 shows a block diagram of the EC mirror glass and associated electronics. Two photosensitive resistors are used to determine the lighting conditions; photoresistor 1 faces the rear window and senses the light level falling on the mirror glass, photoresistor 2 faces the windscreen and senses the ambient light level seen by the driver. A differential amplifier module produces an output dependent upon the difference between these two light levels. When the light level falling on the rear-facing sensor exceeds a certain fraction of the ambient light level falling on the forward-facing sensor, the electronic module applies a bias voltage to the mirror electrodes (b and e). This voltage causes a flow of ions from the electrolyte (c) into the front electrode, which darkens to reduce the amount of light reflected by the mirror. Since the amount of darkening is proportional to the bias voltage, it is dependent, in turn, on the difference in light levels falling on the two sensors.

The change in mirror reflectivity occurs slowly; the dimming time (from 70% to 20% reflectance) is about four seconds, the restoring time (from 10% to 60% reflectance) is around seven seconds. An override input from the reversing light switch disables the dimming function when reverse gear is engaged.

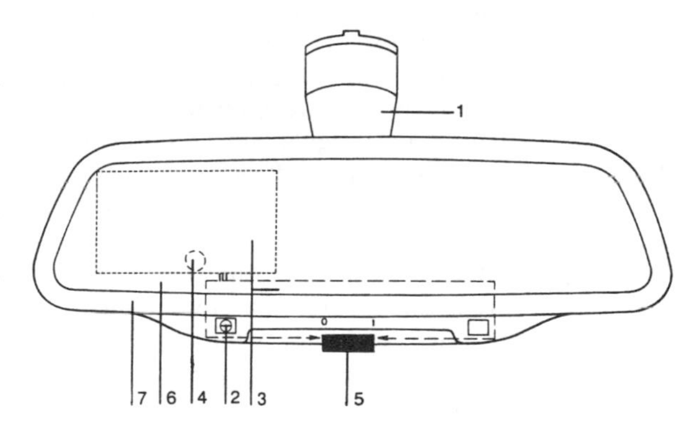

Figure 8.21 Practical arrangement of EC dipping rear-view mirror; [1] mount; [2] photoresistor 1 facing rear window; [3] ECU; [4] photoresistor 2 facing windscreen; [5] ON/OFF switch; [6] EC mirror glass; [7] casing (*BMW*)

The practical arrangement of the electronically dipped mirror is illustrated in Figure 8.21. The sensors and electronic module are contained within the mirror casing and a control is fitted to allow the sensitivity to be varied or the dipping function switched off.

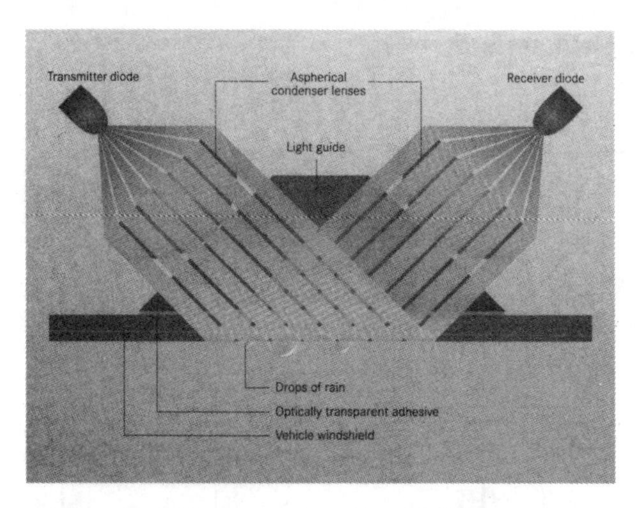

Figure 8.22 Operation of rain sensor (*TEMIC*)

Rain sensor

Although variable windscreen-wiper delay permits the wiper operating interval to be adjusted to suit the amount of rainfall it is often necessary to make frequent adjustments as the weather or road conditions vary. A rain sensor helps in conditions of light rainfall by continuously monitoring the amount of rain striking the windscreen. A microcomputer-based ECU then processes the sensor signal and alters the wiper delay interval to suit the rain intensity.

The rain sensor is mounted inside the car behind the rear-view mirror and operates on the basis of infrared light reflection from the windscreen (Figure 8.22). Light from an infrared transmitter diode is focused through a lens and then coupled to the windscreen via a light guide which is bonded to the glass with a transparent adhesive.

If the windscreen is completely dry then most of the light is reflected from the outer surface of the glass and returns, via the light guide, to a second lens which focuses the rays onto the receiver diode.

If rain falls onto the windscreen then some of the transmitted light is lost from the system by refraction into the raindrops, and so no longer reaches the receiver. This causes a loss of signal which the ECU interprets as a particular level of rainfall, enabling it to adjust the wiper delay interval, as appropriate.

8.6 HEATING AND AIR-CONDITIONING SYSTEMS

All cars marketed in Europe are normally fitted with passenger-compartment heaters which use a heat exchanger to transfer heat from the engine's coolant to the air entering the cabin. The operation of these systems is generally simple, relying on the driver to manually adjust a lever-operated air-flap which blends heated and unheated outside air until a comfortable temperature is achieved.

The volume of air flowing through the cabin also needs to be varied with a multi-speed blower that must be adjusted according to weather conditions and vehicle speed.

Air conditioners are an enhancement to the basic heater unit and incorporate two separate heat exchangers for heating and cooling the incoming air. The vehicle occupants can blend the output from each heat exchanger to obtain cool, warm or dehumidified air inside the cabin. Due to the generally mild summer weather experienced in Northern Europe, automobile air conditioners are only encountered on luxury vehicles or as an optional fitment on some medium-sized vehicles. However, as buyer expectations rise it is likely that the demand for car air-conditioning systems will increase.

Blower motor control

Basic car heaters use a centrifugal air blower (a 'Sirocco fan') that is driven by a permanent magnet dc motor. The motor is energized via a multi-way blower speed switch which provides an 'OFF' position and three or four speeds. The various speeds are obtained by having each switch position place a different value resistor in series with the motor supply, altering the voltage applied to the motor and therefore the motor's power and speed. Maximum speed is obtained by having the last switch position connect the motor directly to the supply. Since the motor draws a heavy current (5–10 A), the resistors must be capable of dissipating a considerable amount of power and so are usually fixed to an aluminium heat-sink to prevent them from overheating.

Luxury vehicles are generally fitted with a variable-speed blower motor that can be adjusted to give any desired speed by using a slider control. The slider is a potentiometer that supplies a variable voltage to a motor control ECU, which, in turn, supplies a proportionate bias voltage to the base of a power transistor that regulates the motor current (Figure 8.23). The ECU is a simple analogue circuit which translates the large variation in potentiometer wiper voltage into a smaller voltage variation suitable for controlling the power transistor.

As the speed control slider is moved towards 'HIGH', the wiper feeds a rising voltage to the motor control ECU which provides a proportionate increase

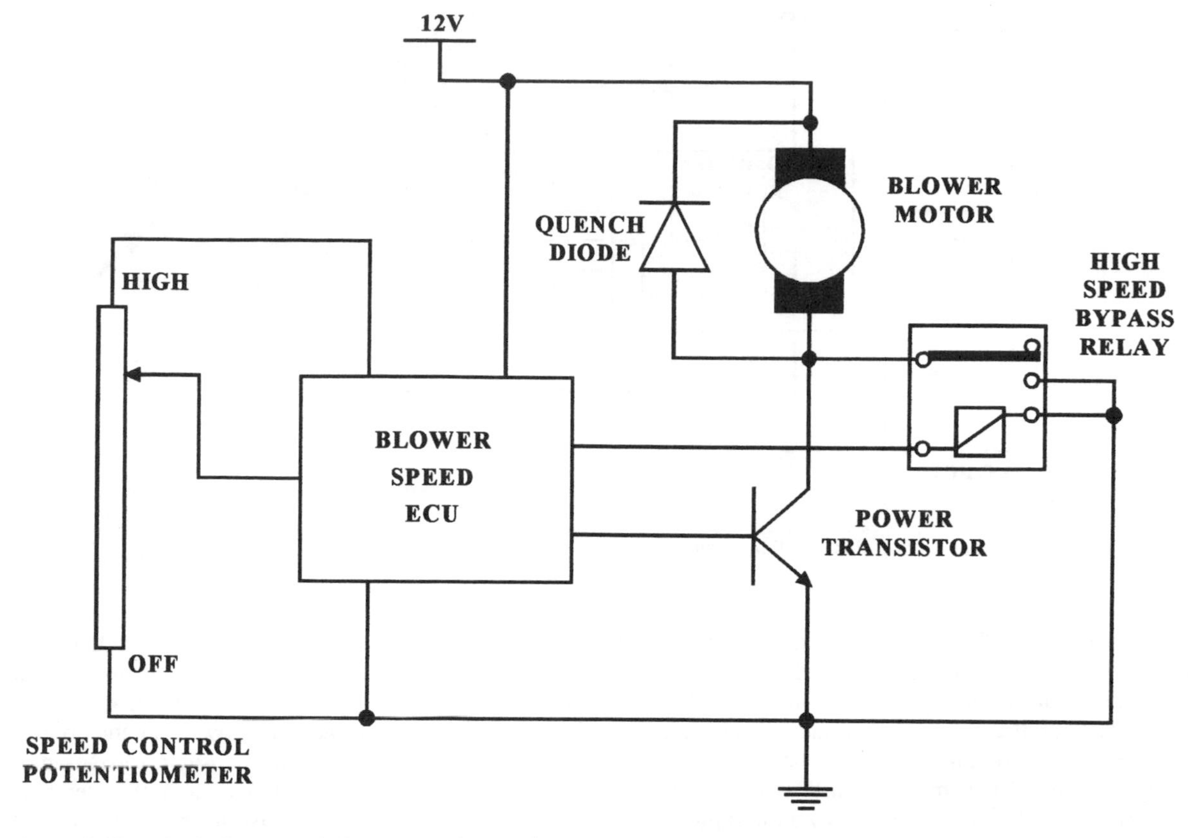

Figure 8.23 Block diagram of blower speed control system

in bias to the power transistor. This reduces the transistor's collector-emitter resistance, thereby increasing the motor's current and therefore its speed. At the 'HIGH' position, the ECU actuates a relay that is wired in parallel with the power transistor, this connects the fan motor negative terminal directly to ground and so gives maximum speed.

A quench diode is connected across the motor to protect the power transistor from induced 'back emf' when the motor is switched off.

Air flap position control

A typical vehicle heating system will contain three air flaps to control the air flow into the passenger compartment, specifically, a hot/cold mixing air flap, a fresh/recirculated air flap and a 'mode' air flap which controls the distribution of air between the various outlets. On the majority of cars the position of each of these flaps is manually controlled via a level-and-cable arrangement, however on luxury vehicles it is now more common for a motor-drive arrangement to be employed. This offers the advantage of precise positioning with minimum effort and also permits an ECU to control the flap positions when an automatic 'climate control' system is fitted.

Air flap position actuation is achieved using one of two methods; servomotor control or stepper motor control.

(a) *Servomotor control* (Figure 8.24). The driver sets the position of each air flap by moving a knob which is coupled to the wiper of a control potentiometer. The wiper thus feeds a continuously variable control voltage to the air flap ECU which actuates the air flap's servomotor (a dc motor with a rotary feedback potentiometer fitted to the output shaft). The ECU acts as a voltage comparator and drives the servomotor in the appropriate direction until the voltages from the control and feedback potentiometers become equal. At this point the flap is in the correct position corresponding to the control knob setting.

(b) *Stepper motor control.* As with servomotor control, the driver commands the air flap position by moving the wiper of a potentiometer which send a

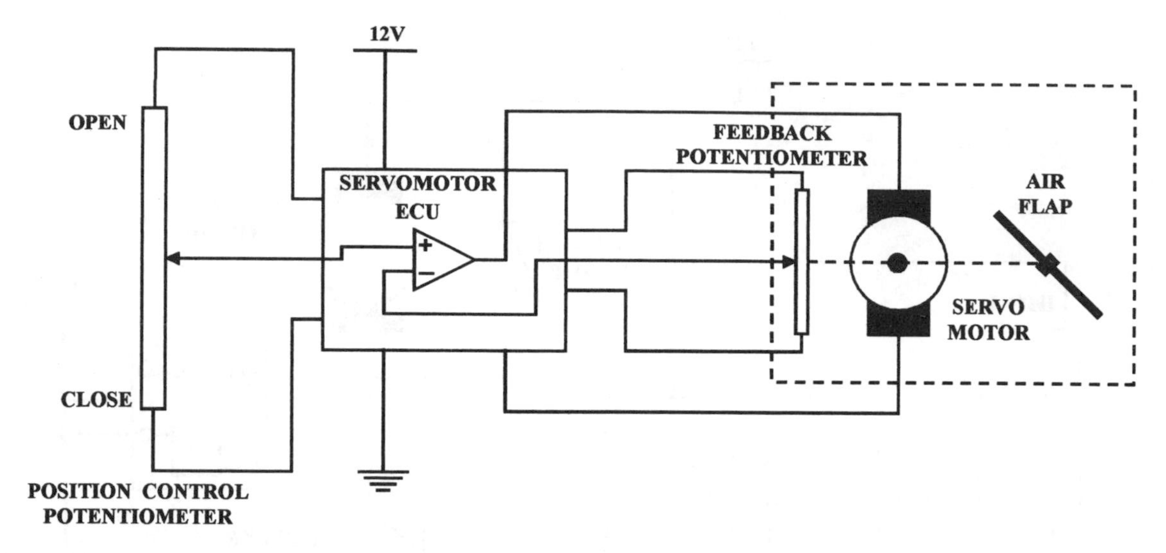

Figure 8.24 Servomotor control of air flap position

voltage to the air flap ECU. Since the stepper motor moves through a precisely defined angle for each energizing current pulse, it is possible for the ECU to keep track of the flap position and to calculate the number of steps required to reach a new position. The stepper motor ECU is more sophisticated than that for servomotor control and may employ a microprocessor.

Air conditioning

Automobile air-conditioning systems operate on the principle of the 'heat pump'; air entering the vehicle is cooled by extracting heat from it and then 'pumping' that heat back to the atmosphere. The pumping medium is a refrigerant compound that normally boils to a vapour at a temperature of about –30°C, but can easily be condensed to a liquid by increasing its pressure with a compressor. By forcing the refrigerant to repeatedly condense and then evaporate, a heat pump circuit is formed.

Older air-conditioning systems use refrigerant R12 (Dichlorofluoromethane, a CFC) whereas recent systems use R134a (Tetrafluoroethane) which is not a CFC and therefore causes less damage to the Earth's ozone layer. R12 and R134a are totally incompatible and must never be mixed, moreover it is not possible to use R134a in a system designed for R12, and vice versa.

Figure 8.25 illustrates the layout of an automobile air-conditioning circuit. The engine-driven compressor (1) increases the pressure of the refrigerant gas, thereby also increasing its temperature (F1). This high pressure gas then flows to a condenser radiator (2) which is mounted at the front of the car, near to the engine coolant-radiator. Here, the hot refrigerant gas is cooled down by ambient air flow (A1) and so condenses to a liquid state (F2). The liquid refrigerant is still at a high pressure and flows into a receiver-drier flask which removes any moisture or debris (10,11). From the receiver-drier the refrigerant passes to the thermostatic expansion valve (4), which regulates the quantity of refrigerant flowing into the evaporator (5) as a function of the temperature at the evaporator outlet pipe. As the refrigerant liquid passes through the expansion valve nozzle, into the evaporator, its pressure falls dramatically and so it cools down very rapidly (F4). Air flowing into the cabin (A2, A3) gives up its heat to the cold refrigerant, which absorbs the heat and so increases in temperature. The hot vapour (F5) is then drawn back into the compressor by suction and the cycle is repeated.

Air conditioner electrical system

In order to support the operation of the refrigeration circuit most automobile air conditioners incorporate a simple electrical control system. At the simplest level this includes two elements.

(a) *Compressor clutch.* So that the refrigeration circuit is only operated when required, the compressor drive-belt pulley is coupled to the drive-shaft with an electromagnetic clutch which engages or disengages in response to various signals (Figure 8.26). The clutch is initially energized as a result of the driver pressing

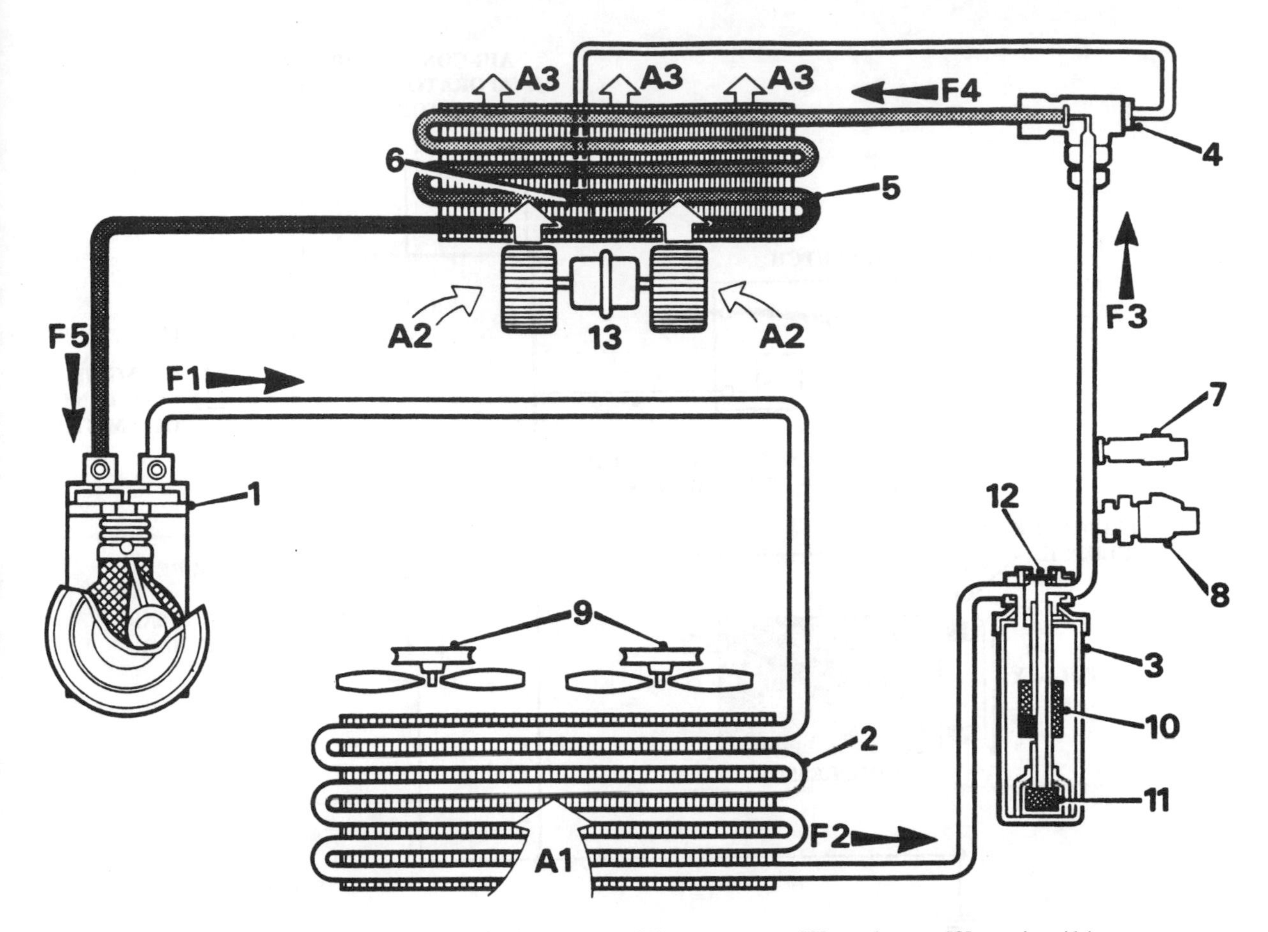

Figure 8.25 Schematic layout of air-conditioning system: [1] compressor; [2] condenser; [3] receiver/drier; [4] thermostatic expansion valve; [5] evaporator; [6] capillary tube; [7] dual pressure switch; [8] high pressure switch; [9] cooling fans to increase air flow; [10] drying agent; [11] filter; [12] sight glass; [13] fan motor; [A1] air flow through condenser; [A2] air flow through fan and evaporator; [A3] air flow to vehicle interior; [F1] high pressure, hot R12 vapour; [F2] high pressure; cooled R12 liquid; [F3] high pressure, cooled R12 liquid with moisture, vapour bubbles and foreign matter removed; [F4] low pressure, low temperature atomized liquid R12; [F5] low pressure, high temperature R12 vapour (*Rover*)

the 'AIR CON' control switch, however it may be de-energized owing to:

(i) The dual pressure switch detecting a refrigerant pressure that is too low (because of a leak) or too high (perhaps because of a blockage). The usual operating pressure at the receiver/drier lies in the range 2–20 bar for R12.

(ii) The control thermistor, mounted on the evaporator, detecting an evaporator temperature of below about 4°C, leading to a risk of frosting and consequent loss of cooling efficiency.

(iii) The engine management ECU demanding full power from the engine, and therefore disengaging heavy loads such as the compressor.

(iv) The engine management ECU detecting that the engine coolant temperature is too high. The compressor is disengaged to help prevent further overheating.

(b) *Condenser cooling fan* Heat is normally removed from the condenser by the air flow through the front of the car. If this flow is inadequate (perhaps because of stationary traffic or a very high ambient temperature) then the refrigerant temperature will rise, leading to a proportionate pressure increase. When the pressure rises above a predetermined limit, the high pressure switch closes (component [8] in Figure 8.25) and the condenser cooling fan is switched on to cool the condenser and reduce the refrigerant pressure.

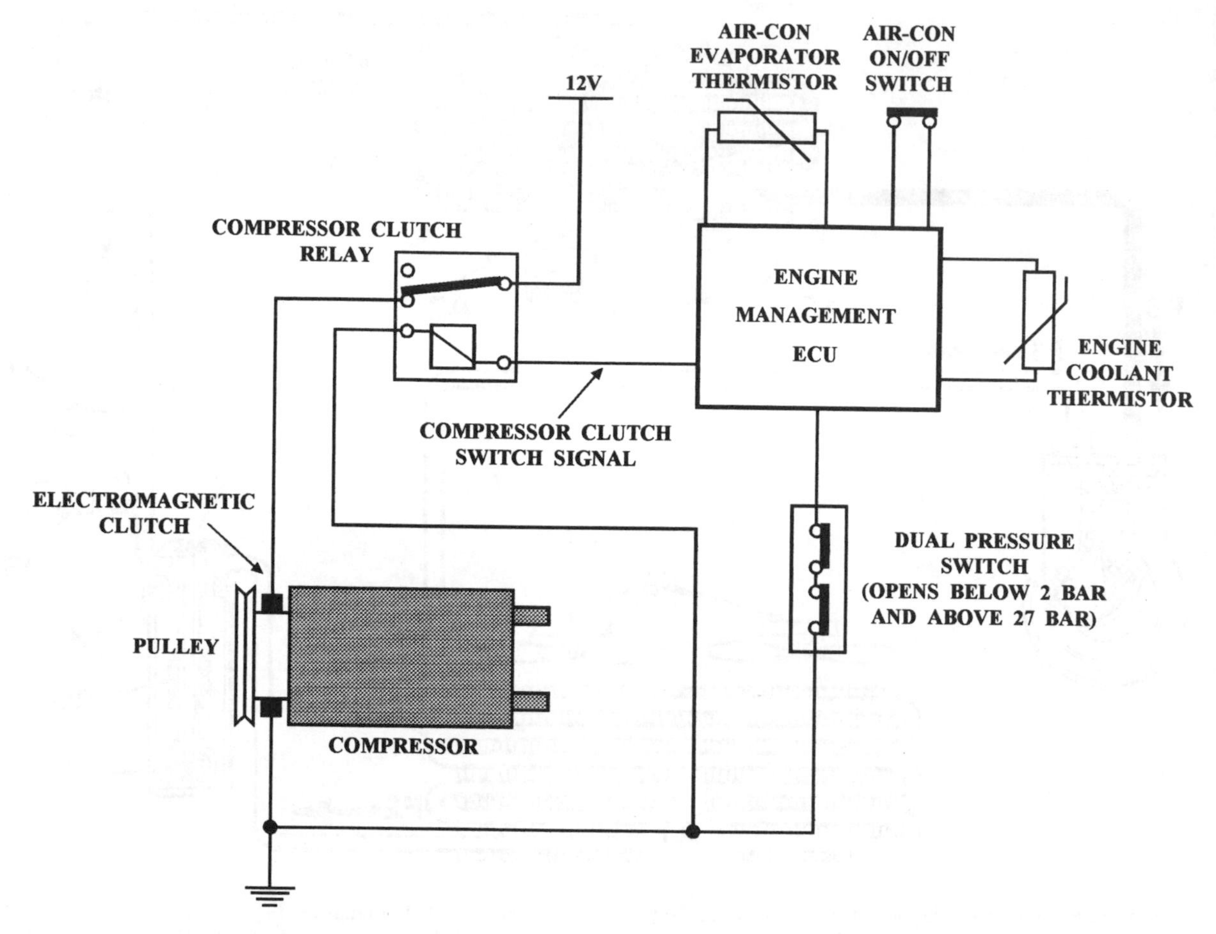

Figure 8.26 Control of air-conditioning compressor's electromagnetic clutch

Electronically controlled air conditioners

Electronically controlled air conditioners are marketed as 'Automatic Temperature Control' or 'Automatic Climate Control' systems and provide automatic regulation of the cabin temperature to a value preset by the occupants. The system comprises a microprocessor-based ECU, connected to sensors which monitor variables such as cabin air temperature, outside air temperature, engine coolant temperature and the level of solar radiation (sunlight) entering the car. In response to these inputs, the ECU automatically adjusts air temperature and distribution so that the cabin environment always remains comfortable.

The system operates by having the driver set the desired cabin temperature on the temperature control switch pack. The ECU adjusts the air temperature by first selecting either fresh or recirculated air. The incoming air is then cooled and dehumidified by the evaporator matrix. The hot/cold mixing air flap is set to feed a proportion of the cooled air through the heater matrix, after which it is mixed with unheated air until the desired temperature is reached. Air is then delivered to the cabin through various vents. According to the temperatures recorded in and around the vehicle, the temperature control ECU automatically provides the following controls:

(i) Air-flow rate. The air-flow rate through the cabin is controlled by varying the blower motor speed.
(ii) Air vent selection. The ECU selects either fresh or recirculated air inlets, and heating (windscreen, floor) or cooling (face) air outlets.
(iii) Temperature control. The ECU modifies the hot/cold mixing air flap to maintain the desired temperature.
(iv) Compressor clutch. The compressor electromagnetic clutch is switched on and off to cycle the

air conditioner, maintaining the desired cooling level.

System components

Figure 8.27 illustrates the components of an electronically controlled air conditioner system.

The driver selects the control temperature setpoint using the switch pack (1), which also contains an LCD display to show the system operating status. The ECU is fitted to the rear of the switch pack and contains an 8-bit microprocessor together with drive circuits for the actuators and analogue interface circuits for the sensor inputs.

A number of ntc thermistors are fitted to measure temperatures, each sensor being connected between a regulated 5 V supply and ground (0 V). The ECU monitors the current passing through each sensor, thereby deriving a temperature value.

An outside air temperature sensor (8) is mounted externally, away from heat sources such as the engine, radiator and exhaust system. It is constructed with a thick casing, giving it a considerable heat capacity so that sudden temperature changes (e.g. from water splash) do not unduly affect the average level measured.

The cabin air temperature sensor (6) is located in the dashboard moulding and is fitted into a housing which contains a small fan. The fan draws a continuous flow of cabin air over the sensor at rate of about 5 litres per minute, thus ensuring an accurate average temperature reading irrespective of outlet vent setting and dashboard temperature.

The evaporator temperature sensor is clipped to the evaporator fins (11). When the fin temperature falls below about 4°C, the temperature control ECU disengages the compressor clutch to prevent further cooling and so avoid possible icing of the evaporator.

The engine coolant temperature sensor measures the temperature of the heater matrix fins (12) and informs the ECU of heated air availability. If the engine coolant is cold, and heating is requested, the ECU keeps the blower on a slow speed and directs air to the screen vents. The compressor is also switched off. This helps with demisting and prevents cold air from being blown into the footwells. Once the coolant temperature rises above about 30°C the other vents are opened and the blower speed is progressively increased.

The solar radiation sensor (7) is mounted on the top surface of the dashboard where it is exposed to direct sunlight. When the amount of sunlight entering the cabin exceeds a predetermined threshold, the ECU can apply a fast acting cooling correction by directing cold air to the face vents. The solar radiation sensor is manufactured either as a thermistor on a light-absorbing backing, or as a photodiode, the latter being preferred since it has a fast response and is relatively insensitive to ambient temperature changes. The diode is mounted in a housing with an optical filter which only admits the sun's infra-red (heating) radiation. A reverse bias is applied by the ECU and the diode's reverse current is monitored as a measure of solar radiation. The reverse current is roughly proportionate to the incident sunlight level.

ECU and temperature control algorithm

The temperature control ECU is shown in block diagram form in Figure 8.28. Analogue data from the various sensors passes through a multiplexer so that many signals can be processed using just one analogue-to-digital converter. Other inputs are of a digital nature and include the control pushbuttons and an engine or compressor speed signal (to indicate the available cooling capacity of the air-conditioning system). The ECU's microprocessor calculates the required temperature of the incoming air and actuates the air flap servomotors and blower motor via the drive circuits and the digital-to-analogue converter, as appropriate. The compressor is also cycled on and off, depending on the external and internal temperatures.

To maintain the required setpoint temperature the cabin must be supplied with air at a temperature, T_{AIR}, calculated as,

$$T_{AIR} = K_{CAB} \times T_{CAB} - K_{EXT} \times T_{EXT} - K_{SUN} \times S_{SUN} - C$$

where T_{CAB} is the cabin air temperature, T_{EXT} is the external air temperature and S_{SUN} is the solar radiation level. K_{CAB}, K_{EXT}, K_{SUN} and C are system constants stored in memory.

When the cabin air temperature, T_{CAB}, differs from the setpoint temperature, T_{SET}, the microprocessor must modify T_{AIR} by an amount ΔT_{AIR} given by,

$$\Delta T_{AIR} = K_{SET} \times T_{SET} - (K_{CAB} \times T_{CAB} - K_{EXT} \times T_{EXT} - K_{SUN} \times S_{SUN} - C)$$

where K_{SET} is a constant associated with T_{SET}. The system thus enters a control loop, with continual modification of the blown air temperature, T_{AIR} to maintain the cabin temperature, T_{CAB} at the setpoint temperature, T_{SET}. Since there is a time lag between changing T_{AIR}, and the resultant change in T_{CAB}, it is important that the controller software incorporates an appropriate delay factor. The actual amount of delay will depend on the blower speed, since a high blower speed will give a smaller delay, and vice versa.

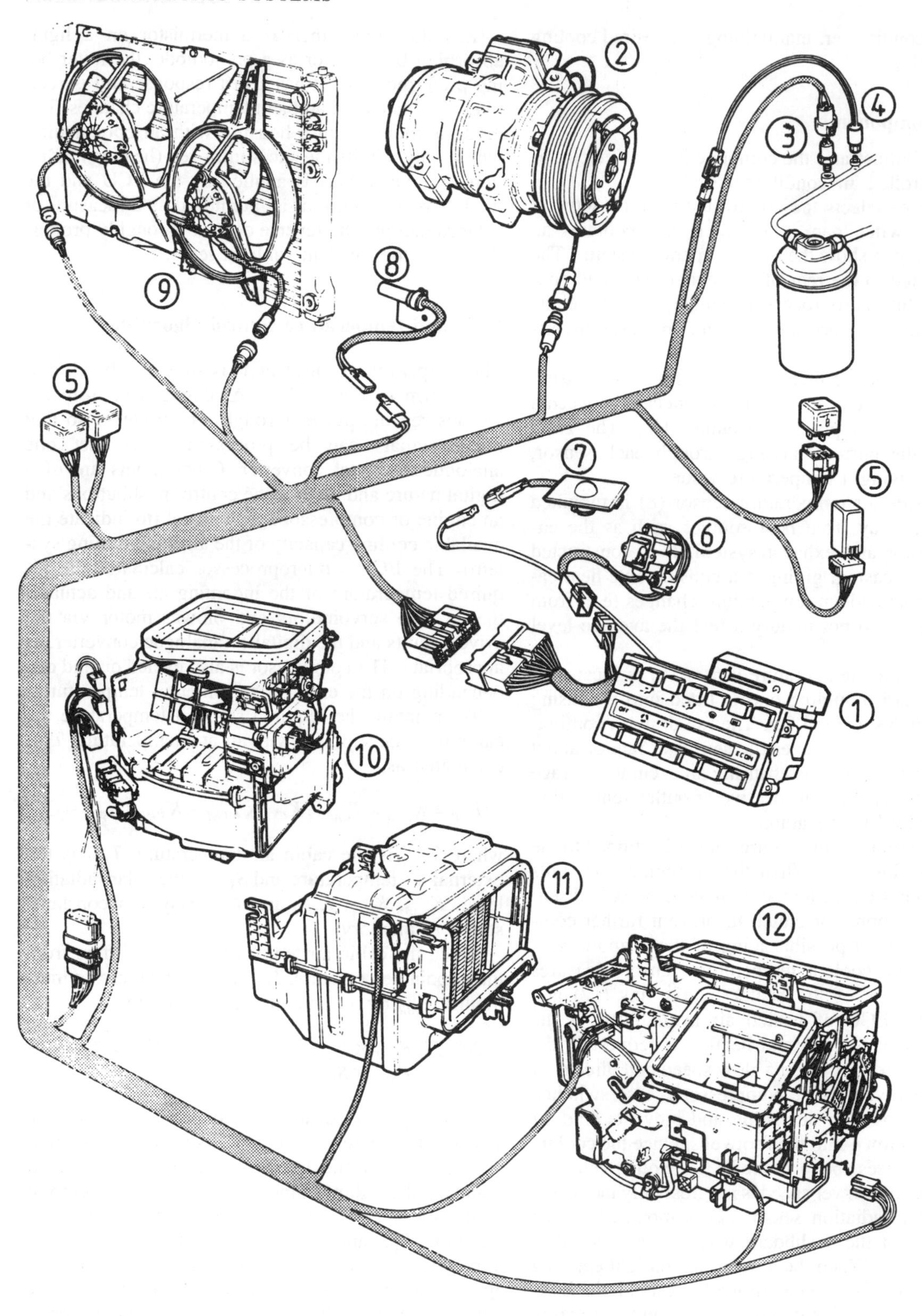
2
4
3
8
9
5
7
5
6
1
10
11
12

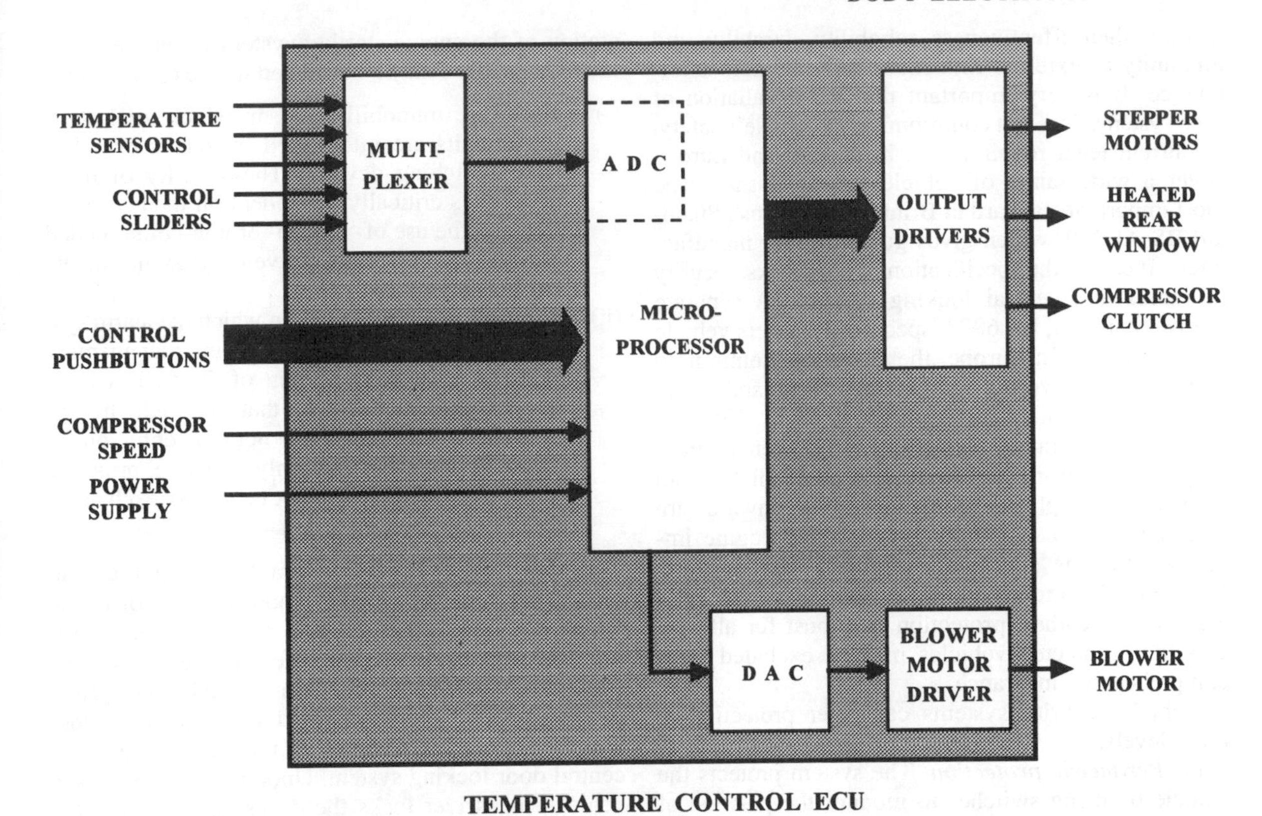

Figure 8.28 Automatic temperature control system ECU

8.7 VEHICLE SECURITY SYSTEMS

It is an unfortunate fact that Britain has one of the worst records for motor vehicle theft of any country in the world; in 1994 alone over half a million vehicles were reported stolen in the UK, and over a third of these were never recovered. Car crime accounts for about 25% of all criminal activity and whilst most of these incidents are associated with the theft of in-car stereo systems and personal effects carried in the vehicle, some involve 'joyriding' in which the vehicle is driven at speed by teenage thieves. Frequently, the inexperience of the driver leads to a road accident. Car security therefore relates not just to the protection of the vehicle and its contents, but also to road safety.

Most car thieves are teenage opportunists and consequently a comparatively simple alarm system can provide adequate security; research has shown that on a typical family saloon car a simple alarm reduces the chances of theft by a factor of 150.

No security system can prevent the theft of a valuable car by really determined professionals; in such cases only a hidden radio beacon and advanced direction finding equipment can help recover the vehicle. Such devices (e.g. the 'Tracker' system) have been extensively used in the USA and a recovery success rate of up to 93% is claimed.

Figure 8.27 Components of the automatic temperature control system (*Rover*): [1] switch pack and ECU; [2] compressor; [3] dual pressure switch; [4] high pressure switch; [5] fan and compressor relays; [6] cabin air temperature sensor; [7] solar radiation sensor; [8] outside air temperature sensor; [9] radiator/condenser with cooling fans; [10] blower assembly; [11] evaporator matrix; [12] heater matrix and air distribution assembly

Electronic alarm systems

Electronic alarm systems are a standard fitment on most new cars and are also available as an aftermarket accessory for older vehicles. A wide variety of systems are in production and generally offer a level of protection that increases with the cost of the product. In assessing alarm systems the primary concerns relate to their effectiveness, reliability, durability and immunity to external influences such as radio interference. It is very important that the installation of such systems does not compromise the vehicle's safety.

Current legal requirements in the UK and Europe cover a wide range of vehicle security issues. The most important standard in Britain is the British Standard BS AU209, which gives guidelines for manufacturers. It covers the specification of door locks, security of radios and central locking systems. A separate British Standard, BS 6803, specifically covers vehicle alarm systems. In Europe, the European Community requirements covering ignition key locks and alarm systems are detailed in the directive 74/61/EEC.

German insurance companies have been particularly keen to bring car theft to a halt. All German market cars supplied from October 1995 onwards are required to have electronically operated engine immobilization as standard equipment. Furthermore, European Community guidlines state that from 1995 electronic anti-theft protection is a must for all new cars; non-protected vehicles may be excluded from comprehensive insurance.

Vehicle security systems can offer protection at three levels.

(a) *Perimetric protection.* The system protects the vehicle by using switches to monitor the position of the opening panels of the vehicle (doors, boot, bonnet, etc.). A siren is activated if tampering is detected. The level of protection may be enhanced by adding sensors to detect movements of the body.

(b) *Volumetric protection.* Volumetric protection systems use ultrasonic, microwave or infra-red sensors to detect unauthorized movements within the passenger compartment of the vehicle. Ultrasonic sensors operate on the Doppler principle whereby any movement inside the vehicle causes a small shift in the received frequency of a 40 kHz ultrasonic transmission. Microwave sensors operate on the same principle, but use a high frequency radio transmission at about 10 GHz. Microwave sensors are less prone to false triggering by air currents and are therefore mostly used on cabriolet vehicles.

Infra-red sensors are mounted within the passenger compartment, on the roof rails at the top of the 'B'-post. Each sensor comprises a transmitter/receiver unit which transmits an invisible 'curtain' of infra-red radiation vertically downwards towards the floor. The receiver constantly monitors the strength of the reflected signal and any sudden disturbance, such as would be caused by an intruder, triggers the alarm system.

(c) *Engine immobilization.* Engine immobilization operates by having the security ECU inhibit the operation of the engine starting system when the alarm is activated. This may be achieved in one of two ways:

(i) Hardware immobilization, in which vital starting circuits are interrupted by relays or solid-state switching devices. The security of these systems is critically dependent on their installation and the use of 'hidden' or non-colour coded wiring is necessary to prevent bypassing of the cut-out devices.

(ii) Software immobilization, in which the alarm system ECU is coupled to the engine management ECU to cause 'scrambling' of the fuel or ignition calibration maps so that although the engine can be cranked, it will not run continuously. These systems can be highly effective, provided that the engine ECU cannot be readily exchanged for a fully functional unit.

The specification of a manufacturer's standard-fit security system will depend upon the type of car to which it is fitted. In all cases perimetric protection is provided, most systems also offer immobilization and some offer volumetric protection. Usually the system can be activated and deactivated using either the door lock key or a remote controller that also operates the central door locking system. Upon leaving the parked vehicle, the driver locks the doors and activates the security system by pressing a button on the remote controller. An indicator LED ('confidence light') flashes rapidly for a moment, informing the driver that the system is armed, and then flashes at a continuous slow rate to deter thieves.

If unauthorized entry is attempted, the security system enters an alarm state, sounding a siren and flashing the indicators or headlights. The engine immobilizer prevents the vehicle from being driven away.

After a period of about 30 seconds the siren is shut off (to prevent nuisance and battery drain due to false alarms) but engine immobilization remains in force until the vehicle owner deactivates it with his key or remote controller.

Alarm system remote controller

The remote controller allows convenient operation of the alarm and central door locking systems. It comprises a small transmitter unit which is carried by the vehicle owner, and a receiver unit which is connected to the door locking and security ECUs.

The transmitter may be enclosed within a key-fob or, in some designs, within the key itself. In order to ensure the smallest possible dimensions the circuit is manufactured using a multilayer printed circuit board and surface-mount integrated circuits. Power is supplied from miniature batteries of the type used in wrist watches.

Arm and disarm information is conveyed from the transmitter to the receiver by means of a complex digital code. The code itself is carried serially, either as a beam of pulsed infra-red (IR) light which is invisible to the human eye, or as an ultra-high frequency radio signal (UHF radio). Infra-red systems tend to suffer from poor range and are very directional, but are not subject to Home Office regulations or Type Approval requirements. UHF systems can give much greater range but are vulnerable to code breaking by 'grabbers' and other sophisticated radio equipment used by criminals. The frequency of the UHF radio signal used depends upon the radio licensing regulations of the territory in which the vehicle is operated. In most of Europe a frequency of 433.9 MHz is used, in France 224 MHz is used, whereas Australia and Italy use 315 MHz. Home Office regulations permit the use of 418 MHz in Britain, although European legislation may lead to the adoption of 433.9 MHz as a pan-European frequency.

To ensure that the transmission link is as secure as possible, many systems operate with 'rolling codes' in which each depression of the transmitter button causes a different set of code data to be sent. The receiver software is matched to the transmitter so that it expects the appropriate code change on each occasion. Generally, a fixed set of codes (say, 25) are rolled through on a cyclic basis. If the transmitter and receiver get out of sequence (perhaps due to inadvertent operation of the transmitter away from the vehicle) then the remote controller will not operate and the system must be re-synchronized by using the door key.

Immobilizer with RF transponder

A further extension of security is the electronic immobilizer with an integrated micro-transponder embedded inside the ignition key. With such a transponder unit in place, and with a corresponding receiver fitted to the car, the ignition lock only operates if the key's code exactly matches that of the lock. The transponder is powered by radio energy transmitted by the receiver at a frequency of 60–150 kHz, so operation is contactless and no batteries are required in the key.

Figure 8.29 illustrates the operation of a transponder system commercialized by Philips and installed by many vehicle manufacturers, including BMW, GM and Volkswagen. The transponder consists of a sophisticated CMOS chip, termed a Programmable Identification Tag (PIT), which is connected to a power supply circuit comprising a pick-up coil and capacitor. When the key is moved to within about 5 cm of the lock, sufficient radio energy is picked up from the lock's coupling coil to energize the PIT, which then enters a 'read' mode and transmits data back to the base station's receiver unit by modulating the current in the pick-up coil. As a result of inductive coupling, the current in the lock's receiver coupling coil is also modulated and the data is extracted. If the code received from the PIT corresponds to the code stored in the base station's memory, then the microcontroller instructs the engine management ECU to permit an engine start sequence. After the engine has been successfully started the receiver unit transmits a unique password back to the transponder and then changes the transponder's code to a new combination, ready for the next time the key is used. In this way a very high level of security is provided, making it almost impossible to duplicate keys.

Security system configuration

Figure 8.30 shows the configuration of a basic vehicle security system which features perimetric and volumetric intrusion protection and engine immobilization. It comprises a microprocessor-based alarm control unit which also controls the central door locking motors; the alarm is armed whenever the car is locked using the radio frequency (RF) remote controller.

Once the alarm is armed, it can be triggered by any of the following actions:

(i) opening of the bonnet, doors or tailgate;
(ii) operation of the driver's interior door lock;
(iii) operation of the ignition switch;
(iv) attempted operation of the starter motor;
(v) disturbance of the microwave signal (volumetric sensor).

When the alarm is triggered it sounds the siren for 30 seconds and disables the engine by inhibiting the starter motor circuit and temporarily altering the fuelling map in the engine ECU.

In order to disarm the system and unlock the doors a valid disarm code must be received from the remote controller.

Additional sensing features

In addition to the basic features of the alarm system described above, many security systems have extra sensors to provide a higher level of vehicle protection.

(a) *Radio and glovebox contacts.* The radio and

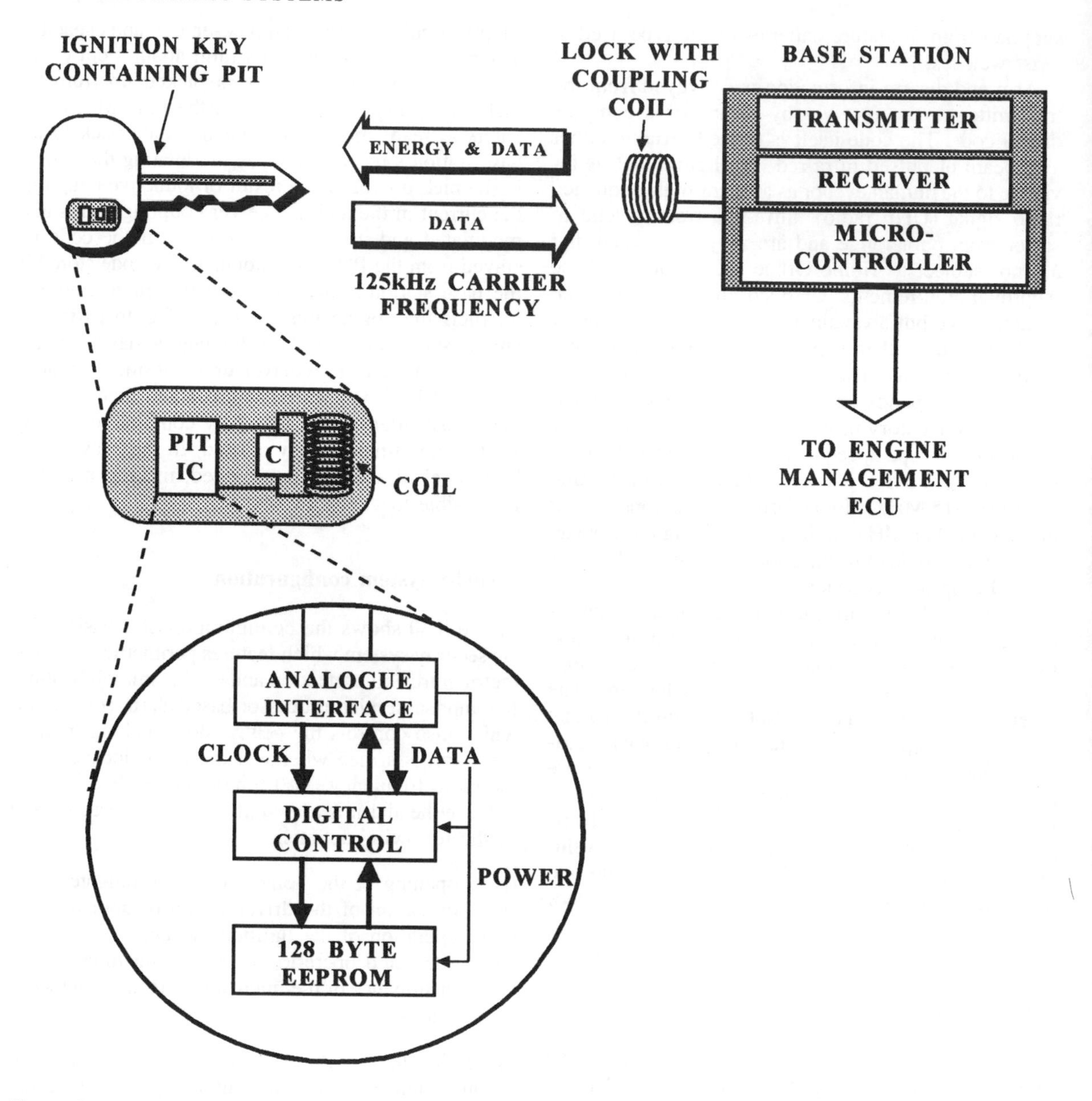

Figure 8.29 Vehicle immobilization using programmable identification tag (PIT)

glovebox may be fitted with switch contacts which cause the alarm to trigger if tampering is detected.

(b) *Glass breakage sensors*. Glass breakage sensors are required to trigger the alarm when a thief attempts to enter the car through the rear or side windows. The front windscreen is not usually fitted with a sensor because it is considered too tough to be broken through.

Two types of sensor may be fitted to the other windows:

(i) Wire loop sensor. A fine wire loop is bonded to the glass and the continuity of the circuit is continually monitored. A breakage triggers the alarm. This technique is most suited to fixed body glass, especially the rear window where the heating element can be used.

(ii) Magnetic sensors. These are used on the moving side windows. Each window has a small magnet bonded to the extreme lower edge of the glass, just below the weatherstrip. The presence

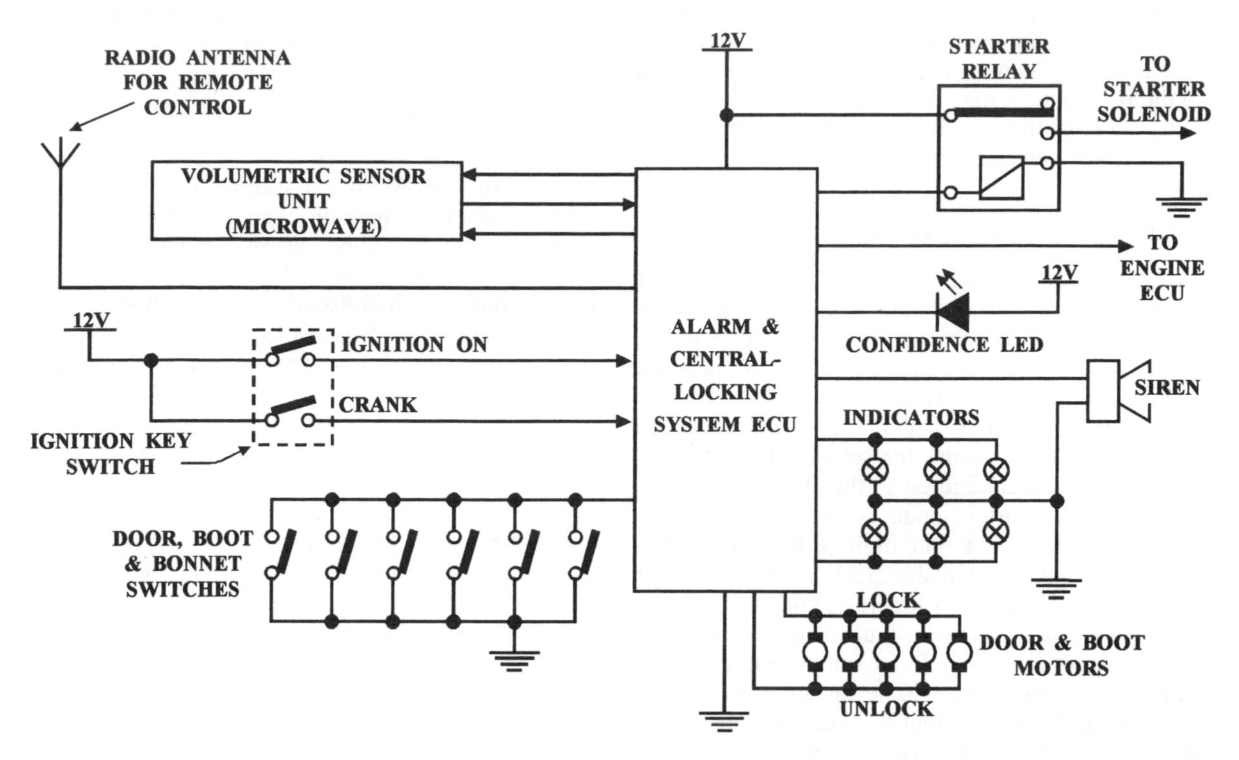

Figure 8.30 Basic vehicle alarm and immobilization system

of the magnetic field is sensed by a small reed switch, which is mounted nearby on the interior of the door casing. With the window glass intact, the reed switch is held closed by the magnetic field and the protection circuit is complete. If the window is broken, the glass shatters and so the magnet falls to the bottom of the door casing, allowing the reed switch to open and the alarm to trigger.

(c) *Tilt sensor*. The tilt sensor is used to trigger the alarm if the vehicle is tilted to an inclination different from that at which it was parked (e.g. if it is jacked up or towed).

One type of sensor operates by having a microprocessor monitor the change in the value of a capacitor which is composed of two metal plates separated by a insulating liquid dielectric. If the sensor is tilted, the plate area covered by the liquid changes and so the sensor capacitance changes. The microprocessor measures the capacitance when the alarm is first armed by applying a high frequency oscillation to the plates and monitoring the resulting current. A trigger signal is issued within the ECU if it detects any change from this value. It is important that the security system features a disable command for this sensor so that it can be turned off when necessary (e.g. during ferry journeys, on car transporters, etc.).

(d) *Vehicle position sensor*. The vehicle position sensor monitors any forward or backward rolling motion of the vehicle, due to towing or pushing, triggering the alarm if it detects any movement.

A change in vehicle position is usually detected using the speedometer transducer pulses; each pulse is recorded and if more than a preset number occur within a given time then the vehicle is judged to be moving and the alarm is triggered.

8.8 IN-CAR ENTERTAINMENT SYSTEMS

Most new cars are now supplied with an in-car entertainment (ICE) system fitted as standard. On the majority of cars this comprises an FM stereo radio-cassette unit and four or six loudspeakers mounted to the front and rear of the passenger compartment. Some luxury vehicles may additionally be fitted with a compact disc (CD) player.

Although the ICE system components may be badged with the car manufacturer's name, it is invariably the case that they are supplied by a specialist car audio manufacturer working to the car maker's specification. In general, when faults occur in the ICE system, the relevant components should be removed from the vehicle and taken to a specialist car audio repair centre.

Security-coded radio-cassette units

Due to the large number of thefts of car radio-cassette units many car audio manufacturers have developed sets which require the entry of a unique four-digit security code every time the power supply is reconnected (e.g. if the unit is stolen).

The security coding circuitry is incorporated into an IC which also contains the set's tuning and amplifier circuits, and configured so that the unit is inoperable until the correct code is entered and the IC reactivated. By using four digits a theoretical 9999 different codes are available, making the set less attractive to a thief. If an incorrect code is entered a delay period of about one minute commences and the user must wait until it has expired before another code can be tried. Each time an incorrect code is entered the delay time doubles (i.e. 1 minute, 2 minutes, 4 minutes, 8 minutes, etc.) so that a thief is deterred from experimenting with all possible codes.

Radio Data System

Radio Data System (RDS) is a pan-European data transmission system that has been developed with the car driver in mind. It functions by having FM radio transmitters send out an additional digital data signal which is 'piggy-backed' onto the the main programme material. The RDS signal is inaudible to the listener, but may be decoded by suitably equipped RDS radios to give an additional level of information and convenience; it is for this reason that RDS is often known as 'Teletext for radio'. Standard radio receivers cannot decode RDS and are unaffected by the RDS signal.

The digital data transmitted by RDS offers three primary features:

(i) Alphanumeric display of the radio station's name on the car radio's display panel.
(ii) Continual automatic retuning of the radio to the strongest receivable signal for the chosen radio station.
(iii) Interruption of cassette playback and increase of radio volume when traffic bulletins are broadcast by the chosen radio station.

Operation of RDS

RDS data contains a series of binary codes transmitted at a rate of 1187.5 bits per second. The codes represent alphanumeric data or switching functions in accordance with the following catagories:

(a) *Programme service (PS) code.* The PS code is a set of data which is decoded to give an eight character alphanumeric display of the radio station name, for example BBC Radio Two is displayed as 'BBC R2'. This facility means that the driver can easily identify each station by its name rather than by having to remember its frequency, a feature that is of great value now that there are so many FM stations across the country.

(b) *Programme identification (PI) code.* The PI code is a set of data, transmitted by every radio station, which contains information giving the station location, nature of the programme material and a station reference number. With this data each FM radio station is uniquely identified.

(c) *Alternative frequencies (AF) code.* The AF code gives a table of alternative frequencies on which the selected radio station is transmitting. Using the AF code, the car radio is able to automatically retune to stronger transmissions as the vehicle travels across the country, thus maintaining reception quality without any intervention on the part of the driver.

(d) *Traffic programme (TP) code.* The TP code is transmitted by a radio station to indicate that it broadcasts regular traffic bulletins. When the RDS car radio receives a TP code it illuminates a 'TRAFFIC PROGRAMME' indication on the display panel.

(e) *Traffic announcement (TA) code.* The TA code is transmitted by a radio station to indicate that a traffic bulletin is being broadcast. When the radio receives the TA code it switches from the cassette player to the radio (if applicable) and adjusts the volume to a preset level.

(f) *Extended other network (EON) codes.* The EON codes contain information about other stations receivable in the area, allowing the car radio tuning presets to be automatically retuned to new frequencies as the car travels accross the country. EON codes can also allow additional information facilities, such as the interruption of a national radio programme to bring traffic news from a local radio station.

9

Reliability and fault diagnosis

9.1 INTRODUCTION

When electronic systems were first introduced onto cars they were usually installed in the relatively benign environment of the passenger compartment. Even so, electronic failures were not unusual and many early systems gained a reputation for unreliability.

Car buyers will no longer tolerate unreliable operation however; the days of drying out 'wet' sparking plugs and cranking the engine with a starting handle on cold mornings are long gone. Consumer expectations are now such that a vehicle must operate efficiently over a life of 100 000 miles with little more than routine servicing and the odd minor repair. When a fault does occur, it must be detected and successfully repaired as quickly as possible to reduce 'vehicle off road' time.

Consumer demand, together with the falling cost of microprocessor controllers, has been a significant factor in stimulating the wide availability of electronic systems such as engine management, ABS, TCS and semi-active suspension systems. In the next decade, electronic control of throttle, transmission, braking, steering and intelligent cruise control will become common fitments. If well designed, these developments should improve the safety of road transport. However there are also disadvantages; increased complexity generally leads to decreased reliability. Manufacturers will try to prevent this, but will find difficulty in doing so throughout a vehicle's life, particularly in the later years when third or fourth owners may not be able to afford the cost of franchised dealer servicing.

It is therefore becoming essential that major safety or emissions-related defects are automatically detected and brought to the attention of the driver, and then easily identified by service technicians, especially during MoT tests.

9.2 RELIABILITY OF AUTOMOBILE ELECTRONIC SYSTEMS

The automotive environment

Whilst most consumer electronic systems operate over a relatively modest 0°–70°C temperature range, automotive systems may have to operate from −40° to 125°C in conditions of high humidity, vibration and rapid voltage variations. It has often been said that automotive electronics must be designed to a military specification, but to transistor radio prices.

Particular reliability problems stem from the extremes of temperature encountered. Temperature requirements for component operation arise from the conditions prevailing at the installation position on the vehicle and the weather conditions in the country where the vehicle is operating. Table 9.1 gives some typical temperature ranges that components must be capable of withstanding during normal usage. It must be remembered that some of these temperatures can change very rapidly when, for example, the engine is started or the air-conditioner is operated.

Shock and vibration is another factor that must be considered; depending on vehicle assembly, repair and driving circumstances, acceleration forces of up to 20 g in the frequency range to 20 Hz may be experienced by the electronic components.

The vehicle electrical environment also affects the electronic systems. Wide voltage fluctuations can occur as a result of load switching and intermittent faults such as loose terminals. Voltage variations are particularly acute when the engine is started. In very cold conditions with a partially discharged battery a nominal 12 V supply may fall as low as 6 V during cranking. At the other extreme a 12 V system may, exceptionally, be required to survive 'jump starting' from a 24 V truck electrical system. Many automotive controllers are therefore specified to operate over a 6–30 V supply range. Sources of electrical noise may cause interference to electronic systems. For example, on spark-ignition engines the coil secondary voltage can rise as high as 50 kV, possibly inducing noise in nearby signal leads and electronic devices. An increasingly important source of electrical noise is radio-frequency interference (RFI) which originates both from on-board systems (microprocessors, mobile telephones, CB radios, etc.) and from off-board sources, such as mobile transmitters (police, taxis, etc.) and fixed television, radio and radar transmitters.

In order to ensure that on-board electronic systems operate reliably in this hostile environment, all vehicle manufacturers conduct both basic electrical tests

Table 9.1 Minimum and maximum temperature requirements encountered at various locations on an automobile

Installation location	Critical components	Non-critical components
Interior – floor	−40 – +85°C	–25 – +70°C
Instrument panel top surface	−40 – +120°C	–25 – +120°C
Instrument panel recess	−40 – +85°C	–25 – +70°C
Rear parcel shelf	−40 – +120°C	–25 – +120°C
Luggage compartment	−40 – +85°C	–25 – +70°C
Engine bay bulkhead	−40 – +125°C	–25 – +125°C
Engine	−40 – +150°C	–25 – +150°C
Exhaust system	−40 – +650°C	–25 – +650°C
Chassis – outside	−40 – +120°C	–25 – +120°C
Insulated chassis surfaces	−40 – +85°C	–25 – +70°C
Chassis surfaces exposed to heat	−40 – +125°C	–25 – +125°C
Parts exposed to hot engine oil	−40 – +170°C	–25 – +150°C

and more sophisticated electromagnetic compatibility (EMC) tests.

Design for reliability

Although there are some basic international standards for the test and evaluation of ECUs, there are as yet no common standards in general use applying to the design of the systems themselves. Each electronic system manufacturer has therefore evolved his own design methodology in the hope that the final product will pass the required tests and prove reliable.

These methodologies generally rely upon making a judgement about the operating environment (temperature, vibration, etc.), selecting suitable components and then validating a large number of prototypes fitted in test vehicles.

Since semiconductor devices are strongly affected by temperature, the thermal environment is generally the most important aspect in conducting a reliability assessment. For example, when designing an engine management ECU to be mounted in the engine compartment, the following time/temperature exposure can be predicted for a 10 year/100 000 mile vehicle lifetime:

40 hours at 125°C
350 hours at 65°C
8 400 hours at 25°C

Table 9.2 Number of failures per million hours as a function of temperature for ECU components

ECU component	25°C	65°C	125°C
TOTAL ECU	14	32	840
Printed circuit board	2	8	170
Intel 8096 microcontroller IC	1	4	110
Injector drive circuit	1	3	66
Knock sensor circuit	1	2	64

Using this information it is then possible for the ECU designer to predict the ECU reliability by using the FIT (Failures In Time) method, where 1 FIT = 1 failure per billion hours of ECU operation.

Component manufactures are asked to supply FIT data for each device used in the ECU circuit and these figures are then summed to arrive at an overall FIT figure for the whole ECU. For an ECU mounted in the engine compartment a reliability figure of about 10 000 FIT (meaning 10 000 component failures per billion hours of operation) might be typical. In percentage terms, this equates to a failure rate per year of less than 0.1%.

To illustrate the influences on ECU failure, typical reliability data for an engine management ECU is presented in Table 9.2. Two features should be noted; (a) the failure rate for the ECU and its major components rises disproportionately with temperature, demonstrating the difficulty of designing ECUs for hot environments, and (b) the least reliable component in the ECU is the printed circuit board, illustrating the enormous importance of manufacturing technology in achieving high reliability. In particular, the quality of the soldered joints is critical.

Electrical testing of ECUs

In addition to standard temperature, shock and vibration tests there are several widely accepted electrical conditions that an automotive circuit needs to survive. These include double voltage supply, reverse battery connection, overvoltage due to alternator regulator failure (16–20 V) and a number of transient voltage conditions, including:

(i) Transient voltages occurring during load dumping when the alternator load suddenly drops due to a battery terminal becoming disconnected. This is usually assessed using an 80 V pulse signal with an energy of 50 J, which decays to a nominal 12 V in about 400 ms.

(ii) Supply disconnection transients occurring when the ignition is switched off.
(iii) Transient voltages occurring when inductive loads such as relays, coils and solenoids are disconnected. These are evaluated using a –100 V, 0.5 J, negative transient and a 240 V, 0.5 J, positive transient which decays to 12 V in about 1.25 ms.
(iv) Transient voltages due to capacitive and inductive coupling between adjacent wires. These are evaluated by laying power and signal leads close to each other and then monitoring the induced noise in the signal leads. Problems can usually be alleviated either by altering the disposition of wiring within a vehicle, or by providing screening around the signal leads.

Since any of these transients can disrupt the operation of electronic circuits it is normal for the ECU designer to suppress them both at their source and at the input to the ECU. The immunity of the ECU is then tested by using a transient generator to inject voltage transients into the vehicle wiring and observing any disturbances.

Electromagnetic compatibility (EMC)

Motor vehicle electromagnetic compatibility was first considered in 1952, when an act of Parliament required the fitting of suppressor resistors to ignition system HT leads in order to prevent interference with domestic radio and TV reception. Since that time, the increased use of electronics for safety and emissions-related systems means that testing for electromagnetic interference has become a major part of the vehicle evaluation process.

In 1989 the European Community responded to the growing problem of electromagnetic compatibility by producing a Directive (number 89/336/EEC) which states that electronic equipment must neither cause, nor suffer from, interference due to electromagnetic disturbances. In due course it is intended that the motor industry will have its own set of EMC standards which will form part of the vehicle Type Approval process.

Many EMC problems arise because cars are mobile and therefore can be exposed to very varied levels of electromagnetic radiation from a wide variety of sources (Table 9.3).

Although fixed radio transmitters can be very powerful, they are usually located some distance from a road and therefore vehicles are not exposed to particularly high field strengths. The greatest problems often arise from mobile transmitters, which although not very powerful, will give rise to very high field strengths around the vehicle on which they are mounted.

Table 9.3 Sources of electromagnetic interference and typical field strengths

Interference source	Field strength
AM broadcast transmitters	10 Vm^{-1}
Mobile radio transmitters	1–100 Vm^{-1}
VHF/UHF broadcast transmitters	1 Vm^{-1}
Radar transmitters	1 000 Vm^{-1}
High-voltage overhead power lines (50 Hz)	10 000 Vm^{-1}

The electromagnetic field strength at the vehicle, E (measured in volts per meter, Vm^{-1}) is related to the transmitter power, P, and the distance in metres of the vehicle from the transmitter antenna, r, by the equation;

$$E = (30P)^{1/2}/r$$

For example a 50 kW broadcast transmitter will generate a field strength of only about 1 Vm^{-1} on a vehicle 1 km away, whereas a mobile transmitter mounted on a police car has a power of only about 25 W, but will generate a field strength of approximately 30 Vm^{-1} on the vehicle. It is therefore very important that mobile transmitting equipment is carefully installed on vehicles.

Vehicle manufacturers normally specify locations for the installation of radio and cellular telephone equipment and problems will not arise if these recommendations are followed. Equipment improperly installed by unqualified persons, could, however, lead to interference with other vehicle systems.

Interference effects and cures

Interference effects can arise in two ways:

(i) Electromagnetic interference can be picked up by the vehicle wiring harness, which acts as an antenna for frequencies in the range 20–200 MHz and converts the field into a current. Typically, 1 mA of interference current is generated for each 1 Vm^{-1} of field strength.
(ii) Electromagnetic interference may couple directly into the ECU via printed circuit board tracks and component leads. This effect is most pronounced at frequencies above about 200 MHz.

The effect of interfering signals on vehicle systems is dependent upon the ECU circuit design. In severe

Figure 9.1 A vehicle undergoing EMC testing whilst running on a rolling-road (*Motor Industry Research Association*)

cases it can lead to temporary fuelling and ignition errors in engine management systems, spurious indications from instruments and trip computers and 'dropping out' on cruise control systems. The solution to EMC problems generally lies in careful design and layout of the ECU printed circuit board, such as using a 'ground plane' layer and fitting components with short leads. In some cases incoming signals may be passed through an input filter to eliminate high-frequency interference before it can enter the ECU's circuitry.

EMC testing

Although automotive electronic systems can be individually evaluated for electromagnetic compatibility it is more usual to perform whole-vehicle testing in which an entire prototype vehicle, with all of its electronic systems installed, is placed in an EMC chamber (Figure 9.1).

An EMC chamber is a large radio-shielded room that has its walls internally lined with large cones of radio-absorbing foam material, enabling tests to be performed under uniform conditions. A rolling-road dynamometer and exhaust-gas extraction facilities are provided so that the engine, transmission, ABS and TCS electronics can be tested under realistic circumstances.

In order to perform an EMC test, the prototype vehicle is carefully positioned within the chamber and a system of transmitters and antennas are used to expose it to electromagnetic radiation over the frequency range 1–2 000 MHz at field strengths up to at least 50 Vm^{-1}. Since it would be unsafe for technicians to enter the chamber during a test, a number of closed-circuit TV cameras and voltage and current probes are used to monitor disturbances. If any problems are observed then the exact nature of the failure must be identified and the relevant circuits or wiring modified.

9.3 AUTOMOBILE ELECTRONIC SYSTEM DIAGNOSTICS

Introduction

The word 'diagnosis' is generally taken to mean the process of identifying the root cause of a problem by examining the symptoms of the problem. Although this is the primary purpose of automotive diagnostics it is also necessary to include the equally important task of simply detecting that a problem exists in the first place.

The excellent reliability of automobile electronic systems, allied to the very high level of technology used, has led to a reduction in the number of simple or 'routine' defects encountered by the service technician, while multiplying the possible causes of failure. This has greatly increased the problem of fault diagnosis on present-day vehicles. The way in which fault diagnosis is performed in a modern repair shop is therefore very different to the 'traditional' approach employed ten or twenty years ago.

Traditional fault diagnosis techniques

Before the widespread application of automotive electronics, car electrical systems were composed of just a few simple independent sub-circuits powered directly from the battery. Such circuits typically consisted of a switch controlling a lamp or motor, perhaps via a relay. Since there were so few electrical components, they were readily identified, even on vehicles that the technician had never encountered before. Simple components could easily be checked using a test-lamp or a multimeter (a voltmeter, ammeter and ohmmeter combined into one unit) and more complicated components, such as flasher units and relays, could be checked by substitution.

This approach had many attractive features, in particular it required only low-cost test equipment and allowed the technician to use his knowledge and experience to optimize the diagnostic process.

In the late 1970s, the arrival of electronic fuelling and ignition systems led to a breakdown of the traditional diagnostic strategy, for three main reasons.

(a) Using the traditional approach, the electronic system was tested by disconnecting the ECU from the other components and then checking them individually. If no fault was found then the ECU was deemed (usually wrongly) to be defective by default. For the car owner this sometimes led to a long repair time and the unnecessary replacement of an expensive component. For the technician it led to a lot of frustration and many unhappy customers.

(b) The interconnection of many sensors and ECUs made it difficult for the technician to hold a mental picture of the functional interactions of all the components. Vehicle manufacturers provided help by distributing service documents giving flow-charts and diagnostic tables, but even then the technician was unlikely to fully understand how the systems interacted. This became a particular problem for multi-franchise and independent repair shops, where technicians serviced a wide range of vehicle types.

(c) The wiring on older cars usually carries just two types of signal; battery voltage and ground. Modern electronic systems augment these circuits with wires carrying complex low-voltage signals between various ECUs and the numerous sensors and actuators. Traditional test-lamp checks are therefore of little value; a multimeter may be of no additional help and may even cause damage if incorrectly used.

During the 1980s, the rapid introduction of engine management electronics produced a requirement for new testing techniques, new test hardware and higher-quality service data. The great diversity of systems also led to the need for rapid access to information relevant to the particular vehicle under repair. The solution to these requirements has been the development of a range of on-board (i.e. incorporated into the ECU) and off-board diagnostic facilities that fall into three categories.

(a) *Off-board diagnostic stations.* These do not connect directly to the ECU and therefore do not rely upon any on-board diagnostic features. Off-board diagnosis is usually restricted to the engine fuelling and ignition systems and therefore off-board testers are generally referred to as *engine testers* or *engine analysers*.

(b) *On-board diagnostic software that provides direct indication of fault codes.* An ECU's software may incorporate self-test routines that can store a 'fault code' when a system fault is detected. The ECU then flashes a lamp or LED in a specific sequence for each stored code. The technician reads the flash-code and interprets its meaning from service literature.

(c) *On-board diagnostic software accessed via an off-board diagnostic tool.* When a vehicle is taken for service a handheld diagnostic tester, usually known as a *scanner* or *scan tool*, can be connected to a diagnostic terminal fitted to certain ECUs. Data and fault codes are then read directly from the ECU's memory and interpreted by the technician. Advanced developments of these tools can be connected to the entire vehicle harness to provide a 'whole-vehicle' diagnostic facility complete with comprehensive repair information.

Off-board diagnostic equipment

Off-board diagnostic testers first became available in the USA, mainly because of the earlier diversification and elaboration of electronic engine management systems there. They are now widespread in Europe, and can be found in most repair shops.

Figure 9.2 illustrates a typical engine tester, in this case manufactured by FKI-Crypton. Similar equipment is available from other manufacturers, including Sun, Bear and Bosch.

The Crypton tester is based around an IBM-compatible host computer with disc-drive and keyboard, which is connected to an engine analyser module. A set of test leads and a timing strobe are plugged into the rear of the analyser module. The tests leads are suspended from a swivel boom, ready for connection to key points such as the ignition coil, HT leads, battery terminals and so on. The computer processes engine data and presents results on a colour monitor and a black-and-white printer. The whole arrangement is mounted on a trolley for easy portability in the workshop.

A set of computer discs is supplied with the tester, giving adjustment data for thousands of different car models and containing instructions to the technician on how to connect the tester to the vehicle and which adaptors to use.

Identifying just the vehicle model is not enough, in addition the engine variant, transmission type, ignition system and options fitted are all relevant to the diagnosis process. Once connected, the tester is able to provide fault diagnosis on major vehicle systems including the starting and charging systems, ignition, compression and induction, fuel system and emissions.

In standard form an engine analyser does not provide diagnostic procedures specific to the ECUs and their associated sensors; optional modules can be purchased to provide this facility.

A key part of the engine testing procedure is exhaust gas measurement. This is usually accomplished using a so-called *four-gas analyser* to measure the concentrations of carbon monoxide (CO), carbon dioxide (CO_2), unburned hydrocarbons (HC) and oxygen (O_2). The gas which is the subject of most legislative control is CO, although MoT inspections also include a limit for HC.

A four-gas analyser is normally supplied with the engine tester, either integrated into the console or as a separate module. The most common technology used for CO, CO_2 and HC measurement is known as non-dispersive infra-red (NDIR) analysis. It involves transmitting an infra-red beam through the exhaust gas and analysing the composition of the received signal.

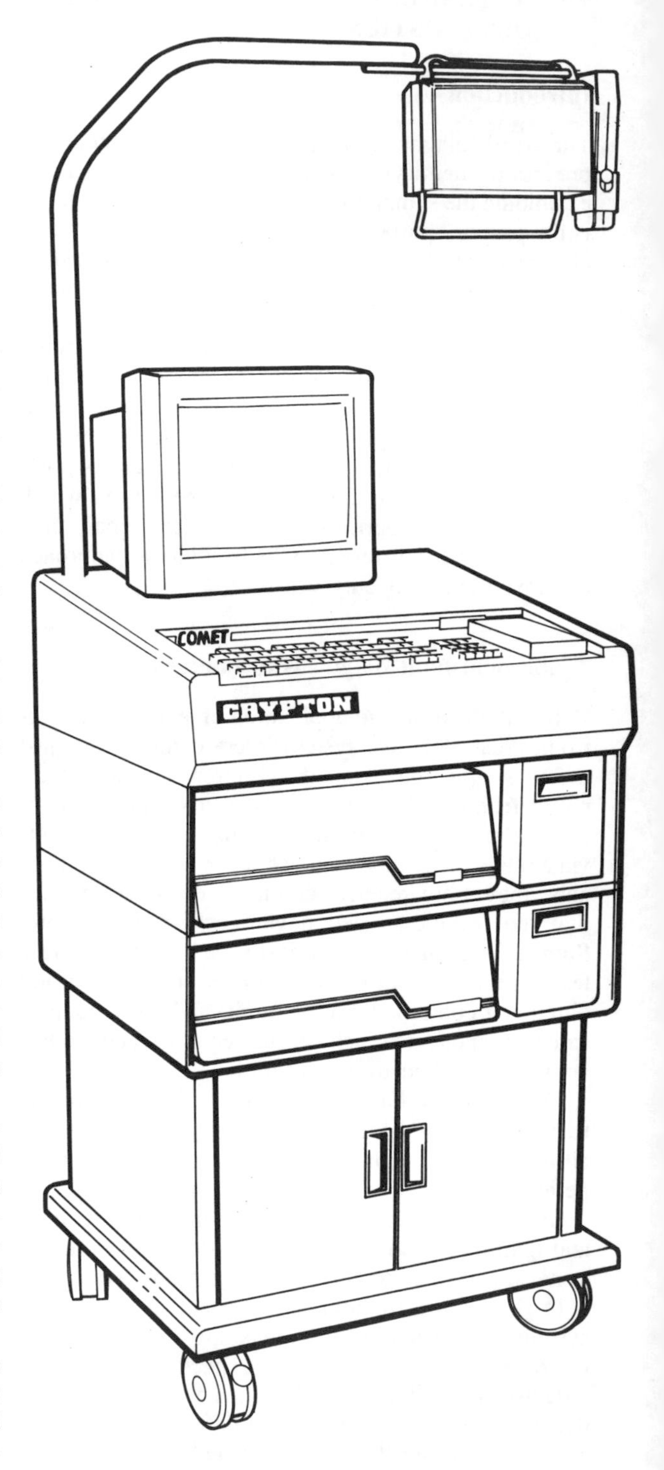

Figure 9.2 Crypton 'Comet' engine tester based on an IBM-compatible computer (*FKI Crypton*)

Since each gas component absorbs radiation at a different characteristic wavelength it is relatively easy to calculate the relative concentrations of the gases.

O_2 is measured using a galvanic cell sensor, similar in construction to an exhaust gas oxygen sensor used for engine air–fuel ratio control. It produces an output voltage proportional to the oxygen concentration in the exhaust gas. Using data from the four-gas analyser it is possible for the tester's software to compute the engine's air–fuel ratio, even when a catalytic converter is fitted in the exhaust system. This information can then be used to determine whether or not the closed-loop fuelling system is functioning correctly.

Using an engine tester

The process of reaching an accurate diagnosis depends on the following steps:

(i) Vehicle identification.
(ii) Correct execution of test procedure.
(iii) Comparison of measurements with vehicle limit data.
(iv) Evaluation of PASS/FAIL results to arrive at a diagnosis.

Before commencing testing, certain preconditions have to be met to ensure accurate results, for example the engine temperature and rpm must be within certain limits and certain ancillary items must be disconnected or disabled. A typical test procedure then involves taking basic data at idle speed, subsequently varying the engine rpm to detect particular faults. For example;

(i) Idle test. Measure idle speed, idle stability per cylinder, exhaust emissions, spark firing voltage, sparkline voltage, spark duration, battery voltage and charge current, coil positive and coil negative voltages, display samples of various electronic system signal waveforms.
(ii) Accelerate sharply. Measure spark firing voltage, sparkline voltage, spark duration, acceleration per cylinder, detect misfires, exhaust emissions, monitor ignition timing variations and battery voltage.
(iii) Release throttle. Measure deceleration per cylinder, exhaust emissions, voltages.

During a test the equipment displays the measured values and automatically compares them against the 'LOW', 'GOOD' and 'HIGH' values which are held on disc for the particular engine variant.

Once all tests have been completed, the technician must review the data and make a diagnosis. Many engine testers aid this process by making a series of suggestions based on comparing the measurements with stored tables of possible combinations. When a characteristic combination is detected, a message informs the operator of the likely nature of the problem. Some testers are even able to make recommendations based on field experience, for example, 'on this model the fuel-pump relay is prone to failure – check first'.

When a diagnosis has been made, fault rectification can be attempted. The tests are then repeated and if the fault persists alternative diagnoses must be considered.

Although off-board diagnostic systems can help cure many common fuelling and ignition problems, they are much less useful in detecting intermittent faults in complex electronic systems. In many instances fault symptoms can appear in other, related systems. Consider the following (real) problem: a customer complains that his vehicle, fitted with a catalytic converter and electronically controlled automatic transmission, occasionally shifts from fourth to third gear, and then back again, when driving at a steady 50 mph. Is the fault in the transmission's electronic controller, or is it an internal hydraulic fault? The answer is neither. After some investigation it is discovered that an intermittently faulty exhaust-gas oxygen sensor is occasionally giving an incorrect signal to the engine ECU, leading to a sudden drastic change in air–fuel ratio. The sudden fall in engine power causes the transmission to momentarily shift to a lower gear.

Intermittent and non-reproducible faults, such as this, can only be detected by constant observation of all relevant parameters whilst the vehicle is in use. This means using on-board diagnostic facilities built into the ECUs.

On-board diagnostics

Many early automotive electronic systems were designed with little thought to the need for fault diagnostics. Therefore when faults occurred on these systems it was necessary to connect diagnostic equipment by inserting adapters into the circuit, generally between the ECU and the wiring harness. The system was thus checked using the connections made available. The disadvantages of this strategy were the large variety of costly wiring adapters required and the fact that the connectors themselves were often responsible for the original fault symptoms. Connecting the diagnostic equipment mysteriously provided a temporary cure to the fault! Most automotive controllers now incorporate a considerable number of fault detection algorithms to monitor functional sequences and continually check each sensor for signal plausibility and consistency. If a malfunction is detected the controller will go into a 'limp-home' mode whereby an appropriate value is substituted for the faulty data. For example, if a coolant temperature sensor were to fail open-circuit, the software might enter a 'limp-home' temperature of 80°C in order to keep the vehicle mobile. The driver is informed of the malfunction via a dashboard warning light and the microprocessor stores a multi-digit 'fault code', indicating the nature of the failure, in non-volatile memory. Fault codes of this type fall into two categories:

(i) 'slow codes', which generate coded lamp flashes.
(ii) 'fast codes', based around the international standard ISO 9141.

Slow codes
When the ECU's self-diagnostic software detects a fault it stores the appropriate fault code in its memory and illuminates a dashboard-mounted light known as a *check-engine light* or *malfunction indicator light* (MIL) to indicate to the driver that the vehicle needs attention. Depending on the electronic system involved, the service technician can then view the slow code in one of three ways:

(i) By looking at a flashing LED mounted on the ECU enclosure.
(ii) By shorting two terminals on a diagnostic connector to enable the ECU to flash the MIL on the dashboard.
(iii) By connecting an LED or analogue voltmeter to a diagnostic connector pin and watching the LED flashing or the meter needle swinging from side to side.

Since slow codes are designed to be visually interpreted by the service technician their frequency is very low (around 1 Hz) and the amount of information conveyed is therefore very limited. The result usually appears in the form of a repeating two-digit flash-code, whose meaning must be looked-up in a service document specific to each ECU.

Typically a long flash (about 1.5 seconds) counts as a 'tens' digit and a short flash (about 0.5 seconds) counts as a 'units' digit, a few seconds pause is inserted between tens and units. For example, if the LED shows two long flashes, a two-second pause, and then four short flashes, the fault code is '24'. Looking in the manufacturer's service data might reveal 'CODE 24 = Vehicle speed sensor fault – open or short circuit in the sensor circuit'.

The technician must then use his skills to determine whether the fault lies in the sensor itself, in the connector, or in the associated wiring. Once the fault has been rectified, the ECU's fault memory must be cleared by momentarily disconnecting the battery or removing the ECU power supply fuse.

Slow codes have the advantages of being simple, reliable and low-cost. Moreover, they can be read without using special equipment, an important consideration when vehicles are used in regions where there are few dealers. The main drawback with flash codes is that they provide very limited information and require the technician to complete the diagnostic process.

Fast codes
In contrast to slow codes, fast codes provide a flood of data via a two-way high-speed serial data connector wired to the ECU. Such data links were originally incorporated for end-of-line testing in the car factory, but they are now widely used for service diagnostics.

Serial links provide an opportunity to communicate with vehicle systems without disturbing any of the wiring harnesses, and most importantly, for a high level of information exchange with a diagnostic tester.

Dedicated serial diagnostic testers are usually supplied by vehicle manufacturers exclusively for use in franchise dealer repair shops. Each tester comes with a set of leads and adapters to fit engine, transmission, ABS, air-conditioning and other ECUs for a specific range of vehicle models from one manufacturer.

Several diagnostic equipment manufacturers, for example Crypton and Bosch, market universal serial diagnostic testers for use in independent repair shops. These testers comprise a standard hand-held scan tool that can be configured to work with a wide range of ECUs from different manufacturers.

The ISO 9141 standard

Since the late 1980s the standardization of serial links has been made possible by the ISO 9141 standard for interchange of digital information. ISO 9141 specifies the way in which an off-board diagnostic tester can communicate with an ECU to access data and fault codes and send instructions. It does not specify compatibility of software and therefore fault codes, diagnostic procedures, etc. still vary from one manufacturer to another.

ISO 9141 specifies that the diagnostic tester shall communicate with the ECU via a one wire ('K' line) or two wire ('K' and 'L' line) data connector. Line K is bidirectional (passes data both ways) whereas line L is unidirectional and is only used when the link is first set up, at all other times it idles at a logic '1'. Vehicle battery voltage and ground terminals are also provided at the connector.

When a diagnostic tester is connected to an ECU's

serial link it must first 'wake up' the ECU. This is a achieved by the transmission of a specific 8-bit address from the tester to the vehicle. The address is sent simultaneously on the K and L lines at a rate of 5 bits per second.

If the address is correct for the ECU under test then it responds by transmitting an 8-bit speed synchronization pattern back to the diagnostic tester. This pattern of alternating '0's and '1's informs the tester of the data rate to be used for the transmission of all subsequent data. After transmission of the speed synchronization pattern, the ECU transmits two 'key words' which inform the diagnostic tester of the form of the subsequent serial communication and the hardware configuration of the K and L lines. When the tester receives the final key word it transmits the logical inversion of this key word back to the ECU to confirm that the initialization process has been successful.

The ECU test procedure following initialization is dependent upon the software provided with the diagnostic tester. Most testers have a basic capability to read stored fault codes from an ECU and decode them into a written statement displayed on an LCD screen.

More sophisticated software can provide a detailed diagnosis of sensors and actuators via an automatic test sequence. Command codes can be issued by the tester to instruct the ECU to perform various functions and operate various actuators. The tester then provides a full display and diagnosis of readings obtained, together with relevant troubleshooting hints.

Universal serial diagnostic testers

Universal serial diagnostic testers are designed to be used by independent repair shops that need a wide-ranging diagnostic capability. They conform to ISO 9141 and can be used to diagnose faults on vehicles from many different manufacturers.

The Crypton 'Check Mate' universal tester (Figure 9.3) comprises a ruggedized microprocessor-based scan tool with a 64 character LCD screen and 6 button keypad. It is normally supplied with a set of connecting leads and a small printer. In order to configure the Check Mate for a particular vehicle model, a *program card* must be pushed into a slot in the rear of the tester. Each card contains all the necessary software to test and diagnose faults on a specific ECU, and provides instructions and prompts to guide the technician through a logical test routine. A wide range of program cards is available to cover all popular engine management systems, plus a few ABS and other systems.

The use of program cards offers a number of advantages, in particular the cost of upgrading and extending the tester's capability, by buying additional cards as they become available, is relatively modest.

Dedicated diagnostic testers

Most vehicle manufacturers supply their franchized dealers with serial diagnostic testers exclusively for use on their own vehicles. Examples of testers of this type include FDS2000 (Ford), TestBook (Rover) and Tech 15 (Vauxhall/Opel). In addition to the facilities offered by the universal testers described above, these dedicated testers have enhanced functionality through the provision of an electronic information library which contains service data, diagrams and repair tips. This information is stored on compact disc (CD-ROM) and a set of updated discs can be sent out to the dealer network every few months.

Ford FDS2000

The facilities offered by a modern diagnostic tester will be illustrated by considering the Ford FDS2000 diagnostic system. Three components make up the FDS2000: a base station, a portable diagnostic unit (PDU) and a vehicle interface adaptor (VIA).

The base station, which is desk or wall mounted, provides a storage area for the PDU and the VIA, and a trickle charger for the batteries in the PDU. Most importantly, the base station provides the PDU with diagnostic data from a 560 MB CD-ROM.

The PDU, illustrated in Figure 9.4 is the heart of the system. It is a tough hand-held unit, specifically designed for the repair shop environment. Data and graphics are presented on a toughened touch-sensitive LCD screen. A programmable serial interface port is provided to support ISO 9141 and a variety of other flash codes.

The VIA is used to connect the PDU to the vehicle wiring loom when the vehicle has no serial link, or when a very detailed analysis is required.

When the FDS2000 is turned on, the first task is to load the PDU with the relevant Vehicle System Test Manual (VSTM) held on the base station's CD-ROM. The operator informs the PDU, via the touch-sensitive screen, which car model is to be worked on, and the appropriate guided diagnostic test sequences and help data is then downloaded to the PDU. CDs are updated approximately every three months as new data and test sequences become available.

Having loaded data from the CD, the PDU is removed from the base station and is now fully portable.

Using the guided diagnostic sequences which the PDU offers, the operator commences the appropriate

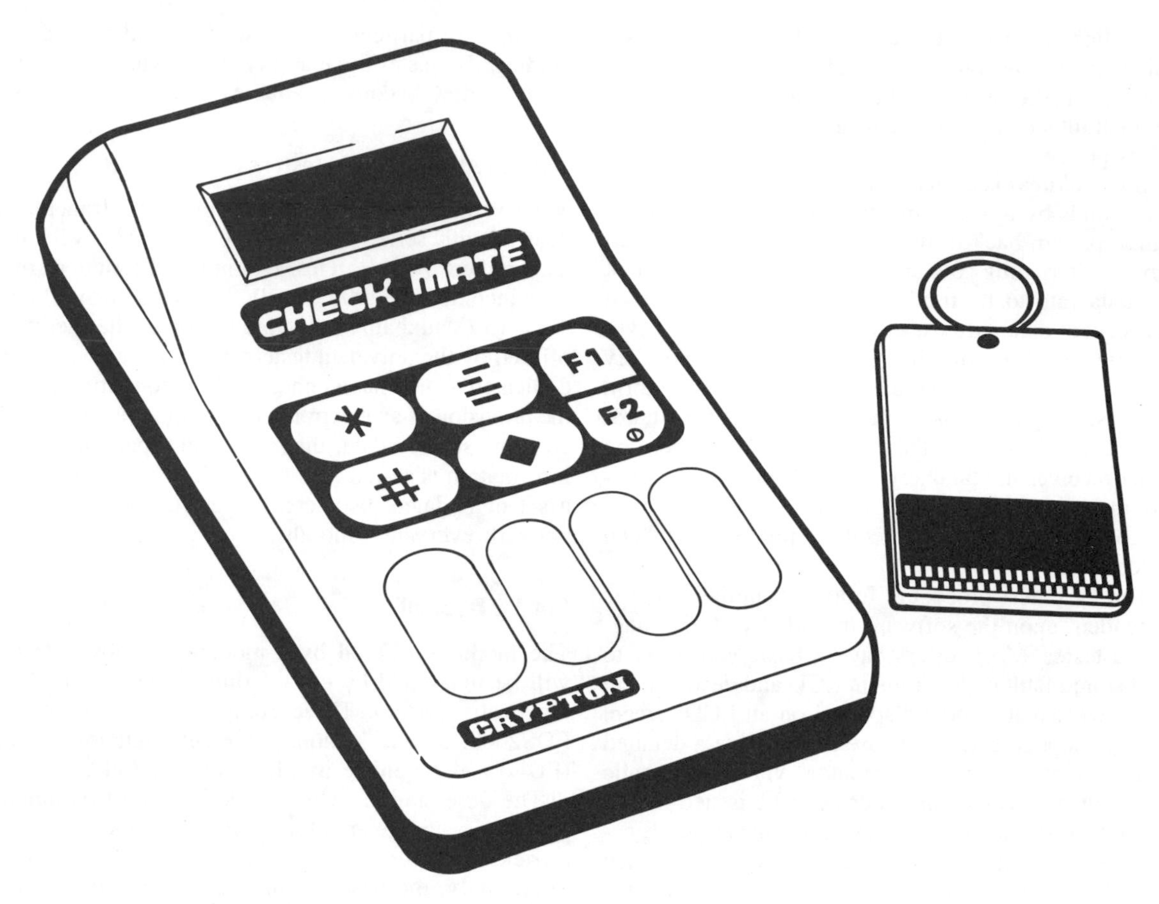

Figure 9.3 Crypton 'Check-Mate' hand-held diagnostic tester with program card. The program card slots into the rear of the tester (*FKI Crypton*)

vehicle tests. Diagrams are presented on the LCD screen to show the technician where, and how, the FDS2000 should be connected to the car's electrical system. Intermittent faults can be automatically detected by driving the car with the PDU linked to the vehicle. A special buffer enables not just the moment of the fault to be recorded, but all data for 30 seconds before it occurs.

When a fault is found, the PDU suggests appropriate remedial work and then checks that the work has been carried out correctly and that no other faults are present. At the end of the procedure there is a disconnection menu to show the technician how to disconnect the equipment.

9.4 CALIFORNIA AIR RESOURCES BOARD OBD II

Whilst for European-market vehicles the level of fault diagnosis is at the discretion of the manufacturer, the situation in the USA is rather different. Since a control system malfunction can give rise to an increase in exhaust emissions, US authorities require that all engine controllers incorporate specific on-board diagnostic (OBD) functions. The most basic monitoring requirements (OBD I) were introduced in 1988 and covered oxygen sensor and EGR failure, and detection of out-of-range sensor signals. More comprehensive requirements (OBD II) were introduced by the California Air Resources Board (CARB) over the period 1994–96 and cover nine specific environmental failure areas which must be displayed to the driver and actively corrected where possible (Table 9.4). It is anticipated that requirements similar to these will be introduced in Europe within the next decade. The general criterion adopted by the CARB is that faults should not cause a vehicle to exceed the relevant emissions certification limits by more than 50%.

Engine misfire is a particularly serious problem in controlling emissions and if sufficiently bad can

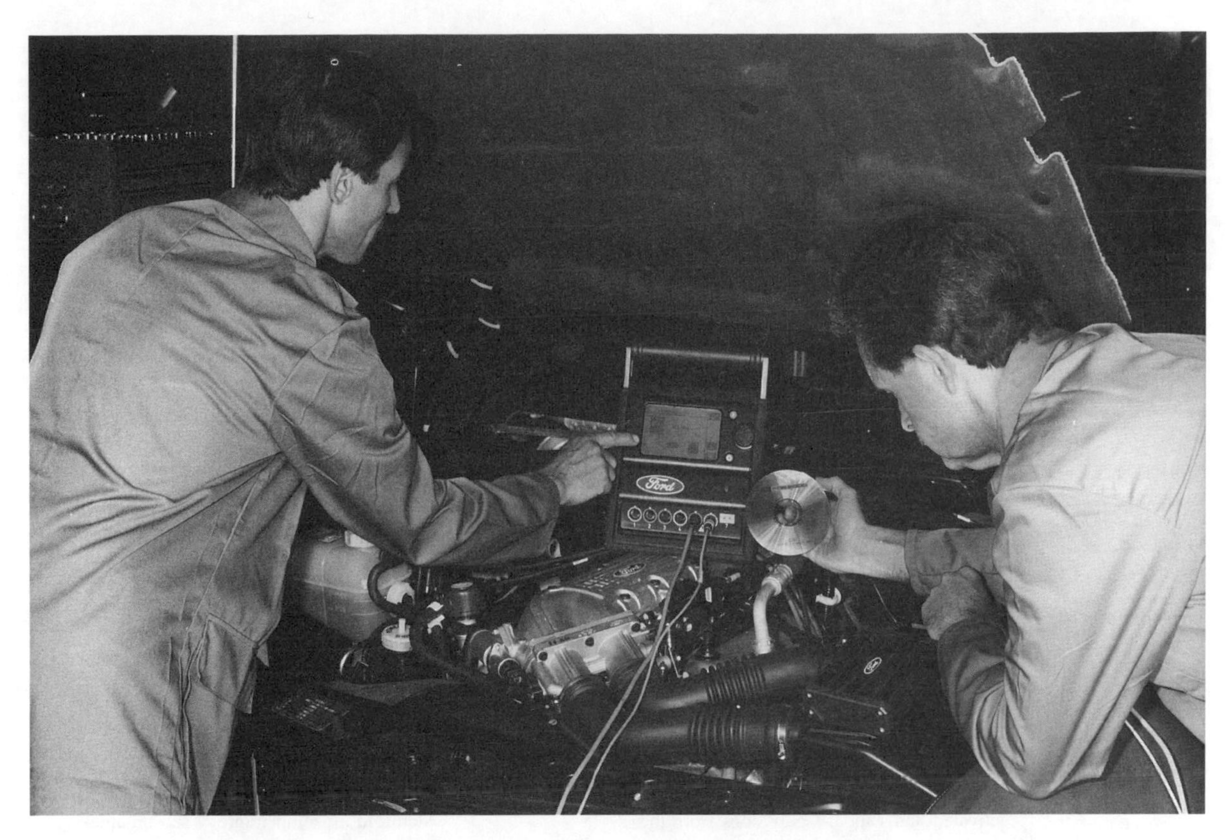

Figure 9.4 Ford FDS2000 Portable Diagnostic Unit (PDU) in use on a vehicle (*Ford*)

permanently damage catalytic convertors. It is for this reason that the CARB propose that during misfire the malfunction indicator light blinks to show the driver that he has a problem that is not only increasing emissions, but could also prove expensive if not attended to. Other faults, including an ineffective catalyst, are indicated by constant illumination of the MIL.

To sell vehicles in California, manufacturers must provide representative vehicles that have covered the required mileage for emissions certification (100 000 miles). Using these vehicles, they must then demonstrate that when the MIL illuminates the emissions do not rise more than 50% above the certification limits.

In addition to the self-diagnostic requirements described above, OBD II makes use of certain Society of Automotive Engineers (SAE) standards for electronic diagnostic facilities, specifically:

SAE Standard J1930 – defines standard electronic system terminology.
SAE Standard J1978 – defines a standard scan tool.
SAE Standard J1979 – defines standard fault codes.
SAE Standard J1962 – defines a standard diagnostic connector – a 16-pin connector mounted beneath the dashboard, which is now a legal requirement for all cars sold in the USA.

A major purpose of OBD II was to bring about the standardization of all fault codes to SAE J1979. Additionally, all vehicle manufacturers must use the same component terminology, the same scan tool and the same scan tool connecting lead. The outcome being that when a car owner is confronted with an illuminated MIL he can drive to the nearest garage, irrespective of franchise, and be assured that the technicians there will be able to diagnose the problem. The system also makes life much easier for technicians in independent repair shops who work on many different makes of car.

OBD II diagnostic trouble codes

The standardized fault codes specified in OBD II are termed *Diagnostic Trouble Codes* (DTCs). DTCs are

Table 9.4 California Air Resources Board (CARB) on-board diagnostics – II (OBD II) monitoring requirements introduced 1994–96

Component/system monitored	Fault criterion	Rationale
Engine misfire	A. Catalyst damage B. 1.5 × Certification emissions limit C. Inspection emissions limit	Misfire increases HC, CO emissions
Catalyst efficiency	Hydrocarbon conversion efficiency 40–50%	Reduced catalyst efficiency increases all emissions
Exhaust gas oxygen sensor (lambda sensor)	1.5 × Certification emissions limit	Emission control system cannot function with defective EGO sensor
Fuel injection system	1.5 × Certification emissions limit	Defective fuel injection system may cause a rise in emissions
Evaporative purge system	No hydrocarbon flow into engine when purge activated and vapour present in canister	Faulty purge system may cause a rise in HC emissions from the fuel supply system
Sensors and actuators	Open circuits or parameters outside manufacturer's specified limits	May cause increase in emissions (already in OBD I)
Air conditioning gas loss (chlorofluorocarbon)	Refrigerant leak	Chlorofluorocarbon refrigerants damage atmosphere
Exhaust gas recirculation	1.5 × Certification limits	EGR system failure will result in increased NO_X emissions
Secondary air system	No secondary airflow detectable in exhaust	Secondary air failure will result in increased emissions during warm-up phase

displayed on the scan tool's LCD screen and consist of two parts: the first part comprising a letter and a numeral, the second part comprising three numerals, for example **P0 111**.

The letter indicates which major vehicle component is at fault, as follows:

Table 9.5 Examples of diagnostic trouble codes to SAE standard J1979

DTC	MIL status	Text displayed on scan tool screen	Malfunction/fault
P0105	ON	P0105 Manifold absolute pressure sensor	Manifold absolute pressure sensor
P0106	ON	P0106 Manifold pressure sensor range/performance	Manifold absolute pressure sensor range/performance
P0107	ON	P0107 Manifold pressure sensor low input	Manifold absolute pressure sensor – signal too low
P0108	ON	P0108 Manifold pressure sensor high input	Manifold absolute pressure sensor – signal too high
P0110	ON	P0110 Manifold air temperature sensor	Temperature sensor, intake pipe
P0112	ON	P0112 Manifold air temperature sensor low input	Temperature sensor, intake pipe – signal too low
P0113	ON	P0113 Manifold air temperature sensor high input	Temperature sensor, intake pipe – signal too high
P0115	ON	P0115 Engine coolant temp. sensor	Temperature sensor, coolant
P0325	OFF	P0325 Knock sensor	Knock sensor
P0443	ON	P0443 Purge control valve circuit	EVAP valve control circuit
P0500	OFF	P0500 Vehicle speed sensor	Speed sensor
P0505	OFF	P0505 Idle system	Idle air control valve

P = Powertrain electronic system
C = Chassis electronic system
B = Body electronic system

In addition, there is also **U** (Undefined) which is held in reserve.

The numeral following the letter indicates whether the fault involves a defect covered by OBD II legislation (in which case a 0 is displayed), or whether the

Table 9.6 SAE J1930 standardized terms and acronyms for automobile electrical/electronic systems

SAE standardized term	SAE standardized acronym
Accelerator pedal	AP
Air cleaner	ACL
Air conditioning	A/C
Charge air cooler (*Intercooler*)	CAC
Continuous fuel injection system	CFI System
Camshaft position sensor	CMP Sensor
Closed throttle position	CTP
Distributor ignition system	DI
Data link connector (*Diagnostic Socket*)	DLC
Diagnostic trouble code	DTC
Diagnostic test mode	DTM
Engine control module (*Engine ECU*)	ECM
Engine coolant temperature sensor	ECT Sensor
Exhaust gas recirculation	EGR
Electronic ignition (*Distributorless Ignition*)	EI
Evaporative emission	EVAP
Fan control (*Electric Cooling Fan*)	FC
Generator (*Alternator*)	GEN
Heated oxygen sensor	HO_2S
Idle air control valve	IAC Valve
Intake air temperature sensor	IAT Sensor
Ignition control module (*Ignition ECU*)	ICM
Knock sensor	KS
Mass air flow sensor	MAF Sensor
Manifold absolute pressure sensor	MAP Sensor
Multi-port fuel injection (*Multipoint Fuel Injection*)	MFI
Malfunction indicator lamp	MIL
On-board diagnostic system	OBD System
Open-loop	OL
Oxygen sensor – unheated	O_2S
Pulsed secondary air injection	PAIR
Park/neutral position switch/sensor	PNP Switch/Sensor
Engine speed sensor	RPM Sensor
Service reminder indicator	SRI
Scan tool	ST
Throttle body	TB
Turbocharger	TC
Transmission control module (*Transmission ECU*)	TCM
Throttle position switch/sensor/potentiometer	TP Switch/Sensor/Pot
Three-way catalytic converter	TWC
Vehicle speed sensor	VSS
Wide-open throttle	WOT

fault code is an additional facility provided by the vehicle manufacturer (in which case a 1 or 2 is displayed).

The first of the subsequent three numerals indicates the subsystem in which the fault has arisen. For powertrain faults, the following codings are used:

P0 1XX – Fuel/air supply
P0 2XX – Fuel/air supply
P0 3XX – Ignition system
P0 4XX – Emission control system
P0 5XX – Engine speed/idle control
P0 6XX – ECU and ECU output signals
P0 7XX – Transmission
P0 8XX – Transmission
P0 9XX – Reserved for SAE
P0 0XX – Reserved for SAE

The last two numerals of the DTC (shown as XX, above) indicate the precise nature of the fault.

Examples of some common DTCs, and the status of the MIL for each code, are given in Table 9.5.

SAE standard J1930 nomenclature

The SAE standard J1930 provides a set of standardized vehicle electrical/electronic system terms, definitions, abbreviations and acronyms. The objective of J1930 is to ensure that all motor manufacturers adopt the same name and abbreviation for the same part, thus making electronic system fault diagnosis easier for technicians who work on many different makes of car. Although the standard is only required for US-market vehicles, the global nature of the car industry has led to the adoption of J1930 by many European and Japanese manufacturers. A example list of SAE J1930 standardized terms and acronyms is given in Table 9.6.

Appendix

SYSTEM OF UNITS

This book uses the International System of Units (also called SI, from the French *Système International des Unités*) which is adhered to by all professional engineers. Table 1 summarizes the six fundamental SI units from which all other units are derived.

In practice it is often necessary to describe quantities that occur in small fractions or large multiples of a unit, a standard set of prefixes are used to denote powers of 10. These prefixes are listed in Table 2.

Table 1 SI units

Quantity	Unit	Symbol
Length	Metre	m
Mass	Kilogram	kg
Time	Second	s
Electric current	Ampere	A
Temperature	Kelvin	K
Luminous intensity	Candela	cd

Table 2 Standard prefixes

Multiple or fraction	Prefix	Symbol	Power of ten
Millionth of a millionth	pico	p	10^{-12}
Billionth	nano	n	10^{-9}
Millionth	micro	μ	10^{-6}
Thousandth	milli	m	10^{-3}
Hundredth	centi	c	10^{-2}
Tenth	deci	d	10^{-1}
Ten	deca	da	10
Thousand	kilo	k	10^{3}
Million	mega	M	10^{6}
Billion	giga	G	10^{9}

Index